PRACTICAL GEOMETRY
AND ENGINEERING
GRAPHICS

by the same author

" The Dimensioning of Engineering Drawings"
" Machine Drawing and Design"
" Perspective ", etc., etc.

EIGHTH EDITION

PRACTICAL GEOMETRY AND ENGINEERING GRAPHICS

•

*A TEXTBOOK FOR ENGINEERING AND
OTHER STUDENTS*

by W. ABBOTT

C.M.G., O.B.E., Ph.D., B.Sc., M.I.Mech.E., M.R.I.

*Formerly Director of Studies, Federation of British
Industries and H.M. Staff Inspector of Engineering,
Ministry of Education*

*This book is available in Portuguese, from Editora Gertum Carneiro,
Rio de Janeiro*

•

LONDON · BLACKIE & SON LIMITED · GLASGOW

BLACKIE & SON LIMITED
450 Edgware Road
London W2 1EG
Bishopbriggs
Glasgow

First Edition 1929
Second Edition 1933
Third Edition 1939
Fourth Edition 1946
Fifth Edition 1951
Sixth Edition 1956
Seventh Edition 1963
Eighth Edition (Paperback) 1971
Reprinted 1972
Reprinted 1974
Reprinted 1977

©

by W. ABBOTT

C.M.G., O.B.E., Ph.D., B.Sc., M.I.Mech.E., M.R.I

First Asian Edition 1961

●

MACHINE DRAWING AND DESIGN
A Textbook of Intermediate Standard for Engineering Students
Eighth Edition

●

THE DIMENSIONING OF ENGINEERING DRAWINGS
This is a companion volume to *Machine Drawing and Design* and
deals particularly with the application of tolerances.

●

TECHNICAL DRAWING
A complete course for use in Secondary and Technical Schools

●

PERSPECTIVE
A comprehensive work for the architect, artist and designer

Printed in Great Britain by
Thomson Litho Ltd, East Kilbride, Scotland

PREFACE

THIS BOOK IS INTENDED TO PROVIDE A COURSE IN PRACTICAL Geometry for engineering students who have already received some instruction in elementary plane geometry, graph plotting, and the use of vectors. It also covers the requirements of Secondary School pupils taking Practical Geometry at the Advanced Level.

The grouping adopted, in which Plane Geometry is dealt with in Part I, and Solid or Descriptive Geometry in Part II, is artificial, and it is the intention that the two parts should be read concurrently. The logical treatment of the subject presents many difficulties and the sequence of the later chapters in both parts is necessarily a compromise; as an illustration, certain of the more easy intersections and developments might with advantage be taken at an earlier stage than that indicated.

In Part I considerable space has been devoted to Engineering Graphics, particularly to the applications of graphical integration. The use of graphical methods of computation is fully justified in most engineering problems of a practical nature—especially where analytical methods would prove laborious —the results obtained being as accurate as the data warrant.

The scope of the course in Descriptive Geometry has been designed to be in keeping with the relative importance of the subject in an engineering curriculum; for, although its educational and practical values have long been recognized, the time allocated to its study is usually limited. The earlier chapters in Part II have been arranged with special regard to the difficulties experienced by students in dealing with three-dimensional geometry for the first time; with these difficulties in view the problems of the straight line and plane have been dealt with after the orthographic and isometric projection of solids, the students being then more accustomed to think in space. In the treatment of the various solids, grouping has been based upon identity of principle rather than similarity of form. A systematic use has been made of auxiliary projections in the solution of problems, typical applications being given in Problems 220, 242, and 323; this powerful method does not seem to have been developed to the extent that it deserves, and by its use many standard problems become merely exercises involving a knowledge of a few fundamental constructions. The numerous pictorial views included are intended to aid visualization, and the student should be encouraged to rely upon this kind of sketch rather than to resort to models.

The subject-matter on each page is self-contained, and arranged so that the text and corresponding diagrams are always adjacent to each other. To facilitate cross reference a consistent notation has been used throughout and suffixes only have been given to the lettering, thus obviating the difficulty of locating a point a_1' among others marked a_1 and a'.

3

In preparing the work the author has been guided by the syllabuses and past examination papers of the various university and other Examinations Boards. The examples included have been limited to about six hundred, and bear directly on the text; most of them are original and of the appropriate standard, and types having little but their difficulty to recommend them have been excluded. The questions set by the examining bodies furnish therefore a further supply from which the student may draw. Answers have been appended to a large number of the examples; in the geometry portion many of the answers have been designed to serve simply as checks upon the accuracy of the constructions involved.

The author's thanks are due to Mr. A. S. Ritchie, B.Sc., for kindly undertaking the reading of the draft and final proofs of the book, to Mr. F. F. P. Bisacre, O.B.E., M.A., B.Sc., for advice and guidance during the preparation of the book and its passage through the press, and to many other friends for useful criticism.

The author will be grateful to receive notification of any errors, ambiguities, or obscurities which may have escaped notice.

Seventh Edition. New material on velocity and acceleration diagrams has been included, and various small improvements made.

W. A.

1963.

ABBREVIATIONS

The following is a list of the abbreviations and symbols used herein:

St. line	Straight line.	O^{ce}	Circumference.
△	Triangle.	Rad.	Radius.
Par^m	Parallelogram.	Diam.	Diameter.
Par^l	Parallel.	+ve	Positive.
Perp.	Perpendicular.	—ve	Negative.
Rt. ∠	Right angle.	Approx.	Approximately.

and other obvious contractions.

The following Greek letters are used:

α (alpha), β (bēta), γ (gamma), δ (delta), θ (thēta), Σ, σ (sigma), φ (phi), π (pi), ψ (psi), ω (omega).

EDITOR'S NOTE ON METRIC WORKING

This book is mainly concerned with geometry, and the constructions given are true regardless of units. However, the exercises do involve drawing to scale and the student will want to do his work in metric units. In the vast majority of examples the difficulty can be got round by the following simple expedient:

Take 1 inch as 2 centimetres.

All numerical data will then convert by simply doubling the figure, and the same will apply to the answers. The drawings will be somewhat smaller than the author intended. (Students who possess larger-than-average drawing-boards may convert at 3 centimetres to the inch.)

The above general rule will be presumed and the exceptions to it are listed as follows:

Page 6
Ex. 4. Scale 100 mm × 0·2 mm

Page 30
Ex. 3. 2·00 m × 1·50 m, scale 1:10

Page 50
Ex. 2. Pitch circles 288 mm and 384 mm dia., addendum 12 mm, dedendum 13 mm, tooth numbers 24 and 32

Page 56
Ex. 1 to 6. Take 2 cm as the unit

Page 60
Ex. 1. Amplitudes 4·5 cm and 3·5 cm
Ex. 2. Amplitudes 3·5 cm and 2·1 cm
Ex. 3. OQ =6 cm, travel 9 cm

Page 74
Section 93. Read centimetre for inch throughout

Ex. 1. Take numerical values from fig. 1; scales $m=4$, $h=2$
Ex. 2. Take

t	0	20	40	60	80	100	120	140 seconds
s	0	22	98	241	402	536	634	707 metres

Ans. 4·2 m/s, −0·055 m/s²
Ex. 3. Take length as 2 m, moment (M) given by $M = 4l^2$ kN m
Ex. 4. Take length as 4 m.

M kN m	·3	2·4	8·2	19·4	38·0	65·9
l m	·67	1·33	2·00	2·67	3·33	4·00

Page 78
Ex. 1. 6 cm radius. Ans. 72 cm³, 254 cm⁴, 3·59 cm
Ex. 2. Data, P, bar, 10 9 8
                V, m³/kg ·194 ·215 ·243

P,	7	6	5	4·5	4·0	3·5
V,	·277	·323	·388	·430	·486	·555

P,	3·0	2·5	2·0
V,	·648	·776	·970

Expansion from 10 bar to 2 bar. Ans. 312, 200 J. Stage pressures 5·90 and 3·45 bar
Ex. 3. Minor radius of cam 31·7 mm, major 41·3 mm; form of all acceleration – time graph correct. Ans. 1·83 m/s, 457 m/s²

Page 80
Ex. 2. Read metres for feet, answers in seconds and metres

Page 82
Ex. 1a. Accept data as given, convert to SI units, give power in kilowatts. Ans. 1180 1360 1380 1180 790.
Ex. 1b. Use actual data, but draw in SI units. Ans. 385 × 10⁶J

Page 84
Ex. 2. Work in inches

Page 86
Problem 105. Initial velocity 120 ft/s
Ex. 5. Accept data as given and express answers in SI units. Ans. 1.02×10^{-4} k m²/N, $.49 \times 10^{-4}$
Ex. p. 80, No. 3. Convert data exactly, then work in metres

Page 92
Ex. 1. For pounds read newtons.
Ex. 2. Take 3 tons as 30 kN, 5 ft. as 1·50 m. Ans. 49·1 and 59·8 kN.

Page 94
Ex. 1. For pounds read newtons, take 1 in. as 2 cm. Ans. 11 N, ·21 N m
Ex. 2. Side 2 cm, forces in N. Ans. 2·45N, +·29 N m
Ex. 3. Side 6 cm, forces in N. Ans. 5·14 N, 3·6 cm, 33·5°

Page 96
All forces in N, distances in m, numerical answers unaffected

Page 98
Ex. 1 and 2. Take 1 in. as 2 cm
Ex. 3. Take feet as metres
Ex. 4. Convert all data at 3 ft. = 1 m, 1 ton = 1000 kg, answers correspond. Note, the answers given for the height of the c.g. are correct but must be treated with caution. For some stability calculations it is necessary to place the load at the point of the jib, regardless of its actual height

Page 100
Ex. 1–4. Forces in newtons, distances in metres, numerical answers unaffected

Page 102
Ex. 1–5. Lengths in metres, loads in kN, numerical answers are then correct

Page 104
Ex. 1–3. Any units
Ex. 4. Take *form* of loading graph as given, span 6 m, ordinate scale 1 cm = 5 kN/m, EI in N m². Ans. 267,000/EI m, 0·1 m to right of centre

Ex. 5. Take form of loading as given, span 10 m, ordinate scale 1 cm = 10 kN/m, EI in N m². Ans. 3.35×10^6/ EI m, ·20 m right of centre

Page 106
Ex. 1. Take 3 ft. as 1 metre, ordinate scale 1 unit = 600 kN/m. Ans. 200 kN, 13·3 m, 2100 kN m, 21·7 m, 940 kN m
Ex. 2. Dimensions 30 m × 12 m × 4·5 m, depth 3 m

Page 108
Ex. 1. Read metres for feet, loads each 1 kN
Ex. 2–3. Any units

Page 110
Ex. 1. Loads 40 kN and 20 kN, scale 1 cm = 5 kN. Ans. 10 times given values
Ex. 2. Span 16 m, rise 0·64 m, loads in newtons. Scale 1 cm = 200 N. Ans. in newtons
Ex. 3. Take 3 ft. as 1 metre, loads 20 and 30 kN. Assume supports free to move laterally. Ans. in kN, 10 times values given
Ex. 4. Take loads as 3 and 5 kN

Page 112
Ex. 1–4. Loads in kN. Ans. in kN.

Page 114
Ex. 1. Loads in N
Ex. 2. Span 13 m, rise 0·975 m

Page 116
Ex. 1–2. Read kN for tons

Page 118
Length 10 m, load 200 kN. Ans. 250 kN m, 75 kN

Page 120
Ex. 1. Read kN for tons

Page 122
Ex. 1. Take elongations as being in centimetres, draw diagram twenty times full size
Ex. 2. Take data for $\delta l/P$ as being in m/N × 10^{-6}, load 40 kN. Ans. $x = .3$ mm, $y = 1.3$ mm. Horizontal force = 68 kN, deflection ·8mm

Page 124

Ex. 1. Panel size 2 m, load 20 kN, f = 80 MN/m^2, E = 200 GN/m^2. Ans. 13·6 mm.

Ex. 2. Panel size 3m, loads at 10 kN = 1 tonf. Ans. 19·2 mm.

Ex. 3. Panel size 3m, f top 80, bottom 100, vertical 60, inclined 80 MN/m^2. E = 200 GN/m^2. Ans. 49·5 mm

Page 136

Ex. 1–10. For ft. read metres, for tons read kN, answers correspond

Page 138

Ex. 1. Work in centimetres, converting lengths at 2 cm to 1 inch. Ans. twice given values

Ex. 2. Initial steam speed 467 m/s, blade speed 150 m/s. Ans. (a) 330 m/s, 135 m/s

Ex. 3. Dimensions OP 0·5 m., OQ 1 m. Ans. 25·4 m/s

Page 140

Ex. 1. Take 1 inch as 2 cm, velocity of Q 3 m/s.
Ans. 3·3 m/s, 55 rad/s, 23·4 rad/s.

Ex. 3. Take 1 inch as 2 cm. Ans. (1) D 1·83, E ·225 m/s. (2) D 2·26, E 1·13 (3) D 2·14, E 3·28 m/s; 9·92 rad/s

Page 146

Ex. 1. Take 1 inch as 2 cm. Ans. 47·3, 41·52, 26·64, 7·68, −21·12, −26·64, −28·32, −28·32 m/s^2

Ex. 3. Take 1 inch as 2 cm. Ans. 36·72, 26·88 m/s^2

Page 184

Ex. 2. Take 3 feet as 1 metre. Scale 1 : 100. Ans. 6·5 m

Page 192

Ex. 4. Take 3 feet as 1 metre. Scale 1 : 20

Page 194

Ex. 1. Take 3 feet as 1 metre. Scale 1 : 10

Page 296

Example. Set off aerofoil shapes for root and tip, taking chord lengths of 3·55 and 1·14 m, and angles of incidence of 5° and 3° as in fig. 3. Determine the aerofoil shape on a plane three quarters along tip to root

CONTENTS

PART I

PLANE GEOMETRY AND ENGINEERING GRAPHICS

PART II

SOLID OR DESCRIPTIVE GEOMETRY

1a. Drawing Equipment.

The following is a suitable equipment for the work in this book:— drawing board, half imperial 23″ × 16″: tee square, 24″ blade: scale, 12″ long, bevelled edges—one graduated in inches and tenths, the other in inches and eighths: protractor—semi-circular type: set-squares, celluloid, 45° 6″ edge, 60° 10″ edge: 5″ compasses (needle-point) with lengthening bar: 4½″ dividers, with fine adjustment: pencil spring-bows: one or two French curves: pencils H, 2H, 3H: rubber: drawing pins, &c.

1b. Accuracy in Drawing.

Assuming perfect equipment, errors in a drawing arise from inaccuracies in the lengths of the lines and the in-exact location of points of intersection. These in turn depend largely upon the thickness of the lines used, and upon the inclinations of the lines to one another.

EXAMPLES

(1) Draw two parallel lines ½″ apart and, using a 2H pencil, draw between them a succession of distinct parallel lines, counting them as they are drawn. With care about 50 lines can be inserted. Assuming that the spaces and lines are equal in width, the thickness of each line is about $\frac{1}{200}$″, or ·005″.

(2) With lines similar to those used in Ex. 1, draw the base BC, fig. 1, 3″ long, and using set-squares complete the △ ABC. Measure BA accurately. Increase the length of BC by $\frac{1}{10}$″ and repeat the construction. Note the increase in length of BA.

Calculate the lengths of BA and compare results.

$$\left(\frac{BA}{\sin 150°} = \frac{BC}{\sin 15°}\right)$$

From these exercises it will be seen that the exact location of the point A requires great care, and that an error in the length of BC gives a magnified error in the length of BA. When the angle BAC is smaller still, errors in the position of A may be considerable without being obvious.

2. Diagonal Scales.

These are of use when the divisions of an ordinary scale become minute, or when distances in three denominations have to be represented.

Fig. 2 shows a diagonal scale reading to eighths and sixty-fourths of an inch. Only the overall length of the scale should be transferred from a ruler, all subdivision being carried out geometrically.

Fig. 3 shows a diagonal scale to measure yards, feet, and inches to a scale of $\frac{1}{30}$.

Both drawings are self-explanatory; the exact widths (AB) of the scales are immaterial.

Note.—These diagonal scales are worthless if the subdivision is inaccurate, and they are included here mainly to serve as tests for accurate draughtsmanship.

EXAMPLES

(3) Using set-squares (not protractor) draw lines about 4″ long radiating from a common point P at 15° to one another. Draw a circle, centre P, rad. 3½″; compare the lengths of the chords of the intercepted arcs.

(4) Construct a diagonal scale 4″ long to read to $\frac{1}{100}$″. Mark on it 2·87″ and 3·46″.

(5) Construct a diagonal scale to measure feet, inches, and quarters of an inch, to a scale of ⅛. Mark on it 3′—9¾″.

(6) * The figure shows a deflection diagram for a structure (see p.125). Using the diagonal scale of Ex. 4, construct the figure and measure OA and AB.

NOTE.—Construction lines are either horizontal, vertical, or inclined at 45°. Start at O; set off *a* and *d* and obtain P; set off *b* and *c* and obtain Q; and so on. Ans. 11·30″, 1·92″.

⁂ A question number in **bold type** indicates that a corresponding diagram is given opposite.

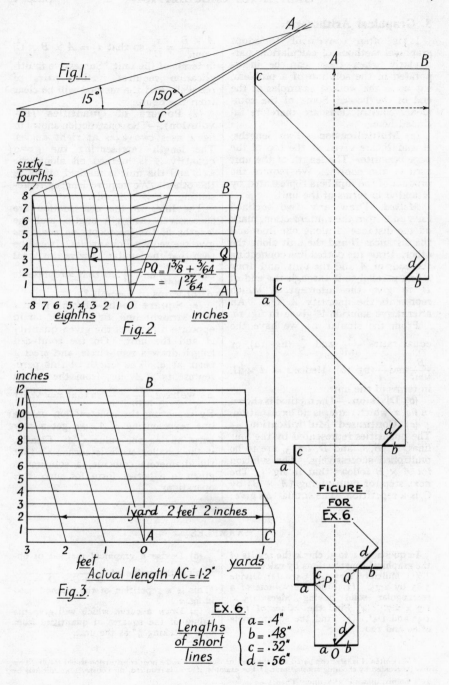

Fig.1.

15°   150°

B   C

A

sixty-fourths

8
7
6
5
4
3
2
1

B

P   Q

$PQ = 1\frac{3}{8}'' + \frac{3}{64}''$
$= 1\frac{27}{64}''$

A

8 7 6 5 4 3 2 1 0   inches   1

eighths

Fig.2.

inches

12
11
10
9
8
7
6
5
4
3
2
1

B

1 yard 2 feet 2 inches

A

3   2   1   0   yards   1

feet

Actual length AC = 1·2″

Fig.3.

A   B

c

d
b

c
a

d
b

c
a

**FIGURE
FOR
Ex.6.**

d
b

c P Q
a

d
a O b

Ex.6.

**Lengths
of short
lines**
$\begin{cases} a = .4'' \\ b = .48'' \\ c = .32'' \\ d = .56'' \end{cases}$

3. Graphical Arithmetic.

It is often convenient to adopt graphical methods of calculation, particularly when these can be incorporated in the solution of a problem, e.g. as in the worked examples at the end of the book. Some of the commoner constructions are therefore included here.

(a) **Multiplication.**—Two lengths, A and B, are given at the top of the page opposite. The length of the unit used is also shown. We require the product of the numbers represented by A and B in terms of the unit.

Method.—Draw any two inclined lines and, from their intersection, mark off the distance A along one line, and the distances B and the unit along the other. Draw the dotted line connecting the ends of A and the unit, and draw a line parallel to it through the end of B to give the intercept x, which represents the quantity $A \times B$. An alternative solution is given in fig. 1b.

From the similar $\triangle$s we have the equal ratios $\dfrac{A}{unit}$ and $\dfrac{x}{B}$ (fig. 1a), or

$\dfrac{B}{unit}$ and $\dfrac{x}{A}$ (fig. 1b). Hence $x = A \times B$, in terms of the unit.

(b) **Division.**—The method is shown in fig. 2, which requires no explanation.

(c) **Continued Multiplication.** — The quantities represented by the four lines, A, B, C and D, fig. 3, are to be multiplied successively. The solution for $A \times B$ follows that in fig. 1. The next step, of multiplying $(A \times B)$ by C, is a repetition. The similar $\triangle$s give:

$\dfrac{A \times B}{unit} = \dfrac{x_1}{C}$, so that $x_1 = A \times B \times C$, in terms of the unit. Successive multiplication requires further pairs of parallels, and the method will be clear from the figure.

(d) **Powers of Quantities (Involution).**—The construction shown in fig. 4 uses two axes at right angles. The length representing the given quantity A is marked off along one axis and the unit is marked off along the other. We require lengths representing A^2, A^3, A^4, etc.

The lines in the construction are successively at right angles, and the lengths of the intercepts on the axes give the required quantities. It will be seen that the similar $\triangle$s give the equal ratios $\dfrac{A}{unit}$ and $\dfrac{x}{A}$, so that x represents A^2 in terms of the unit.*

(e) **Square Roots.**—From a point a in a straight line, fig. 5, mark off in opposite directions the given quantity A and the unit. On the combined length draw a semi-circle, and erect a perp. at a. The length of this perp. represents $\sqrt{A}$, for, from the dotted $\triangle$s, we have $\dfrac{x}{unit} = \dfrac{A}{x}$, so that $x = \sqrt{A}$.† By repeating the construction on the line representing $\sqrt{A}$, we get a new perp. having the value $\sqrt[4]{A}$. Clearly, if the smaller circle is turned so that the diameters coincide, a solution is given in a better form for successive operations.

EXAMPLES

In questions 1 to 5, check the results of the graphical constructions by calculation.
(1) Multiply 4·75 by 3·5. (2) Divide 5·85 by 2·75. (3) Find the volume of a rectangular solid having edges 1·5 × 1·9 × 2·7. (4) Find the values of 1·4², 1·4³ and 1·4⁴. (5) Find the square roots of 80 and 120.

(6) Devise a graphical method of obtaining the value of $\dfrac{A}{B \times C \times D \times \ldots}$. (This is a repetition of the method used in fig. 2.)
(7) Draw a curve which will give the values of the squares of quantities from 0 to 4, taking $\frac{1}{4}''$ as the unit.

---

* Very often A is large compared with the unit (or vice versa) and a poor construction would result. This may be avoided by choosing another unit. † For example, if $\sqrt{90}$ is required, the construction unit may be 25 × normal unit, i.e. 5^2 × unit. Then $\sqrt{90} = \dfrac{x}{5}$.

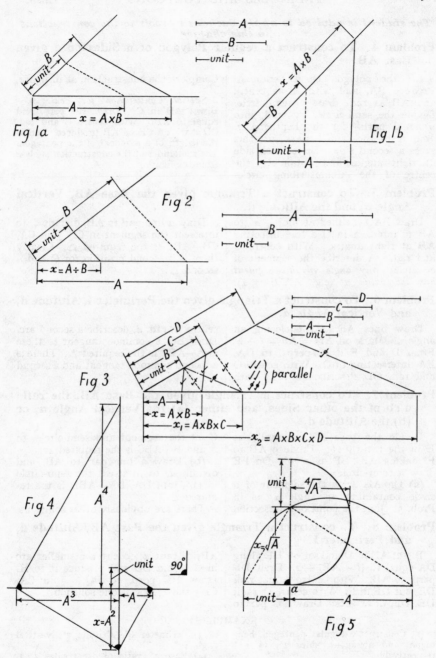

Fig 1a

$x = A \times B$

Fig 1b

$x = A \times B$

unit

A

Fig 2

unit

B

$x = A \div B$

A

Fig 3

D

C

B

unit

A

$x = A \times B$

$x_1 = A \times B \times C$

$x_2 = A \times B \times C \times D$

$\left.\begin{matrix} // \\ /// \\ //// \end{matrix}\right\}$ parallel

Fig 4

A^4

A^3

A

unit

$x = A^2$

90°

Fig 5

unit

$\sqrt[4]{A}$

$x_3 \sqrt{A}$

unit

A

α

The student is advised to apply Euclidean proofs to the constructions in this chapter

Problem 4. To construct a regular Polygon of n Sides on a given Base AB.

Let the polygon be a heptagon. Produce BA, and with A as centre and AB as rad. draw a semi-circle. Divide the semi-circle, by trial, into n equal parts—for the heptagon 7. Join A to the *second* mark, 2; this gives a second side. Bisect each side at right angles and obtain O, the centre of the circumscribing circle.

Complete the construction as in figure.

Special Construction for Pentagon.—Bisect AB in C. Draw BD perp. and equal to AB. With centre C and rad. CD mark off CE on AB produced. AE is the length of a diagonal of the pentagon. The remainder of the construction is clear from the figure.

Problem 5. To construct a Triangle given the Base AB, Vertical Angle α, and the Altitude d.

Draw BAO inclined at $(90° - α)$ to AB to intersect in O a line bisecting AB at right angles. With centre O and rad. OA describe the segment of a circle: *any angle in this segment has the value* α (e.g. $\angle ADB = α$).

Draw a line par^l to AB, distant d, to intersect the segment in C. Join CA, CB—ABC is the required $\triangle$. Clearly there is a second position for C and a second $\triangle$.

Problem 6. To construct a Triangle given the Perimeter 1, Altitude d, and Vertical Angle α.

Draw lines AF, AG enclosing an angle α. Mark off $AD = AE = l \div 2$. From D and E draw perps. to DA, EA, intersecting in O. With centre O and rad. OD describe an arc; with

centre A, rad. d, describe a second arc. Draw BC, a common tangent to these arcs—ABC is the required $\triangle$. There is a second common tangent and a second $\triangle$.

Problem 7. To construct a Triangle given the Base AB, the ratio a : b of the other Sides, and either (a) the Vertical Angle α, or (b) the Altitude d.

Divide the base AB internally at E in the ratio $a:b$. Produce AB to F making AF : BF as $a:b$. On FE draw a semi-circle.

(a) On AB draw the segment of a circle containing an angle α, as in Prob. 5. Join the point of intersection

C, of the segment and semi-circle, to A and B—ABC is the required $\triangle$.

(b) Draw a line par^l to AB, and distant d, to intersect the semi-circle in D. Join DA, DB—ABD is the required $\triangle$

There are obviously other solutions.

Problem 8. To construct a Triangle given the Base AB, Altitude d, and Perimeter 1.

Bisect AB in D and set off DE along DA equal to $(l - AB) \div 2$. Draw DF perp. to AB. With centre A and rad. DE, cut DF in F. With centre D, radii DE, DF, draw arcs. Draw PQ par^l to

AB, distant d, to cut the smaller arc in G. Join DG and produce it to H. Draw HC perp. to PQ. Join CA, CB—the $\triangle$ABC is one solution.

EXAMPLES

(1) Construct a regular pentagon, heptagon, and nonagon; sides 2″, 1½″, 1″ respectively.

Construct triangles to the following data. Measure the sides.

(2) Base 2″, vertical angle 30°, altitude 2¾″.

(3) Perimeter 8″, altitude 2″, vertical angle 45°.

(4) Base 2″, ratio of other sides 5 : 3, (a) vertical angle 30°; (b) altitude 1¾″.

(5) Base 3″, altitude 1¼″, perimeter 8″.

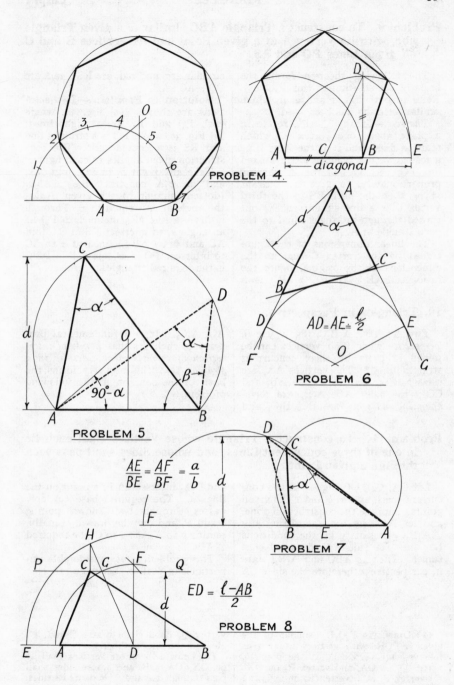

PROBLEM 4.

diagonal

$AD = AE \doteq \dfrac{\ell}{2}$

PROBLEM 6

PROBLEM 5

$90°-\alpha$

$\dfrac{AE}{BE} = \dfrac{AF}{BF} = \dfrac{a}{b}$

PROBLEM 7

$ED = \dfrac{\ell - AB}{2}$

PROBLEM 8

Problem 9. To construct a Triangle ABC similar to a given Triangle abc, with its Vertex A at a given Point and its Vertices B and C on given Lines PQ and RS.

The following theorem forms the basis of the solution to this Problem. Refer to both figs. 1 and 2, in which similar lettering has been used.

Theorem.—If any △ABC rotate in a plane about one vertex A, which remains fixed, and if at the same time a second vertex B move along a fixed line PQ, the sides of the △ varying proportionately to give a constant *shape*, then the path RS of the third vertex C is a line similar to PQ but turned through an angle equal to the vertex angle at A.

The linear dimensions of the figure traced by the vertex C bear to the dimensions of the original figure the ratio AC : AB. In fig. 2, PQ is a circular arc, and rad. arc RS : rad. arc PQ :: AC : AB.

Solution to Problem.—Two solutions are shown: (1) Fig. 3a, where both PQ and RS are straight lines; (2) Fig. 3b, where PQ is a straight line and RS is a circular arc. The construction given applies to each figure.

Take any point b_1 in PQ, join b_1A, and on b_1A construct the △Ab_1c_1 (dotted) similar to the given △abc; the angle at A is equal to α. Through c_1 draw a line MC, inclined to PQ at an angle α, to intersect RS in C. Join AC and draw AB inclined at α to AC to intersect PQ in B. Join BC—ABC is the required triangle.

10. Triangles in Perspective.

Two △s ABC, $A_1B_1C_1$ (fig. 4) are *in perspective* when their vertices can be joined in pairs by three concurrent straight lines.* The vertices AA_1 are *corresponding vertices*, as are BB_1 and CC_1; the sides AB, A_1B_1 are *corresponding sides*, as are BC, B_1C_1 and AC, A_1C_1. It is a fundamental property of such △s that *points of intersection of corresponding sides lie in a straight line*: this property forms the basis of the solution to the following problem.

Problem 11. To construct a Triangle whose Vertices shall each lie in one of three concurrent Lines, and whose Sides shall pass each through a given Point.

Let OA, OB, OC (fig. 4) be the concurrent lines, and a, b and c the given points. Suppose the construction done, and let ABC be the △. Draw any △$A_1B_1C_1$ to satisfy all the conditions but one—the side A_1C_1 does not contain b. The △s ABC and $A_1B_1C_1$ are in perspective; therefore the sides CA and C_1A_1 intersect in P, a point on the line ca. The required base CA now passes through two known points, b and P, and may be drawn. Finally, joining a to A, and c to C the required △ABC is given.

This problem is of considerable importance in Statics. See page 116.

EXAMPLES

(1) Draw △s $P_1Q_1R_1$, similar to the given △PQR, with vertices situated as follows: (a) vertex P_1 at the point b, vertex Q_1 on OA, and vertex R_1 on OX; (b) vertex P_1 at a, vertex R_1 on OX, and vertex Q_1 on a circle centre O rad. 8". Measure the sides of the △s.

(2) Draw a △ whose vertices shall lie on OX, OA, OB, and whose sides shall pass through a, b and c. Measure its sides.

* The lines OA, OB, OC may meet in a point at infinity, i.e. they may be parallel.

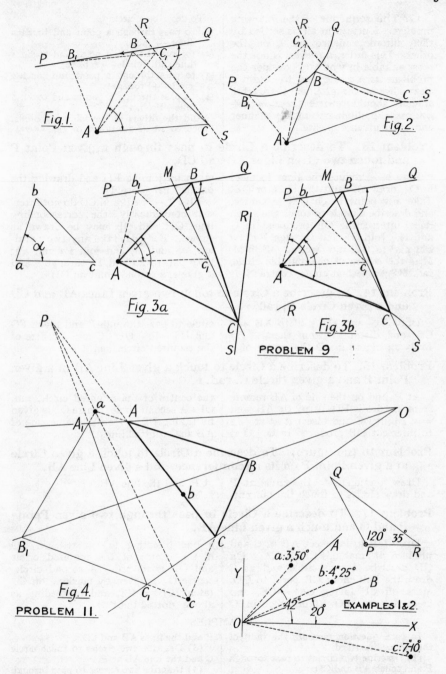

Fig.1.

Fig.2.

Fig.3a

Fig.3b

PROBLEM 9

Fig.4.

PROBLEM 11.

EXAMPLES 1 & 2.

$a : 3'', 50°$
$b : 4'', 25°$
$c : 7'', 10°$

45°
20°
120° 35°

12. The sequence of the following problems, fourteen in all, is settled by their interdependence one upon the other. This order, however, is not the most suitable in which to consider the problems *as a whole*, and to obtain a clear idea of the ground covered the student should rewrite them *in outline*, as a preliminary, in the manner and order given opposite.

To describe a circle:
(1) to pass through a point and touch 2 lines—Prob. 13;
(2) to pass through a point and touch 1 line and 1 circle—Prob. 19;
(3) to pass through a point and touch 2 circles—Prob. 21;
(4) to pass through 2 points and touch 1 line—Prob. 17.

Add the others in the following order: Probs. 20, 18, 22, 14, 23, 15, 16, 24, 25a, 25b.

Problem 13. To describe a Circle to pass through a given Point P and touch two given Lines AB and CD.

Fig. 1.—Produce the lines to meet in O; draw OP and the bisector OE. Take any point F, on OE, as centre, and describe a circle to touch the given lines, intersecting OP produced in G and H. Join GF and through P draw PM par^1 to GF, intersecting OE in M. M is the centre of one suitable circle, rad. MP; a second circle, centre N, is given by joining FH and drawing the parallel PN (dotted).

Fig. 2.—If AB and CD do not intersect conveniently, the corresponding lines OH and OE may be drawn as shown. For OH, take any two parallels, *ab* and *cd*, and join P*a* and P*b*; lines par^1 to P*a* and P*b* through *c* and *d* give *e*, a second point on OH.

Problem 14. To describe a Circle to touch two given Lines AB and CD and a given Circle E, rad. r.

Draw FG and HJ par^1 to AB and CD, and distant *r* from them. By Prob. 13 determine O the centre of a circle to pass through E and touch FG and HJ. Join OE. O is the centre of the required circle, rad. OK.

Problem 15. To describe a Circle to touch a given Line AB in a given Point P and a given Circle C, rad. r.

At P, and on the side of AB remote from C, draw PD perp. to AB and = *r*. Join DC and bisect it at rt. ∠s to intersect DP produced in O. O is the centre of the required circle, rad. OP. A second circle, centre O$_1$, is given by taking PE = *r* on the same side of AB as C (dotted lines).

Problem 16 (no figure). To describe a Circle to touch a given Circle at a given Point P on its circumference, and a given Line AB.

Draw a tangent to the circle at P and draw circles to touch the tangent, at P, and the line AB.

Problem 17. To describe a Circle to pass through two given Points P and Q and touch a given Line AB.

Draw QP to intersect AB in C and produce it making CD = CP. On QD describe a semi-circle and at C draw the half-chord CE perp. to QD. Mark off CF (along CA) = CE, and erect a perp. FO to intersect in O a line bisecting PQ at rt. ∠s. O is the centre of the required circle and OF is its rad. A second circle, centre O$_1$, is given by marking off CG (along CB) = CE, and proceeding as above (dotted lines).

EXAMPLES

In each question measure the radii of the circles described.
(1) Describe two circles to pass through P and touch AB and CD.
(2) Describe a circle ¾″ rad. about P. Then describe two circles to touch circle P and the lines AB and CD.
(3) Describe two circles to touch circle C and the line AB at P.
(4) Describe two circles to pass through P and Q and touch the line AB.

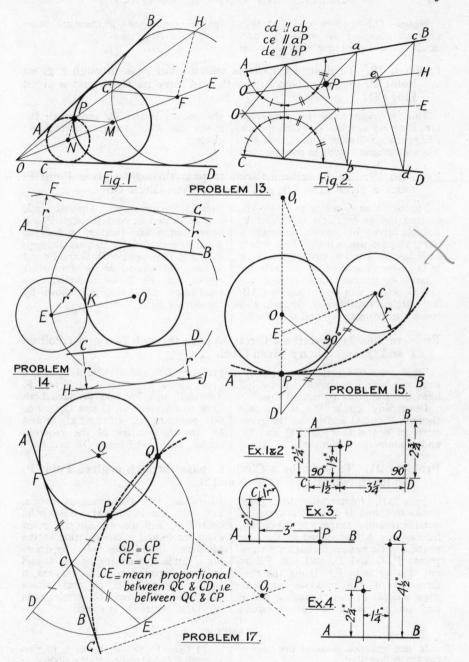

PROBLEM 13.

Fig. 1

Fig. 2

cd ∥ ab
ce ∥ aP
de ∥ bP

PROBLEM 14.

PROBLEM 15.

PROBLEM 17.

CD = CP
CF = CE
CE = mean proportional
between QC & CD, i.e.
between QC & CP.

Ex. 1 & 2

Ex. 3.

Ex. 4.

Note.—The student will find that extreme accuracy is necessary, particularly in registering the points of intersection of lines, in these and other similar constructions.

Problem 18. To describe a Circle which shall pass through a given Point P, touch a given Line AB, and have its Centre on a given Line CD.

This problem may be converted into Prob. 17 as follows. From P draw PEQ perp. to CD and make QE = PE. Then determine a circle to pass through the points P and Q, and touch the given line AB. This is the required circle.

Problem 19. To describe a Circle to pass through a given Point P, touch a given Line AB, and touch a given Circle C.

Through C draw a line perp. to AB, cutting the given circle in D and F, and the given line in E. Describe a circle to pass through the three points D, E and P: join PF, cutting this circle in G. Now determine, as in Prob. 17, a circle to contain the two points P and G, and touch the given line AB. This circle, centre O rad. OP, will also touch the given circle, and is one solution to the problem. Another circle can be drawn to pass through P and G and touch AB; further, by drawing a construction circle to pass through F, E and P, instead of D, E and P, and joining PD instead of PF, two other circles may be drawn to satisfy the conditions. One of these is shown by the dotted circle, centre O_1.

Problem 20. To describe a Circle to pass through two given Points, P and Q, and touch a given Circle A.

There are two solutions to the problem, and both are shown, one by full lines and the other by dotted lines.

Draw any circle PQCB to pass through P and Q and to cut the given circle in B and C. Join PQ and BC and produce them to meet in D. Draw tangents DE and DF from D to the given circle, touching it at E and F. Join AE and FA and produce both lines to intersect, in O and O_1, a line GH bisecting PQ at rt. ∠s. O and O_1 are the centres of the required circles, their radii being OE and O_1F.

Problem 21. To describe a Circle to pass through a given Point P, and touch two given Circles A and B.

Join AB, intersecting the given circles in C and D (also H). Draw an outside common tangent to the circles, intersecting AB produced in E. Describe a circle to pass through the three points P, C and D, and join PE to cut this circle in F. Using the construction of the previous problem, draw a circle to pass through P and F and touch one of the given circles (there are two such tangent circles, but only one is shown): this circle, centre O, will also touch the other given circle and is one solution to the problem. *Dotted lines only*: by drawing a circle to pass through P, C and H, and a crossed common tangent, a similar construction to that above gives two other suitable circles, one of which, centre O_1, is shown.

EXAMPLES

In each question measure the radii of the circles described.

(1) Describe two circles, each to pass through P and touch both the line AB and the circle C.

(2) Describe two circles each to pass through P and Q and touch the circle C.

(3) Describe two circles, corresponding to those in the figure of Prob. 21, to pass through P and touch the circles A and B.

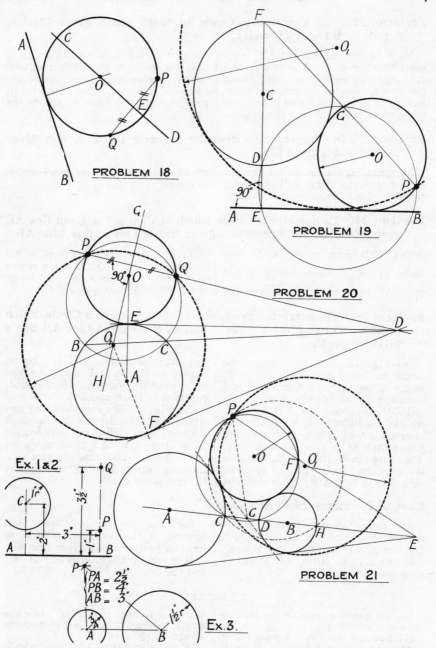

PROBLEM 18

PROBLEM 19

PROBLEM 20

PROBLEM 21

Ex. 1 & 2.

Ex. 3.

PA = 2½"
PB = 4"
AB = 3"

Problem 22. To describe a Circle to touch three given Circles, A rad. r_1, B rad. r_2, C rad. r_3.

With centre A and rad. $(r_1 - r_3)$ describe a circle ; with centre B and rad. $(r_2 - r_3)$ describe a circle. Using the construction of the previous problem, describe a circle centre O to touch these two circles and pass through the centre C. Join OB (or OA or OC): O is the centre of one circle, rad. OD, which touches the three circles externally. Seven other circles, in general, may be drawn to satisfy the conditions.

Problem 23 (no figure). To describe a Circle to touch two given Circles and a given Line.

By regarding the line as the circumference of a circle of infinitely large rad., the above construction may be readily modified to give the required circle, or circles.

Problem 24. To describe a Circle which shall touch a given line AC at the Point C, and intercept a given length l on a given Line AB.

Draw OD perp. to AC at C, and make $CD = \dfrac{l}{2}$. With centre A and rad. AD describe an arc intersecting AB in E. At E draw EO perp. to AB to intersect DO in O. O is the centre of the required circle and OC is its rad. ; FG is the given length l.

Problem 25. To determine the Locus of the centre of a Circle which touches (a) two given Circles A and B, (b) a given Line AB and a given Circle C.

(a) Fig. 1.—Join AB cutting the circles at D and E. Bisect DE at F, and with centre F mark off any equal distances F1 along AB on each side of F. With centres A and B and radii A1 and B1 describe arcs as shown to intersect in the points 1, 1 ; obtain other points such as 2, 2 in the same way. The curve drawn through these points is the required locus. It is a hyperbola with foci at A and B. Four different loci may be drawn for the circles shown, as the point F may be taken in three other positions, viz. the mid-points of GH, GE and DH.

(b) Fig. 2.—By regarding the line AB as the circumference of a circle of infinitely large radius the construction given above may be applied, as shown in the figure. The locus is a parabola. Another locus is given by taking F at the mid-point of EG.

Uses of Problems 25(a) and 25(b).

The above loci may be used if it is required to describe a circle to touch (a) two given circles, or (b) a given line and a given circle, *and to satisfy some other condition.*

For example (dotted lines both figures), if the required circle is to have its centre O on a given line MN, O is given by the intersection of MN and the loci.

EXAMPLES

In each question measure the radii of the circles described.

(1) Describe one circle, corresponding to that in the figure of Prob. 22, to touch the three given circles.

(2) Describe a circle to touch circles A and B and the line MN.

(3) A line AC is 3″ long and a second line AB is inclined to it at 35°. Describe a circle to touch AC at C and to intercept a length of 1½″ along AB.

(4) Describe a circle to touch circles C and D and to have its centre on MN.

(5) Describe a circle to touch AB and the circle D and to have its centre on MN.

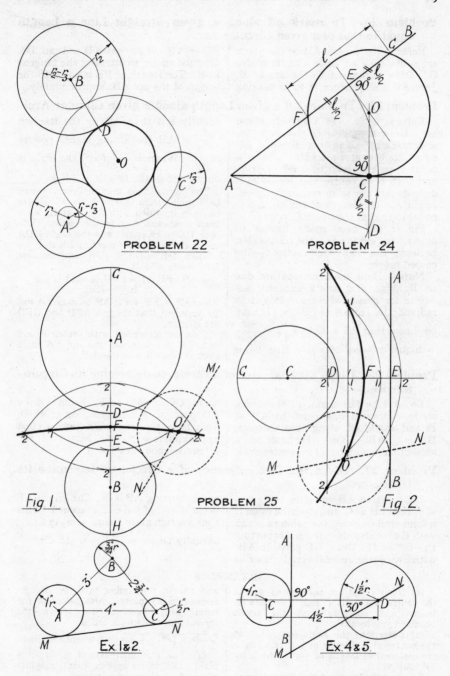

PROBLEM 22

PROBLEM 24

Fig 1.

PROBLEM 25

Fig. 2.

Ex. 1 & 2.

Ex. 4 & 5.

Problem 26. To mark off along a given Straight Line a Length equal to that of a given Circular Arc.

Refer to fig. 1. Let AB be the given arc subtending an angle α at its centre O. Draw a tangent to the arc at B. Join AB and produce it to D making BD $= \frac{1}{2}$AB. With centre D and rad. DA describe an arc to intersect the tangent in E. The length of BE is equal to the length of the arc AB, approximately.

Problem 27. To mark off a given Length along a given Circular Arc.

Refer to fig. 2. Let AB be the given arc. Draw a tangent to the arc at B, and mark off along it a distance BE equal to the given length. Take a point F on BE, such that BF $= \frac{1}{4}$BE, and with F as centre and FE as rad. describe an arc to intersect the given arc in A. The length of the arc AB is equal to that of the line BE, approx.

Fig. 2 has been made similar to fig. 1, to emphasize the connexion between the two constructions—discussed below.

Note.—Both constructions are due to Rankine and are sufficiently accurate for graphical work. If α is in radians, the arc AB = OA . α. It may be shown that EB = OA . $\alpha \left\{ 1 - \dfrac{\alpha^4}{1080} \right.$ — higher powers of $\alpha \Big\}$, so that EB is slightly less than AB by the fraction $\dfrac{\alpha^4}{1080}$ of AB (neglecting higher powers of α). When $\alpha = \dfrac{\pi}{2}$ (90°) the error is only ·0056 of the length of AB.

The construction used in Prob. 27 may be deduced from that of Prob. 26. Refer to fig. 3 in which both figs. 1 and 2 have been combined, to a smaller scale. Suppose DE to be joined and the angle EDA bisected, the bisector meeting BE in F.

Then DE = DA = 3DB, and because DF bisects $\angle$EDB,

EF : FB :: DE : DB (Euc. VI, 3);
$\therefore$ BF = $\frac{1}{4}$BE.

Also, AF = EF, for if AF be joined it will be apparent that the $\triangle$s AFD and EFD are similar;
$\therefore$ an arc described with centre F and rad. FE will intersect the arc BA in A such that arc BA = line BE.

Problem 28. Given the Diameter of a Circle, to determine its Circumference.

Let AB be the diam. of the given circle. Draw a line perp. to AB at B and mark off along it a distance BE = 3 . AB. On AB describe a semi-circle, and from O its centre draw a rad. OC at 30° to AB. From C draw CD perp. to AB intersecting it at D. Join DE: the length DE gives a close approximation to the length of the circumference. See Ex. 3.

Problem 29. Given the Circumference of a Circle, to determine its Diameter.

Draw a line AB equal to the given $\bigcirc^{ce}$, bisect it at C, and describe upon it a semi-circle. Using the same rad. and with B as centre, describe an arc cutting the $\bigcirc^{ce}$ in D. Draw DE perp. to AB; with E as centre, and ED as rad., describe an arc cutting AB in F. The length AF is the diam. of the circle, approx. This construction gives Diam. = 0·317 Circ. ; actually Diam. = $\dfrac{\text{Circ.}}{\pi}$ = 0·318 Circ.

EXAMPLES

(1) Rectify the arc subtending 50° at the centre of a circle 4″ rad. Calculate its length and compare with the result obtained graphically.

(2) Determine the angle subtended at the centre of a circular arc, 4″ rad., by a circumferential length of 3·5″. Check by calculation.

(3) Find the length of the circumference of a circle, 3″ diam., using both Probs. 26 and 28: in the former take the circumference in six parts. Also calculate the length which Prob. 28 should give—the formula for the length DE is

$\sqrt{9d^2 + \tfrac{1}{4}d^2(1 + \cos 30)^2}$, where d = diam.

(4) The circumference of a circle is 6·28″. Obtain its approx. diam. graphically and check by calculation.

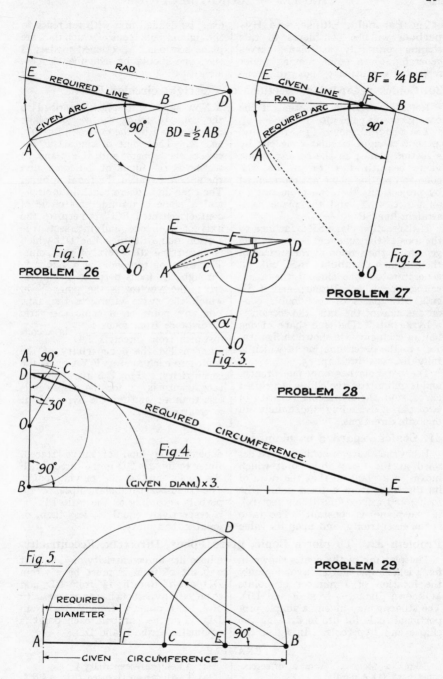

Fig.1.
PROBLEM 26

Fig.2.
PROBLEM 27

Fig.3.

PROBLEM 28

Fig.4.

PROBLEM 29

Fig.5.

The **Parabola, Ellipse, and Hyperbola** will be considered in this chapter primarily as plane curves generated by a point moving under restricted conditions; they will, however, be defined first with reference to the right circular cone of which they are plane sections.* Sectioned models of the cone should be examined if possible.

30. Conics regarded as Sections of the right circular Cone.

Refer first to the definition of the cone given on page 170

The pictorial views, figs. 1, 2, and 3, show a right circular cone cut by a section plane; in the corresponding views beneath, figs. 1a, 2a, 3a, the cone and section plane are represented by projections, the cone projecting as an isosceles $\triangle$, and the plane as a straight line SP.

If the section plane SP is inclined to the axis of the cone and is par[1] to one generator, the section at the surface of the cone is a **parabola**; if SP cuts all generators on one side of the apex, the section is an **ellipse**; and if SP cuts both parts of the double cone on one side of the axis, the section is a **hyperbola**. The true shape of one half of each curve is shown in figs. 1a, 2a, 3a—the determination of which is fully discussed in Chap. 15.

The parabola has only one branch, and is unlimited; the ellipse is a closed curve, and the hyperbola consists of two parts extending indefinitely in opposite directions.

Now suppose a sphere inscribed to the cone to touch the section plane SP, as shown by the circles in figs. 1a, 2a, 3a. The point of contact F between the sphere and the plane of section is the **focus** of the conic, and the sphere is called the **focal sphere.** The cone and sphere touch in a circle, and a plane containing this circle of contact, lettered PCC, is perp. to the axis of the cone and intersects the section plane SP in a line DD which is called the **directrix** of the conic. The **axis** of the conic is a line AA$_1$ through the focus perp. to the directrix; the **vertex** is the point V in which the curve intersects the axis. For any point on a conic the ratio $\dfrac{\text{distance from focus F}}{\text{distance from directrix DD}}$ is a constant called the **eccentricity** of the conic; in each figure VF/VA gives the eccentricity. For the parabola the eccentricity is 1; for the ellipse it is less than 1; and for the hyperbola it is greater than 1.

31. Conics regarded as plane Loci.

From the above, a conic may be defined as the locus of a point which moves in a plane so that the ratio of its distances from a fixed point (the focus) and a fixed line (the directrix), in the plane, is constant. The ratio is the eccentricity, and upon its value depends the kind of curve traced. Refer to fig. 4: DD is the directrix, F the focus, and P the moving point. The conic is a parabola, ellipse, or hyperbola according as the ratio PF/PD is respectively equal to, less than, or greater than unity.

Problem 32. To plot a Conic, given Focus, Directrix, Eccentricity.

The plotting of the conics shown in fig. 4 is a simple matter involving only the location of a number of points at known distances from F and DD. The student may invent a simple proportional scale for use in drawing the ellipse and hyperbola. In fig. 4 the ellipse has an eccentricity of $\frac{2}{3}$, so that a point 2″ from F must be 3″ from DD, one 1″ from F, 1½″ from DD, and so on; the hyperbola has an eccentricity of $\frac{3}{2}$, i.e. a point 3″ from F is 2″ from DD. For the parabola each point is equidistant from F and DD.

EXAMPLES

Taking a focus 2″ from a directrix, construct: (1) A parabola.

(2) An ellipse, eccentricity $\frac{2}{3}$.
(3) Two hyperbolas, eccentricities $\frac{3}{2}$ & $\frac{9}{8}$.

* The curves were studied first as sections of the cone by the ancient Greeks. Long afterwards Pappus discovered that they could be defined as plane loci.

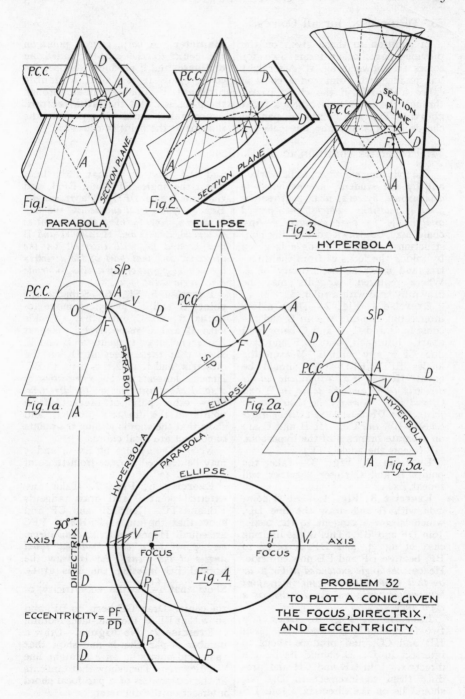

Fig 1. PARABOLA

Fig 2. ELLIPSE

Fig 3. HYPERBOLA

Fig 1a.

Fig 2a.

Fig 3a.

Fig 4.

ECCENTRICITY = $\dfrac{PF}{PD}$

PROBLEM 32

TO PLOT A CONIC, GIVEN
THE FOCUS, DIRECTRIX,
AND ECCENTRICITY.

33. Terms used for all Conics.

In addition to those given on the previous page, other terms used for conics are shown in fig. 1. A straight line joining two points on a conic is called a **chord**; if the chord passes through the focus it is called a **focal chord**. The mid-points of parallel chords lie in a straight line called a **diameter**. A perp. from a point on the conic to the axis is called an **ordinate**, and if produced to meet the conic again it is called a **double ordinate**; the double ordinate through the focus is called the **latus rectum**. A **normal** to a conic is a perp. to the tangent at the point of contact.

34. Properties common to all Conics.

The following exercises, to be worked out by the student, are designed *to demonstrate by graphical construction certain relations which are proved analytically in pure geometry.* Any conic may be drawn to test the constructions; a suitable curve is given by taking the focus 2″ from the directrix and using an eccentricity of $\frac{8}{9}$. Where required, tangents may be drawn by trial with a ruler.

Exercise 1, Fig. 2.—Draw a line intersecting the directrix in D and the conic in B and C, B and C being well apart. Join each point to F and produce CF to any point E. Measure the angles BFD, EFD—they should be equal. Hence: *if a straight line cut the directrix in D and the conic in B and C, and if D, B and C be joined to the focus, then* DF *bisects the exterior angle between* BF *and* CF. (If B and C are on separate branches of the hyperbola, DF bisects the angle BFC).

Exercise 2, Fig. 3.—Take the points B and C closer together and repeat Ex. 1.

Exercise 3, Fig. 4.—Let C coincide with B and draw the line BD, which is now a tangent to the conic. Join DF and BF. This is the limiting case of fig. 3, and DF is perp. to BF, because BF and FE now coincide. Hence: *the angle subtended at the focus by that part of the tangent intercepted between the conic and the directrix is a right angle.*

Exercise 4, Fig. 5.—Draw any two focal chords BC and GH. Join HB and CG, and produce them to intersect in D: D should lie on the directrix. Join GB and CH and produce them to intersect in D_1: D_1 should lie on the directrix. Join DF and D_1F, and show that these lines bisect the angles BFG and BFH, and that the angle DFD_1 is a rt. angle. Hence: *if two focal chords BC and GH be drawn, lines joining the extremities G and B, and C and H, also H and B and C and G, will intersect on the directrix, and that part of the directrix between the points of intersection subtends a rt. angle at the focus.*

Exercise 5, Fig. 6.—Suppose the points G and H, fig. 5, to move towards B and C until they coincide with B and C respectively. Repeat Ex. 4; i.e. draw tangents at B and C. Show that these meet in D_1 on the directrix, and that D_1F is perp. to BC. Hence: *tangents at the extremities of a focal chord intersect on the directrix.* Draw other pairs of tangents from points on the directrix (dotted), and show that the chords joining the points of contact are focal chords.

Note.—The results of Ex. 4 and 5 may be deduced at once from those of Ex. 1 and 3.

Exercise 6, Fig. 7.—Take any external point P and draw tangents PB and PC. Join BF and CF and show that the angles PFB and PFC are equal. Hence: *tangents drawn from any point to a conic subtend equal angles at the focus.* At B draw the normal BG meeting the axis at G. Show that $\dfrac{FG}{FB}$ = the eccentricity of the conic. Draw GE perp. to BF, and show that BE = the semi-latus rectum.

Exercise 7. No Figure.—Draw a number of par^l chords and show that their mid-points lie on a straight line—a diameter. Then show that tangents at the extremities of a par^l focal chord intersect on the diameter.

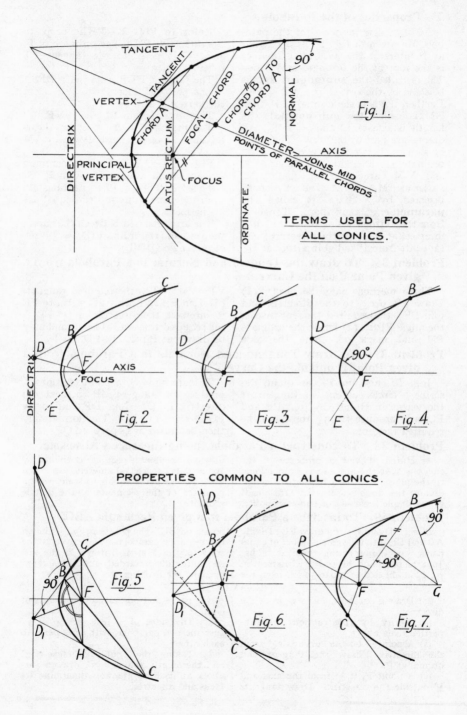

Fig. 1.

TANGENT

TANGENT

VERTEX

CHORD A

FOCAL CHORD

CHORD B // TO CHORD A

NORMAL

90°

DIRECTRIX

PRINCIPAL VERTEX

LATUS RECTUM

FOCUS

DIAMETER—JOINS MID POINTS OF PARALLEL CHORDS

AXIS

ORDINATE.

TERMS USED FOR ALL CONICS.

DIRECTRIX

C

B

D

F

AXIS

FOCUS

E

Fig. 2

C

B

D

F

E

Fig. 3

B

D

90°

F

E

Fig. 4

PROPERTIES COMMON TO ALL CONICS.

D

C

B

90°

F

D₁

H

C

Fig. 5

D

B

F

D₁

C

Fig. 6.

P

B

90°

E

90°

F

C

G

Fig. 7.

35. Properties of the Parabola.

Additional terms used for the parabola are given in **fig. 1**. If a tangent at P intersect the axis at T, and M is the foot of the ordinate at P, then **TM** is called the **subtangent** and is bisected by the vertex; i.e. VT = VM. Further, if PN is the normal at P, then NM is called the **sub-normal**; its length is constant, and = 2VF. All diams. are par[l] to the axis. That part of a diam. between its vertex and an ordinate is called an **abscissa**: VM and V_1M_1 are abscissæ, and V_1M_1 varies as $PM_1 \times RM_1$.* The focal chord bisected by a diam. is called the **parameter** of that diam. Tangents from the extremities of all focal chords intersect at rt. ∠s on the directrix; e.g. tangents from P and Q intersect at D.

Refer to Fig. 2.—PB is perp. to the directrix, PM is the ordinate of P, and FC is perp. to the tangent PT from P. Then:—

The tangent PT bisects the ∠BPF.

The perp. FC meets a tangent from the vertex V at C on PT.

The latus rectum $LL_1 = 4 \cdot VF$.

The ordinate[2] = the product (latus rectum × abscissa), i.e. $PM^2 = LL_1 \cdot VM = 4VF \cdot VM$.

Fig. 3.—PQ and PR are common tangents from an external point P.

Then: ∠PFQ = ∠PFR; i.e. tangents from the same point subtend equal ∠s at the focus.

The area enclosed between the curve, the axis, and the ordinate QM = $\frac{2}{3}$ (area of rectangle QMVL).

Problem 36. To draw the Tangent and Normal to a Parabola from a given Point P on the Curve.

Three methods may be used: **(1)** Draw PB perp. to the directrix and join PF; the required tangent bisects the angle FPB. **(2)** Draw the ordinate PM and mark off along the axis VT=VM; PT is the required tangent. **(3)** Join FP and draw FD perp. to FP to intersect the directrix in D; PD is the required tangent. The normal may be drawn at P perp. to PD.

Problem 37. To draw Tangents and Normals to a Parabola from a given Point P outside the Curve.

Join PF and on PF as diam. describe a circle cutting the tangent at the vertex in B and C. Join PB and PC, and produce them; these are the required tangents.

To determine Q, a point of contact, produce the tangent CP to meet the directrix in D. Join DF, and draw FQ at rt. ∠s to FD. The normal may then be drawn at Q perp. to QP.

Problem 38. To construct a Parabola, using measured Abscissæ.

As stated above,* a property of the curve is that the abscissa is proportional to the product of the parts into which it divides the double ordinate. This leads to the simple construction shown opposite, in which the double ordinate is divided by 10 equally spaced diameters, and the abscissæ are made equal, to scale, to the products of the segments of the double ordinate.

Problem 39. To inscribe a Parabola to a given Rectangle ABCD.

Bisect AD and BC and draw FE. Divide AB and BE into the same number of equal parts, say 5, numbering them as in fig. Join the points on AB to F, to intersect in p_1, p_2 ... lines drawn par[l] to EF from the points on BE. The points p_1, p_2 ... lie on the required parabola. The curve is the trajectory of a stone thrown in the air i.e. uniformly retarded and accelerated motion; see Problem 90.

EXAMPLES

(1) Draw a parabola, focus 2″ from the directrix:

(a) Verify by construction the properties given above.

(b) A point P on the curve is $2\frac{1}{2}$″ from the directrix: draw the tangent and normal at P.

(c) A point P_1 is $\frac{1}{2}$″ from the axis and $\frac{1}{4}$″ outside the directrix. Draw tangents to the curve and normals at the points of contact.

(2) The sides of a rect. are 4″ × 3″. Inscribe two parabolas with axes perp. to each other.

(3) Taking the double ordinate = 4″ and the length of axis = 3″, draw a parabola as in Prob. 38 and determine its focus and directrix.

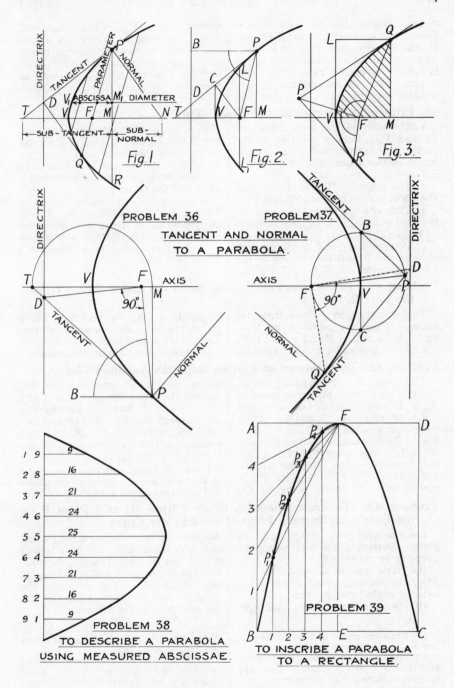

Fig 1.

Fig. 2.

Fig. 3.

PROBLEM 36

PROBLEM 37.

TANGENT AND NORMAL
TO A PARABOLA.

DIRECTRIX

TANGENT

T   V   F   AXIS   M

90°

TANGENT

NORMAL

B   P

TANGENT

DIRECTRIX

AXIS   F   V   D

90°

NORMAL

TANGENT

PROBLEM 38

TO DESCRIBE A PARABOLA
USING MEASURED ABSCISSAE.

PROBLEM 39

TO INSCRIBE A PARABOLA
TO A RECTANGLE.

The ellipse cuts the axis in two points, and is called a **Central Conic.** It has two foci and two directrices. That part of the axis within the curve is called the **major axis** (VV_1), and a perp. to it at the **centre** C, terminated by the curve, is called the **minor axis** (BB_1). Circles described about the axes as diams. are called **major and minor auxiliary circles.** All diams. pass through the centre and are bisected by it; tangents at the extremity of any diam. are parallel.

40. Principal properties of the Ellipse. Fig. 1.

1. The sum of the focal distances PF and PF_1, from a point P on the curve, is constant, and equal to the length of the major axis VV_1. Hence $FB = VC$.

2. The tangent PT and the normal PN bisect the ∠s between PF and PF_1.

3. Tangents QR and QS, from an external point Q, are equally inclined to the focal distances QF and QF_1.

4. Perps. FT and F_1T_1 to the tangent at any point P meet the tangent on the auxiliary circle, and $BC^2 = FT \times F_1T_1$.

5. A circle (dotted) containing the foci F and F_1, and any point P on the curve, cuts the minor axis in points T_2 and N, which are the points of intersection of the tangent and normal, at P, and the minor axis.

6. PM is the ordinate of P. Then $PM^2 : VM \times V_1M :: BC^2 : VC^2$. If MP be produced to meet the auxiliary circle in D, then $DM^2 = VM \times V_1M$ and the ratio may be written PM : DM :: BC : VC.

Problem 41. To construct an Ellipse, given the major Axis VV_1, and the Foci F, F_1.

This construction follows from (1) above. Take any point A on VV_1, fig. 2. With centres F and F_1 and radii VA and V_1A describe arcs intersecting in points A_1, A_2, A_3, and A_4. These points lie on the required ellipse. Take other points such as A and plot a succession of points on the curve.

Problem 42. To construct an Ellipse, using the Auxiliary Circles.

The ellipse and its auxiliary circles in fig. 1 have been partly redrawn in fig. 3. P is any point on the curve, and PM is its ordinate. Join DC, cutting the minor auxiliary circle in G. Join PG. Then because GC = BC and VC = DC, the ratio PM : DM :: BC : VC, from (6) above, may be written PM : DM :: GC : DC; therefore PG is par¹ to MC. To obtain points on the ellipse, such as P, draw any rad. DC cutting the auxiliary circles in G and D; through D draw DP par¹ to BC, and through G draw GP par¹ to VC— the lines intersect in P. Other points obtained in this way are shown for the lower half of the curve.

NOTE.—Tangents to the auxiliary circles at D and G, and a tangent to the ellipse at P, intersect on the axes at T_2 and T_3.

Problem 43. To draw Tangents to an Ellipse (i) at a Point P on the Curve (ii) from a Point Q outside the Curve.

Let the axes VV_1, BB_1, fig. 4, be given. Determine the foci, using the property at (1) above.

(i) Draw the focal distances PF, PF_1; the required tangent bisects the exterior angle between them.

(ii) With centre Q and rad. QF, describe an arc. With centre F_1 and rad. VV_1 describe an arc to intersect it in A and B. Join AF_1, BF_1 to intersect the ellipse in R and S: these are the points of contact between the tangents and the ellipse, and QS and QR are the required tangents.

EXAMPLES

(1) Draw any ellipse and verify the properties (1) to (6) given above.

(2) The foci of an ellipse are $3\frac{1}{2}''$ apart and the major axis is $5''$ long. Draw the ellipse and measure the minor axis.

(3) The major and minor axes of an ellipse are $5''$ and $4''$. Describe the ellipse and determine its foci.

(4) A point P is $6''$ from one focus and $2\frac{3}{4}''$ from the other, in Ex. 2. Draw tangents to the ellipse from P and measure their intercepted lengths.

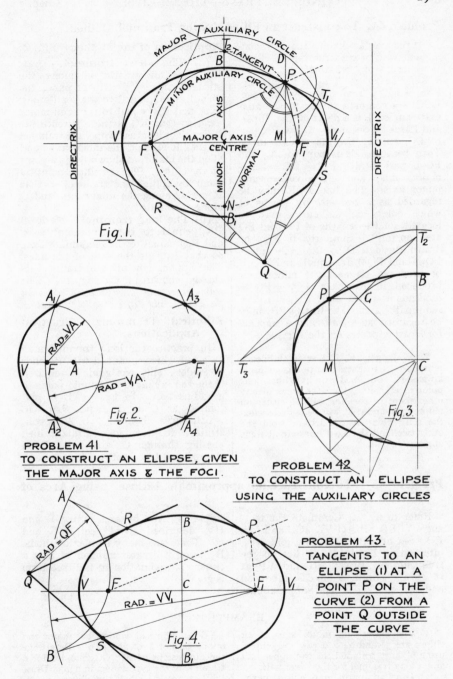

Fig. 1.

AUXILIARY CIRCLE
MAJOR
MINOR AUXILIARY CIRCLE
TANGENT
AXIS
MAJOR C AXIS
CENTRE
MINOR
NORMAL
DIRECTRIX
DIRECTRIX

Fig. 2.

RAD=VA
RAD = V₁A

PROBLEM 41
TO CONSTRUCT AN ELLIPSE, GIVEN
THE MAJOR AXIS & THE FOCI.

Fig. 3

PROBLEM 42
TO CONSTRUCT AN ELLIPSE
USING THE AUXILIARY CIRCLES

Fig. 4.

RAD = QF
RAD. = VV₁

PROBLEM 43.
TANGENTS TO AN
ELLIPSE (1) AT A
POINT P ON THE
CURVE (2) FROM A
POINT Q OUTSIDE
THE CURVE.

Problem 44. To construct an Ellipse by the Trammel Method.

If the axes are known the most convenient way of constructing an ellipse is by means of a trammel.

Refer to fig. 1. This is a reproduction of fig. 3 from the previous page. DGC is a common radius to the auxiliary circles, P is a point on the ellipse, and PM is its ordinate. Draw PR par¹ to DC, cutting the major axis in Q. Then because DM is par¹ to BC, and PG is par¹ to MC, PR = DC = semi-major axis, and PQ = GC = semi-minor axis. The line PQR may be regarded as a *trammel* by means of which points on the curve may be located, for the lengths of PR and PQ are constant. Similarly if PQ₁ be drawn through P so that the angles PQ₁M and PQM are equal, and if Q₁P be produced to meet the minor axis produced in R₁, then PQ₁ = PQ = semi-minor axis, and PR₁ = PR = semi-major axis. The line Q₁PR₁ may be regarded as an *alternative trammel* for locating points on the curve.

Note.—In general, if a straight line be moved so as to have two points along it always in two fixed straight lines, one point in each line, then the locus of any other point in the line will be either an ellipse or a circle: it will be a circle when the third point is equidistant from the other two, and the fixed lines are at right angles.

Application of the Method. Fig. 2.

(a) The short trammel.—Draw the axes and mark off along the straight edge of a strip of paper the distances PQ, equal to the semi-minor axis, and PR, equal to the semi-major axis, so that RQ is the difference between their lengths. Apply the trammel so that R falls on the minor axis, and Q on the major, as shown—P is a point on the ellipse. Plot the ellipse point by point by sliding the trammel so that R moves along the minor axis and Q along the major.

(b) The long trammel.—Mark off PR₁, equal to the semi-major axis, and P₁Q₁ equal to the semi-minor axis, so that R₁Q₁ is the sum of the axes. The application of the trammel is clearly shown in the figure. *This trammel should always be used when the axes do not differ greatly in length.*

Elliptical Trammels. Practical Applications.

In practice the long trammel may be a rod which is moved with its ends in guides; only one guide is required if the rod is hinged at its mid-point to a radius bar. In fig. 3 AB is pin-jointed at C to a radius bar OC which turns about O. If A is constrained to move in a straight line, passing through O, any point P on the rod will trace out an elliptical arc.

Problem 45. To construct an approximate Ellipse, using Arcs of Circles.

Refer to fig. 4. Complete the rectangle VCBD. Bisect VD at E and join EB. Set off CB₁ = CB and join DB₁, intersecting EB in P. P lies on the true ellipse. Bisect PB and VP at rt. ∠s and obtain G on BB₁ produced and H on VC. The approximate ellipse is given by arcs radii GB and HV, described about centres G and H. There are many similar methods. By using a greater number of arcs a closer approximation to the true form is given.

EXAMPLES

(1) The major and minor axes of an ellipse are 5″ and 3½″. Draw the ellipse, using a short trammel for the upper half and a long trammel for the lower half.

(2) Draw an approximate ellipse on the same lines as axes (Ex. 1) and compare the two curves.

(3) An elliptical window has major and minor axes of 8′ 0″ and 6′ 0″. Determine the dimensions of AC, OC, and PC for a trammel of the type shown in fig. 3. Draw the locus of P for half the ellipse. Scale: 1″ = 1 foot.

PROBLEM 44
TO CONSTRUCT AN ELLIPSE
BY THE TRAMMEL METHOD.

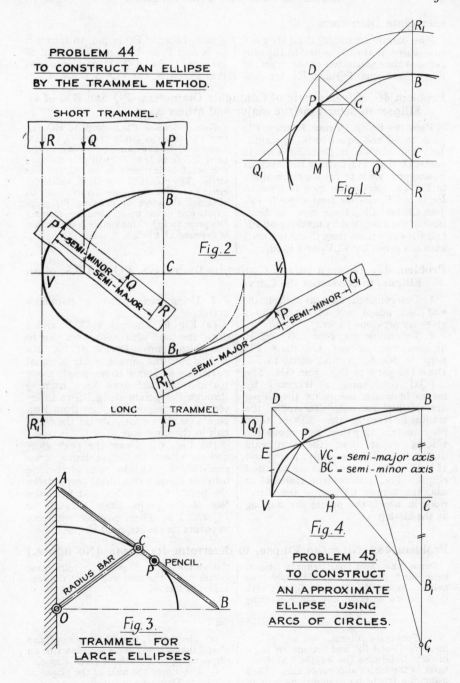

SHORT TRAMMEL

Fig. 2

LONG TRAMMEL

Fig. 3.
TRAMMEL FOR
LARGE ELLIPSES.

Fig. 1.

Fig. 4.

VC = semi-major axis
BC = semi-minor axis

PROBLEM 45.
TO CONSTRUCT
AN APPROXIMATE
ELLIPSE USING
ARCS OF CIRCLES.

Conjugate Diameters.

Two diams. are conjugate if they are so situated that each is par¹ to the tangents at the extremities of the other; in fig. 1 the diams. PQ and RS are conjugate, because PQ is par¹ to tangents at R and S, or because RS is par¹ to tangents at P and Q. It may be proved that $PC^2 + SC^2 = VC^2 + BC^2$.

Problem 46. Given a pair of Conjugate Diameters, PQ and RS, of an Ellipse, to determine the major and minor Axes.

Refer to fig. 2. From P draw PD perp. to SC and equal to SC in length. Join DC and on it describe a circle centre O. Join PO, cutting the circle in E, and produce it to cut the circle again in F. *The semi-minor axis is equal in length to* PE, *and the semi-major to* PF. Join CE and CF; these give the *directions* of the axes, and by marking off PE and PF along them from C, the required axes are given by VCV₁ and BCB₁.

Note.—Because PD is perp. to HC, the line PD and the circle centre O will intersect SC in the same point H; also a perp. DG from D to PQ will pass through the point of intersection of PQ and the circle. The points H and G, in which the circle cuts the conjugate diams., may be obtained therefore by drawing PD perp. to SC and equal to SC, and by drawing DG perp. to PQ. This simple construction is required at 2 below.

Problem 47. Given a pair of Conjugate Diameters, PQ and RS, of an Ellipse, to construct the Curve.

1. Determine the axes, as in Prob. 46, and then adopt any of the methods given on previous pages.

2. Determine the points G and H as described above; i.e. draw PD perp. to SC, fig. 3, and equal to SC. Draw DG perp. to PQ. Join GH. The △PGH constitutes a trammel by means of which points on the ellipse are quickly located. To apply the trammel, transfer the △PGH to a piece of tracing paper, and move the tracing paper over the conjugate diams. so that G lies always in PQ and H in RS; the locus of P is the required ellipse. Fig. 4 shows the trammel in use, the arrows indicating the directions in which the points are moving at the instant.

3. Using circumscribing parallelograms.

(a) Fig. 5. Through P, Q, R, and S draw the parᵐ *abcd* with sides par¹ to the given diams. Divide RC, SC, R*a* and S*d* into the same number of equal parts, say 6. Join P to the points along R*a* and S*d*, and draw lines from Q through the points along RS to intersect the corresponding lines from P in points on the curve. Obtain the other half in the same way.

(b) Fig. 6. Draw the parᵐ *abcd*, and on adjacent sides as diams. draw semi-circles. Divide each semi-circle into six equal parts, drop perps. from the points to the sides, and from the feet of the perps. draw parallels to PQ and RS. These parallels intersect in points on the required ellipse.

Problem 48. Given an Ellipse, to determine its Axes. (No figure.)

Draw any two par¹ chords, bisect them, and draw a diam. through the points. Bisect the diam. at C, and with C as centre describe any circle, cutting the ellipse in four points. Join these points to C; the axes will bisect the four angles so formed.

EXAMPLES

(1) Conjugate diams. of an ellipse measure 8″ and 6½″ and include an angle of 60°. Determine the lengths and directions of the major and minor axes. Then using the triangular trammel method of figs. 3 and 4 plot the ellipse.

(2) The sides of a parᵐ measure 5″ and 4″ and the included angle 50°. Inscribe an ellipse using both methods given above.

(3) Determine the axes of the ellipse in (2), using Prob. 48.

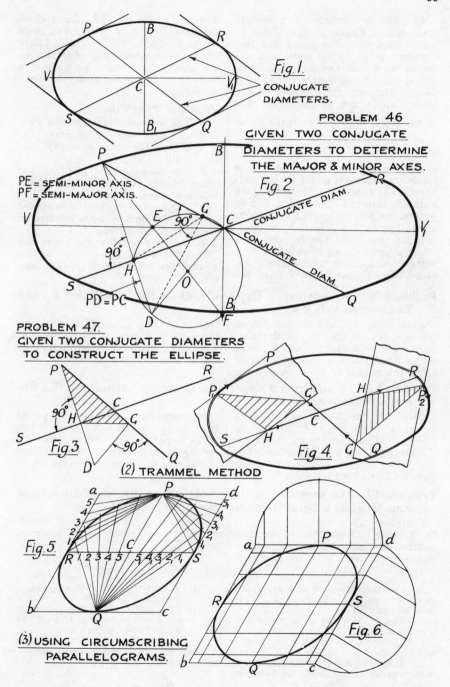

Fig. 1.
CONJUGATE
DIAMETERS.

PROBLEM 46

GIVEN TWO CONJUGATE
DIAMETERS TO DETERMINE
THE MAJOR & MINOR AXES.

Fig. 2.

PE = SEMI-MINOR AXIS.
PF = SEMI-MAJOR AXIS.

90°

CONJUGATE DIAM.
CONJUGATE DIAM.

90°

PD = PC

PROBLEM 47.
GIVEN TWO CONJUGATE DIAMETERS
TO CONSTRUCT THE ELLIPSE.

90°

Fig. 3.

90°

Fig. 4.

(2) TRAMMEL METHOD

Fig. 5.

Fig. 6.

(3) USING CIRCUMSCRIBING
PARALLELOGRAMS.

49. The hyperbola is a central conic and is similar to the ellipse in many respects. The curve has two branches, given by the sections of the double cone, and two foci and directrices; each branch has the same eccentricity. The focus and directrix for one branch apply also to the other: i.e. in fig. 1, $\dfrac{VF}{VD} = \dfrac{V_1F}{V_1D}$ = eccentricity, or $\dfrac{V_1F_1}{V_1D_1} = \dfrac{VF_1}{VD_1}$ = eccentricity. That part of the axis between the vertices VV_1 is called the **transverse axis**; its midpoint is the **centre** C. The **conjugate axis** BB_1 bisects VV_1 at rt. ∠s, and its extremities BB_1 are located by describing an arc, rad. FC, centre V, to intersect the perp. in B and B_1. St. lines which approach the curves indefinitely without actually touching them are called **asymptotes** *; they form the diagonals of the rect. LL_1LL_1 of which the axes are centre lines. The circle centre C diam. VV_1 is an auxiliary circle. It intersects each directrix on the asymptotes and a perp. from a focus to an asymptote meets the circle tangentially.

Principal Properties.

1. The *difference* between the focal distances of a point is constant and is equal to the transverse axis, i.e. $PF_1 - PF = VV_1$.

2. The tangent at a point on the curve bisects the angle between the focal distances, i.e. $\angle FPT = \angle F_1PT$. The length of a tangent intercepted between the asymptotes is bisected at the point of contact, i.e. $GH = GJ$.

3. If Q is a point on the curve and QR and QS are drawn par^l to the asymptotes, then $QR \times QS$ = a constant.

Problem 50. To construct a Hyperbola, given the Foci, F and F_1, and Transverse Axis VV_1.

Refer to fig. 1. Take any point A on the axis, outside the foci. With centre F_1 and rad. V_1A describe an arc; with centre F and rad. VA describe a second arc to intersect it in A_1, a point on the required curve. The same radii give three other points on the curve, and a series of points may be plotted by taking other points such as A.

Problem 51. To construct a Hyperbola, given the Asymptotes CL, CL_1 and a Point Q on the Curve.

Refer first to **(3)** above. Through Q, fig. 2, draw QS and QR par^l to CL, CL_1. Draw any radial line from C cutting QS in I and RQ produced in I_1; through these points draw lines par^l to CL, CL_1, to intersect in (1) on the required curve. Obtain other points (2), (3), (4), in the same way. When LC is perp. to CL_1 the curve is called a *rectangular hyperbola*—a curve common in gas expansions and compressions.

Problem 52. To construct a Hyperbola, given the semi-transverse Axis CV_1 and a Point O on the Curve, ordinate AO.

Complete the rectangle V_1AOG, fig. 3, divide GO and AO into the same number of equal parts, and join these points to V_1 and V. The points of intersection (1), (2) . . . lie on the required curve, which is symmetrical about V_1A.

EXAMPLES

(1) Plot both branches of a hyperbola, given eccentricity = $\frac{3}{2}$, and focus $1\frac{1}{2}''$ from directrix. Mark the second directrix and focus. Draw the asymptotes and measure the angle between them.

Use the hyperbola in Ex. 1 for Exs. 2 and 3.

(2) Take any point P on one branch and draw the tangent and normal to the curve.

(3) Take a few points such as Q, fig. 1, and confirm that $QR \times QS$ = constant.

(4) Plot a rectangular hyperbola, given that a point on the curve is $1''$ from one asymptote and $2''$ from the other.

(5) Take the following dimensions for fig. 3 and plot the complete curve. $CV_1 = 2\frac{1}{2}''$, $V_1A = 2''$, $AO = 3\frac{1}{2}''$.

---

* Refer to p. 266 where the asymptotes are discussed in relation to the cone.

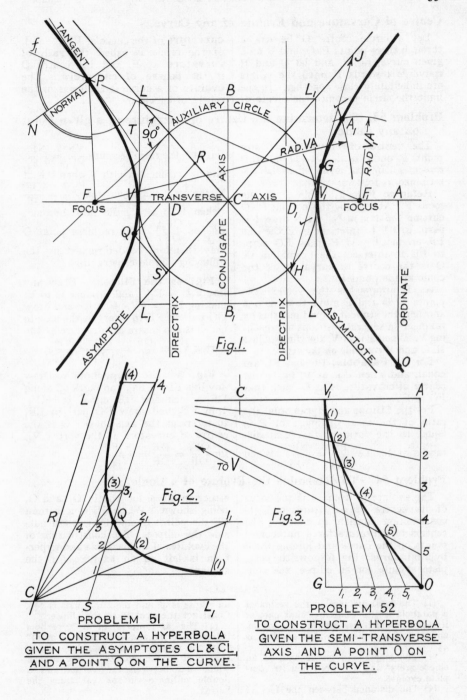

Fig. I.

Fig. 2.

PROBLEM 51
TO CONSTRUCT A HYPERBOLA
GIVEN THE ASYMPTOTES CL & CL₁
AND A POINT Q ON THE CURVE.

Fig. 3.

PROBLEM 52
TO CONSTRUCT A HYPERBOLA
GIVEN THE SEMI-TRANSVERSE
AXIS AND A POINT O ON
THE CURVE.

Centre of Curvature and Evolute of any Curve.

Let a circle centre O be drawn through three points P, Q, and R on a given curve, fig. 1, and let Q and R converge towards P until the points are indefinitely close together. In the limit the circle becomes the **circle of** curvature of the curve at P; the rad. of the circle is called the **rad. of curvature** at P, and the centre O is the **centre of curvature.** The **evolute** of a curve is the locus of the centre of curvature.

Problem 53. To determine the Centre of Curvature at a given Point on any Conic.

The centre of curvature at any point P on a conic, *except when* P *coincides with the vertex,* may be determined as follows:

Refer to Fig. 2. Join P· to the focus F. At P draw the normal PNO cutting the axis in N. At N draw NE perp. to NP to intersect in E the line PF produced. At E draw EO perp. to PE to intersect the normal in O. O is the centre of curvature of the conic at the point P.

As P approaches the vertex, the points N, O and E move towards one another; the student should test this by taking P in several positions approaching V. When P is at V, the three points N, O and E coincide *on the axis.*

For the **Parabola,** PF and FE are equal, and when P is at V, fig. 3, the centre of curvature is at O, such that FO = FV.

For the **Ellipse** and **Hyperbola,** the ratio of the focal distances of P is equal to the ratio of the focal distances of N: i.e. $\dfrac{PF}{PF_1} = \dfrac{NF}{NF_1}$. When

P is at V the ratio becomes $\dfrac{VF}{VF_1} = \dfrac{NF}{NF_1}$; and as O coincides with N when P is at V, the ratio can be written $\dfrac{VF}{VF_1} = \dfrac{OF}{OF_1}$ (when P is at V_1 the ratio becomes $\dfrac{V_1F_1}{V_1F} = \dfrac{OF_1}{OF}$). The required point O may therefore be determined by the following simple construction.

Fig. 4: the Ellipse. — Draw any line $F_1D = F_1V$, and produce it to E, making DE = FV. Join EF and draw DO parl to EF to intersect the axis in O. O is the centre of curvature at the vertex V, for $\dfrac{OF}{OF_1} = \dfrac{DE}{DF_1} = \dfrac{VF}{VF_1}$.

Fig. 5: the Hyperbola. — Draw any line FD = FV_1 and mark off along DF a distance DE equal to F_1V_1. Join EF_1 and draw DO parl to EF_1 to intersect the axis in O. O is the centre of curvature at the vertex V_1, for $\dfrac{OF_1}{OF} = \dfrac{DE}{DF} = \dfrac{V_1F_1}{V_1F}$.

Problem 54. To determine the Evolute of a Conic.

The evolute of a curve is the locus of its centre of curvature, and its construction consists in plotting the centres of curvature for a number of points on the curve and joining them by a fair curve. Fig. 6 shows the complete evolute for an ellipse, the construction lines for points O_1 and O_2 being shown. When CB is less than CF the points O_3 and O_4 will fall outside the ellipse. The construction of the evolutes of the parabola and hyperbola is left as an exercise for the student.

EXAMPLES

(1) The focal distance of the vertex of a parabola is ¼″. Construct that part of its evolute which lies between the vertex and a double ordinate whose abscissa is 4½″.

(2) The major and minor axes of an ellipse are 5″ and 4″. Construct its complete evolute.

(3) The distance between the foci of an ellipse is 4″ and the minor axis is 2½″. Construct the evolute to the ellipse.

(4) The asymptotes of a hyperbola intersect at 70° and at a distance of 1¼″ from the vertices of the branches. Draw that part of the evolute to one branch which lies between the vertex and a double ordinate distant 4½″ from the vertex.

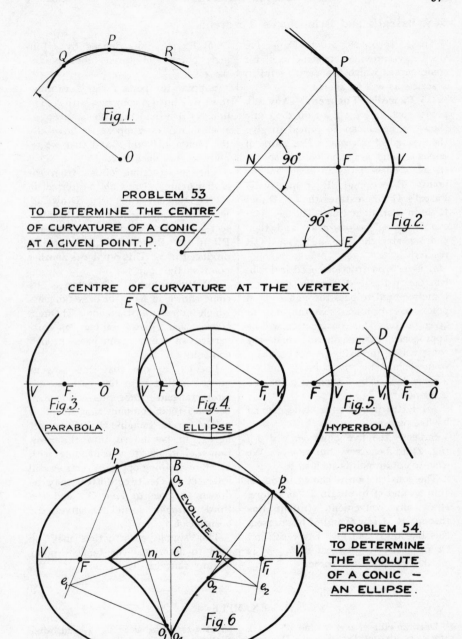

Fig. 1.

Fig. 2.

PROBLEM 53
TO DETERMINE THE CENTRE
OF CURVATURE OF A CONIC
AT A GIVEN POINT. P.

CENTRE OF CURVATURE AT THE VERTEX.

Fig. 3.
PARABOLA.

Fig. 4.
ELLIPSE

Fig. 5.
HYPERBOLA

EVOLUTE

PROBLEM 54.
TO DETERMINE
THE EVOLUTE
OF A CONIC —
AN ELLIPSE.

Fig. 6

54a. Pascal's and Brianchon's Theorems.

These theorems from Projective Geometry provide constructions for conic curves which are very useful in practice, as will be shown.

(1) Pascal's Theorem.—Any six points such as *a b c d e f* on the conic shown in fig. 1 can be joined to give the hexagon * *abcfeda*. The pairs of opposite sides are then *ab*, *fe*; *bc*, *ed*; *cf*, *ad*. These pairs intersect in the points P, Q and R respectively. Pascal's Theorem states that P, Q and R lie in a straight line. The hexagon can be taken as *abfedca*; and then another straight line such as PQR results.

A useful construction, used in Problem 54*b* on the next page, is that for a conic required to pass through a given point and to touch two tangents at given points. This construction will be approached by considering first the construction for extra points on (*a*) a conic defined by *five* points; and (*b*) a conic defined by a tangent point and three points.

(*a*) Extra points on a conic defined by five points.

Suppose the five points *a c d e f* (fig. 1) to be given, but not *b*. We wish to obtain points such as *b*.

The construction is shown in fig. 2. Join *ad* and *cf* to obtain R, as before. Draw any convenient straight line through R. Draw *fe* and *ed* to intersect this line in P and Q. Draw *a*P and Q*c* produced to intersect in b_1; b_1 is another point on the conic.

(*b*) Extra points defined by a tangent point and three other points, fig. 3.

Suppose the point *f*, fig. 1, to move towards *e* until it coincides with *e*. The line *e*(*f*)P would then be a tangent at *e*, as in fig. 3. Suppose we have also the points *a*, *b* and *d* and that we require a point such as *c*.

The construction follows from the earlier figures. Draw *ab* to intersect in P the tangent from *e*. Draw *ad*. Draw any convenient line PR to intersect *ad* in R. Draw *ed* to intersect PR in Q. Draw *b*Q to intersect *e*R produced in *c*. This point *c* is another point on the conic.

(2) Brianchon's Theorem.—The conic shown in fig. 4 has six tangents which intersect. Brianchon's Theorem states that lines joining opposite vertices of the hexagon pass through one point O.

Let us suppose that one tangent moves until it is in line with an adjacent tangent; their intersection then becomes their common tangent point on the curve. Imagine the tangent *ab*, fig. 4, to be moved into the position a_1b_1 in fig. 5, *a* coinciding with a_1. Lines joining opposite vertices still intersect in O; two, dotted, may be drawn at once to give O; and the third, through *c* and O, gives the position of a_1.

This simple construction may be used to determine the tangent point for any tangent.

EXAMPLE

Draw an ellipse, axes 8″ and 5″. Mark any six points and check Pascal's Theorem. Obtain other points on the curve. Draw any six tangents, as in fig. 4, and, after checking Brianchon's Theorem, find the positions of the tangent points, as in fig. 5.

---

* A hexagon is regarded as a figure given by joining six coplanar points no three of which lie in a straight line.

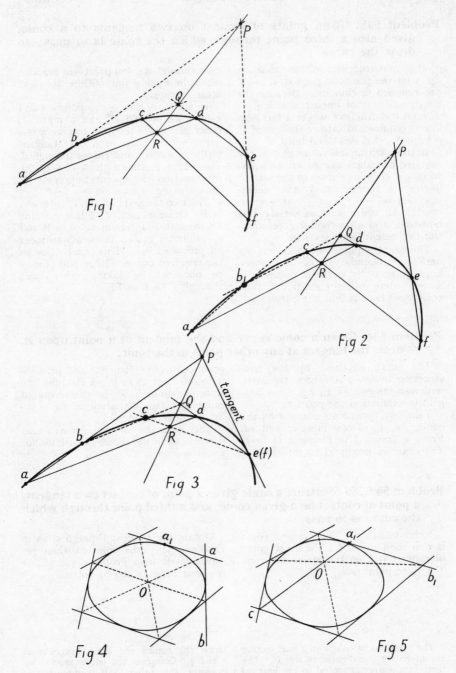

Problem 54b. Given points of contact on two tangents to a conic, given also a third point through which the conic is to pass; to draw the curve.

This construction follows that in fig. 3 on the previous page if *a* and *b* are assumed to coincide. Because the construction is of importance, e.g. in aircraft construction, where a fair outline is required to satisfy the specified conditions, it is described fully.

In conic lofting it is assumed that the required boundary curves will be satisfactory if they are curves of the second degree, i.e. if they satisfy the general equation $ax^2 + by^2 + cxy + px + gy =$ const. All conic sections satisfy this equation, and all second-degree equations represent conic sections.

The usual requirements are that an outline or boundary curve will meet lines tangentially at given points and satisfy some other conditions: that considered here is that the curve shall

pass through a given point—in practice often to enable the outline to clear some feature.

In fig. 1 are shown two points *a* and *e* on lines *ab* and *ef*, and a point *d*; other points are required to give a fair curve which meets *ab* and *ef* tangentially at *a* and *e* and passes through *d*. The curve is shown dotted, and the construction for points upon it is shown in fig. 2.

The two tangents *ab* and *ef* intersect in P. Draw *ad* and *ed*. Draw any line PS through P, intersecting *ad* in R and *ed* produced in Q. Draw *aQ* to meet *eR* produced in *c*. This point *c* lies on the required curve. Other points may be obtained by taking other lines through P such as PS₁.

Problem 54c. Given a conic curve and the tangent at a point upon it, to draw the tangent at any other point in the conic.

The curve obtained by the construction in fig. 2 has been repeated, with one tangent *ab*, in fig. 3. A tangent is required at the point *e*.

Take any two points *c* and *d* on the conic, fig. 4. Draw lines *ec* and *ed*. From *a* draw *ad* to intersect *ec* in R and draw *ac* produced to intersect *ed*

produced in Q. Join RQ and produce it to intersect *ab* in P. A straight line from *e* through P is the required tangent. The similarity of fig. 4 to fig. 2 will be noted.

The accuracy of the construction depends upon the choice of position for *c* and *d*.

Problem 54d. To construct a conic given a point of contact on a tangent, a point of contact on a given conic, and a third point through which the curve is to pass.

Refer to fig. 5. The tangent point is *e* on the tangent *ef*; the other point of contact is *c* on the given conic *ac*. The third point is *d*.

Obtain the tangent through *c*, as in Problem 54*c*; the solution then resolves itself into Problem 54*b*. The required conic is shown dotted.

EXAMPLE

The positions of points on a half section are given by coordinates in fig. 6. The three conics are tangential to the vertical at *p*, and pass respectively through *q*, *r* and *s*. They meet the horizontals tangentially.

Draw the conics and draw tangents at *r* and *q*. Complete the whole section by drawing the other half symmetrically about the vertical. (Construction lines for two points are shown dotted.)

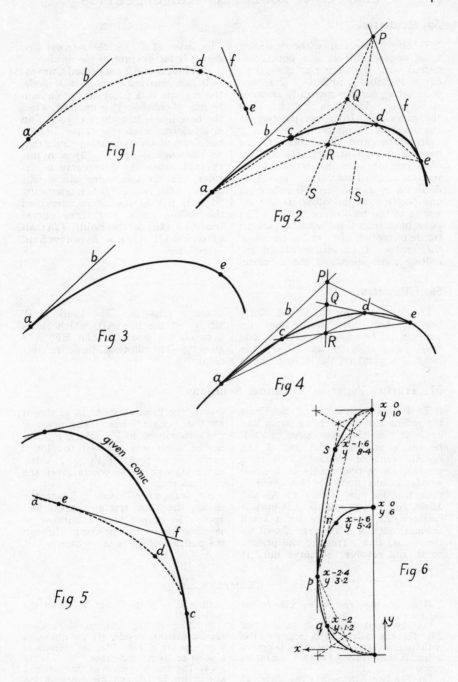

41

Fig 1

Fig 2

Fig 3

Fig 4

Fig 5

given conic

Fig 6

55. Roulettes.

If one curve roll without sliding upon another curve, any point connected with the rolling or generating curve describes a roulette. The base or directing curve is generally assumed to be fixed. The locus of P, fig. 1, as the curve rolls from one position to the other, is a roulette.

Involute and Evolute.—If the generating line is straight, the roulette of a point in the line is called the *involute* of the base curve. The locus of P, fig. 2, as the line AB rolls from one position to the other, is the involute of the base curve shown. The point of contact C is the instantaneous centre of motion of P, and is therefore the centre of curvature of an infinitely short length of the involute.

The locus of C, i.e. *the original base curve, is the* **evolute** *of the involute.*

Cycloidal and Trochoidal Curves. —If the generating curve is a circle, the roulette of a point on the circumference is called: **(1) a cycloid,** when the base line is straight, fig. 3; **(2) an epicycloid,** when the base line is circular and the generating circle rolls on the *outside*, fig. 4; **(3) a hypocycloid,** when the base curve is circular and the generating circle rolls on the *inside*, fig. 4. If the generating point is not on the circumference of the rolling circle, the three curves become: **(1a) a trochoid; (2a) an epitrochoid; (3a) a hypotrochoid** —see later.

56. Glissettes.

If a line, straight or curved, slides between two other fixed lines (or between (1) two fixed points; (2) one fixed point and one fixed line), the locus of a point on the sliding line is

called a glissette. The locus of P, fig. 5, on the line AB, which slides between the lines CD and EF, is a glissette—an elliptical curve in this case.

57. Tracing Paper or Trammel Solutions.

To draw any Roulette, fig. 6. Draw the rolling curve on tracing paper and apply it over the base curve. Let P be the generating point and A the initial point of contact. Using a pricker, i.e. a fine needle in a small wooden handle, mark the position of P; then transfer the pricker to A, and allow the curve to roll about it until it *overlaps* the fixed curve by a small amount, cutting it at B, as shown by the dotted line. Transfer the pricker to B, and revolve the curve until it

touches the fixed curve at B, as shown by the " chain " line. The point P has now moved to P_1. Mark P_1, and in the same way proceed to plot a succession of points $P_1 P_2 \ldots$. A fair curve through these points gives the required roulette.

To draw any Glissette. Draw the sliding line on tracing paper and apply the trammel in a number of suitable positions, pricking through the position of the point in each.

EXAMPLES

(Use a tracing-paper strip for the following.)

(1) The parabola CD rolls on the line AB. Plot the roulette of its focus F. (The roulette is a catenary, the curve taken by a flexible cord hanging between horizontal supports.)

(2) The line AB touches the ellipse at the end C of its latus rectum, and rolls

until it is in contact with V_1. Plot the roulette of P.

(3) A circle $1\frac{1}{2}''$ dia. revolves once around (*a*) the outside, (*b*) the inside, of a circular arc $3''$ rad. Plot the roulette of a point on its circumference.

(4) The line AB moves so that one end always lies in DE and the other on the circle centre O. Plot the locus of C.

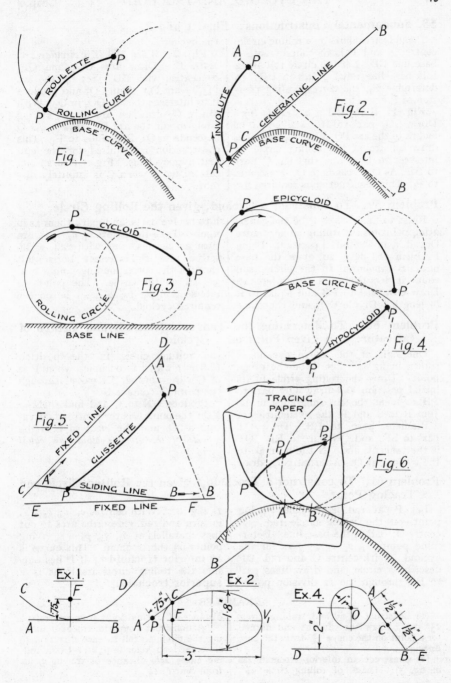

Fig. 1.

ROULETTE
ROLLING CURVE
BASE CURVE

Fig. 2.

INVOLUTE
GENERATING LINE
BASE CURVE

Fig. 3.

CYCLOID
ROLLING CIRCLE
BASE LINE

Fig. 4.

EPICYCLOID
BASE CIRCLE
HYPOCYCLOID

Fig. 5.

FIXED LINE
GLISSETTE
SLIDING LINE
FIXED LINE

Fig. 6.

TRACING PAPER

Ex. 1.

Ex. 2.

Ex. 4.

58. Fundamental Constructions. Figs. 1 and 2.

P and Q are points on a rolling circle, centre O, and P is in contact with a base line BB. Let the circle roll along this base line until Q lies in BB: to determine P_1, the corresponding position of P.

Fig. 1. *When BB is a straight line.* Draw OO_1 par¹ to BB and equal in length to the arc PQ. With O_1 as centre draw an arc of the rolling circle to intersect in P_1 a line through Q par¹ to BB. As Q descends to Q_1, P ascends to P_1. This construction is required for the cycloid.

Fig. 2. *When BB is a circular arc, centre C.* Draw arcs OO_1 and QP_1 concentric with BB. Set off the arc $PQ_1 =$ arc PQ. Join CQ_1 and produce it to intersect the arc OO_1 in O_1. With centre O_1 draw an arc of the rolling circle to cut the arc QP_1 in P_1. As Q descends to Q_1, P ascends to P_1. This construction is required for the epi- and hypocycloids. Fig. 1 is a special case of fig. 2 when C is infinitely distant.

Problem 59. To construct a Cycloid, given the Rolling Circle.

Refer to fig. 3. Let P be the generating point on the rolling circle, centre O, rad. r, in the initial position. Using Problem 28, page 20 draw the base line PB tangential to the circle, and equal in length to its circumference. Draw OO_1 par¹ to PB and equal to it in length. Divide OO_1 and the circle into twelve parts, and mark them as in figure. Take the points $c_1, c_2, c_3 \ldots$ in turn as centres, and with rad. r describe arcs to intersect horizontals through the corresponding points 1, 2, 3 . . . on the circle. The points of intersection $p_1, p_2, p_3 \ldots$ lie on the required cycloid.

Problem 60. To determine the Tangent, Normal, and Centre of Curvature at a given Point on a Cycloid.

One half of the above cycloid is shown in fig. 4. Let Q be the given point. Draw the rolling circle in its initial position and draw QR par¹ to PB to meet the circumference in R. Join R to S and P, the extremities of a diameter perp. to PB. Draw QT par¹ to SR, and QN par¹ to RP; QT is the tangent, and QN the normal, to the cycloid at Q. Alternatively, draw the rolling circle in the position (dotted) which it occupies when P is at Q: the normal QN passes through the point of contact N.

Produce QN to C and make CN = NQ; the point C is the centre of curvature at Q, and the locus of C is the *evolute of the given cycloid—an equal cycloid*

Problem 61. To construct a Trochoid,* given the Rolling Circle and Tracing Point.

Let P, at rad. r, be the tracing point, on the rolling circle rad. R centre O: BB is the base line. Determine points $c_1, c_2, c_3 \ldots$ as for the cycloid. With centre O and rad. OP describe a circle, and draw lines par¹ to BB through the 12 division points 1, 2, 3 . . . . With centres $c_1, c_2, c_3 \ldots$ in turn and rad. r describe arcs to cut these parallels at $p_1, p_2, p_3 \ldots$, giving points on the trochoid. This curve is an **inferior trochoid** †; if P lies outside the rolling circle its locus is a **superior trochoid.**‡

EXAMPLES

(1) Construct a cycloid, rolling circle $2\frac{1}{2}''$ dia. Draw the tangent and normal for a point on the curve $1\frac{3}{4}''$ from the base line.

(2) Construct an inferior trochoid (as in fig. 5). Diam. of rolling circle $2\frac{1}{2}''$, distance of tracing point from centre, $1''$.

(3) Construct a superior trochoid allowing the circle to roll in each direction to show the loop (refer to p. 49. Diam. rolling circle $2\frac{1}{2}''$, distance of tracing point from centre, $1\frac{1}{2}''$.

* The Trochoid is of importance in Naval Architecture. Wave profiles approximate to trochoids having the inverted form of fig. 5. † Or prolate cycloid. ‡ Or curtate cycloid.

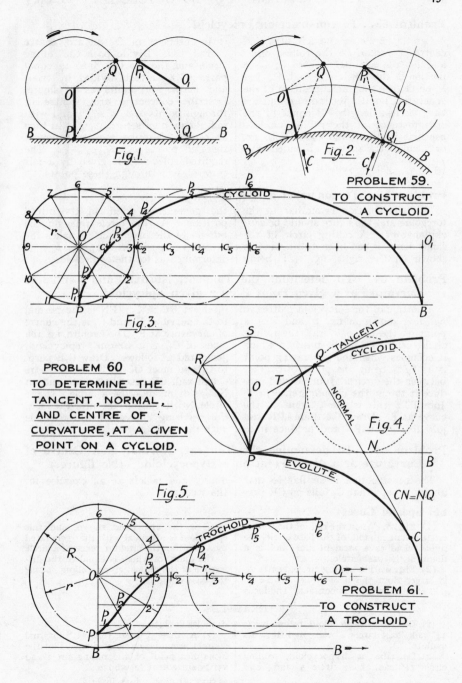

Fig.1.

Fig.2.

CYCLOID

Fig.3.

TANGENT

CYCLOID

NORMAL

Fig.4.

EVOLUTE

$CN=NQ$

Fig.5.

TROCHOID

Problem 62. To construct an Epicycloid.

Let the circle at the left of fig. 1, centre O, rad. r, be the rolling circle and let the arc BB, centre C, be a portion of the base circle, rad. R. Let P, on OC, be the initial position of the generating point. Mark off around the base circle a length PP equal to the circumference of the rolling circle— use first Prob. 28 and then Prob. 27; or calculate the angle α subtended at C, $\left(\alpha = \dfrac{r}{R} \cdot 360°\right)$. Join CP and produce it to intersect in O_1 an arc, centre C, rad. CO. Divide both the rolling circle and the arc OO_1 into 12 equal parts (OO_1 is best divided by trial) and mark the points as in figure. Describe concentric arcs, centre C, through points 1, 2, 3 . . ., and with rad. r and points c_1, c_2, c_3 . . . in turn as centres describe arcs to intersect them in points p_1, p_2, p_3 . . . . The required epicycloid is given by a fair curve drawn through these points.

Problem 63. To construct a Hypocycloid.

This construction is similar to the foregoing and the figure should be self-explanatory. A rolling circle of the same diam. r has been taken, and is shown on the right, fig. 1, P being the generating point in its initial position. As the circle rolls on the *interior* of the base circle, P takes the successive positions, p_1, p_2, p_3 . . . determined as for the epicycloid.

Problem 64. To determine the Tangent, Normal, and Centre of Curvature at a given Point P on a given Epicycloid.

A portion of the epicycloid plotted in fig. 1 is shown in fig. 2, and P is a point on the curve. Draw the rolling circle, centre O, in the position which it occupies when the generating point is at P, N being the point of contact between the circles. Join CO and produce it to cut the circumference in T. Join PT: this is the tangent to the curve at P. Draw the diam. PR and join RC. Join PN and produce it to intersect RC in C_1; PN is the normal to the curve at P, and C_1 is the centre of curvature at P. The evolute, i.e. the locus of C_1, is a *similar* * epicycloid, generated as follows. Draw C_1E perp. to C_1P to meet OC in E; with centre C and rad. CE describe an arc and on NE as diam. describe a circle. If the circle NC_1E roll on the arc, rad. CE, the resulting epicycloid will coincide with the evolute.

Problem 65. To determine the Tangent, Normal, and Centre of Curvature at a given Point on a Hypocycloid. (No figure.)

The construction is similar to that above, except that C_1 falls on CR pro- duced, and is left as an exercise for the student.

66. Special Cases.

(1) **Fig. 3.** When the rad. of the rolling circle is half the rad. of the base circle, the hypocycloid is a straight line and is a diam. of the base circle.

(2) **Fig. 4.** The rolling circle, centre A, is larger than the base circle, centre B, and they have *internal* contact. The locus of P is the epicycloid shown, and this epicycloid is identical with that generated by a point on the circumference of a circle C, the diam. of which is equal to the difference of the diams. given, rolling *outside* the base circle B.

EXAMPLES

(1) Describe an epicycloid, rolling circle $1\frac{1}{4}''$ rad., base circle $4''$ rad., and draw its evolute.

(2) Describe a hypocycloid, rolling circle $1\frac{1}{2}''$ rad., base circle $5''$ rad., and draw its evolute.

(3) A circle $5''$ diam. rolls with internal contact around a circle $4''$ diam. Plot the locus of a point on the rolling circle and verify statement (2) above.

* Two epicycloids are similar when the ratio $\dfrac{\text{rolling circle rad. of 1st}}{\text{rolling circle rad. of 2nd}} = \dfrac{\text{base circle rad. of 1st}}{\text{base circle rad. of 2nd}}$.

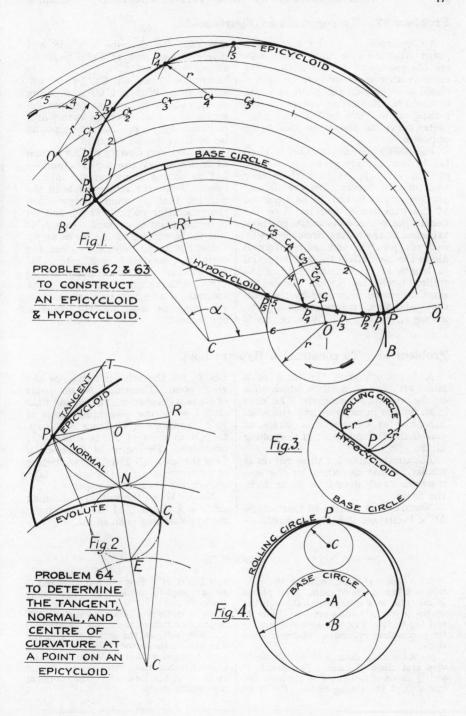

THE EPICYCLOID AND HYPOCYCLOID

EPICYCLOID

BASE CIRCLE

HYPOCYCLOID

Fig. 1.

PROBLEMS 62 & 63
TO CONSTRUCT
AN EPICYCLOID
& HYPOCYCLOID.

TANGENT

EPICYCLOID

NORMAL

EVOLUTE

Fig. 2

PROBLEM 64
TO DETERMINE
THE TANGENT,
NORMAL, AND
CENTRE OF
CURVATURE AT
A POINT ON AN
EPICYCLOID.

ROLLING CIRCLE

HYPOCYCLOID

BASE CIRCLE

Fig. 3.

ROLLING CIRCLE

BASE CIRCLE

Fig. 4.

Problem 67. To construct an Epitrochoid.

An epitrochoid is the locus of a point attached to a circle which rolls on the *outside* of a base circle. The locus is an *inferior* or *superior* epitrochoid according as the point is *within* or *outside* the circumference of the rolling circle. One half of each complete curve is shown in the figure opposite.

The following construction is similar to that already given for the epicycloid and applies to both inferior and superior curves. Let the circular arc BB, centre C, rad. R, be a portion of the base circle, and let P be the generating point in its initial position, attached to the rolling circle, centre O, rad. r. Join OC and set off from B along the base circle the arc BB equal in length to the circumference of the rolling circle, i.e. $= 2\pi r$. Draw the arc OO, centre C, limited by the radii CBO, and divide it into 12 equal parts, giving $c_1, c_2, c_3 \ldots$. With centre O

and rad. OP describe a circle and divide it also into 12 equal parts as shown, numbered 1, 2, 3 . . .; with centre C and rad. C1, C2, C3 . . . describe arcs. With rad. OP and points c_1, c_2, c_3 . . . in turn as centres describe arcs to intersect the former arcs in points p_1, p_2, p_3 . . .; these points lie on the required trochoid.

Normal and Centre of Curvature at a Point on the Epitrochoid.— Let p_4 be the point, on the inferior curve. Draw the rolling circle in the position that it would occupy when the generating point is at p_4, and let N be the point of contact. Join p_4N: this is the normal to the curve at p_4. Draw NR perp. to p_4N to meet the diam. p_4c_4 produced in R. Join RC and produce p_4N to intersect RC in C_1: C_1 is the centre of curvature of the epitrochoid at p_4. A similar construction may be used for the superior epitrochoid.

Problem 68. To construct a Hypotrochoid.

A hypotrochoid is the locus of a point attached to a circle which rolls on the *inside* of a base circle. The locus is an *inferior* or *superior* hypotrochoid, according as the point is *within* or *outside* the circumference of the rolling circle.

The construction for these curves is similar to that described for the epitrochoids, and should be clear from the figure.

Normal and Centre of Curvature at a Point on the Hypocycloid.—

Let p_4, on the inferior curve, be the given point. Determine N, the point of contact between the rolling and base circles when the generating point is at p_4, and join p_4N. Draw NR perp. to p_4N to meet in R the diam. p_4c_4 produced. The point of intersection C_1 of the lines p_4N and CR is the centre of curvature at p_4: in the figure C_1 falls outside the limit of the paper.

Note. When the rad. of the rolling circle $= \frac{1}{2}$ the rad. of the base circle, the hypotrochoid is an ellipse.

EXAMPLES

(1) A circle 2″ diam. rolls on the outside of a base circle 6″ diam. Two points P and Q lie on a straight line through the centre O of the rolling circle. OP is $\frac{3}{4}$″ and OQ is $1\frac{1}{2}$″. Plot the loci of P and Q for a complete revolution of the rolling circle.

(2) A circle $2\frac{1}{2}$″ diam. rolls on the inside of a base circle 8″ diam. Two points P and Q lie on a straight line through the centre O of the rolling circle. OP is 1″

and OQ is $1\frac{1}{2}$″. Plot the loci of P and Q for a complete revolution of the rolling circle.

(3) A circle A 4″ diam. rolls along XY in the direction of the arrow. A circle B 2″ diam. rolls on the inside of A in the direction of the arrow. Both circles move with uniform velocity. When the point C reaches XY the circles are in contact at C. Plot the locus of O, the centre of the smaller circle.

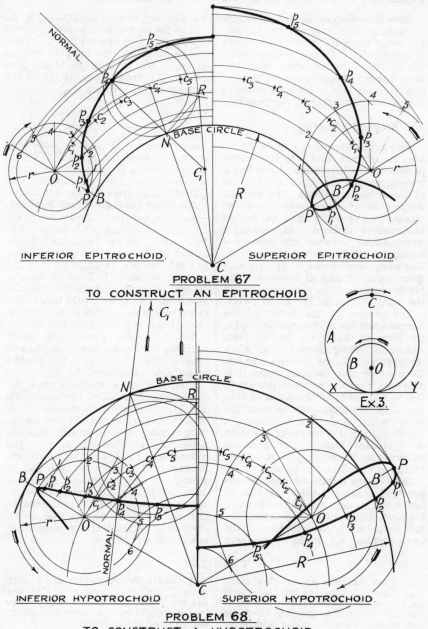

NORMAL

P_5

P_4

c_5
c_4
c_3
c_2
P_3

c_1

R

N

BASE CIRCLE

C_1

R

r

O

P_1
P_2
P
B

6 3 4 3 2 1

INFERIOR EPITROCHOID.

P_5

c_5
c_4
c_3

P_4

c_2

4 3 5

c_1
P_3

r

O

B P_2
P P_1

SUPERIOR EPITROCHOID.

PROBLEM 67
TO CONSTRUCT AN EPITROCHOID

C_1

C

A

B O

X Y

Ex.3.

BASE CIRCLE

N

R

2

3

c_5
c_4
c_3

2

c_1
1

c_2
3
4

c_4
c_3

c_5

c_2

B P_1
P_2
P_3
c_1
P_4
P_5

B P
p_1
p_2
p_3

r

O

5

5

6

NORMAL

C

R

P_4
P_5

6

INFERIOR HYPOTROCHOID

SUPERIOR HYPOTROCHOID.

PROBLEM 68.
TO CONSTRUCT A HYPOTROCHOID.

Problem 69. To draw an Involute of a given Circle.

Determine graphically, as in Problem 28, page 20, the length of the circumference of the given base circle, centre O, and divide it into, say, 12 equal parts. Divide the circumference into the same number of equal parts, 1, 2, 3 . . ., and at these points draw tangents to the circle. Mark off along the tangents,

successively, lengths equal to $\frac{1}{12}$, $\frac{2}{12}$, $\frac{3}{12}$, . . . of the circumference, giving points p_1, p_2, p_3 . . . on the required involute.

The normal to the involute at Q is given by the tangent QN to the base circle; QT drawn perp. to QN is the tangent to the involute at Q.

70. Toothed Gearing.

Two plain wheels, A and B, fig. 2, are in contact and revolve. about parallel axes, the motion of one being transmitted to the other by friction at the rubbing surfaces. To prevent slipping at P when power is to be transmitted, grooves may be cut in the rubbing surfaces, and projecting strips added between the grooves— forming the **gear teeth** shown in fig. 3. The imaginary circles in fig. 3 corresponding to A and B are called the **pitch circles** of the gear wheels, and the point P is called the **pitch point.**

The profiles of the teeth will be correct when the motion transmitted is the same as that given by the plain wheels in rolling contact; i.e. when the angular velocity ratio is constant. To produce this constant ratio it is necessary that the common normal to the teeth profiles at the point of contact should always pass through the pitch point, and this condition is satisfied if the profiles are of **involute** or **cycloidal** form. For practical reasons the involute type is almost exclusively used at the present time.

In involute forms the path of the point of contact, called the **line of action,** is a straight line passing through the pitch point, and is tangential to the base circles of the involutes forming the teeth profiles: e.g. in fig. 5 the teeth

have contact at Q, and the locus of Q coincides with the normal QN and passes through P, the pitch point; the involute base circles are tangential to QN and concentric with the pitch circles.

To draw Involute Teeth Profiles. —Let the two pitch circles be given, fig. 5, P being the pitch point. The base circles of the involutes are obtained as follows. Through P draw a line NN making an angle θ* with the common tangent TT. Draw circles concentric with the pitch circles and tangential to NN: these are the base circles of the involutes and NN is the common normal at the point of contact Q.

Using tracing paper, plot short portions of involute curves to the two base circles: these give the profiles of suitable wheel teeth.

The part of a tooth beyond the pitch surface is called the addendum (fig. 4), and that within, the dedendum or root. Measured radially, the dedendum is made a little greater than the addendum to give clearance. The overall height and width of a tooth are proportioned from considerations of strength.† The addendum and dedendum circles give the limits of the involute curve required. If the dedendum circle is smaller than the base circle, the outline of the tooth between them is a straight radial line.

EXAMPLES

(1) Draw the involute of a circle 2″ diam.

(2) The pitch circles of two gear wheels are 11″ and 15″ diam. If $\theta = 17°$ determine the shape of a complete tooth for each wheel, full size, taking addendum

$= \frac{1}{2}″$, dedendum $\frac{9}{16}″$, width along pitch circle $\frac{3}{4}″$ (i.e. circular pitch $= 1\frac{1}{2}″$). Cut out the teeth in cardboard, using strips long enough to include the wheel centres, and plot the line of action.

---

* The angle of $14\frac{1}{2}°$ was at one time largely used for θ, but 20° is now the standard for a number of gear-tooth systems. When $\theta = 14\frac{1}{2}°$ interference occurs between pinions having less than 32 teeth and a rack.

† For a full discussion refer to works on Machine Design.

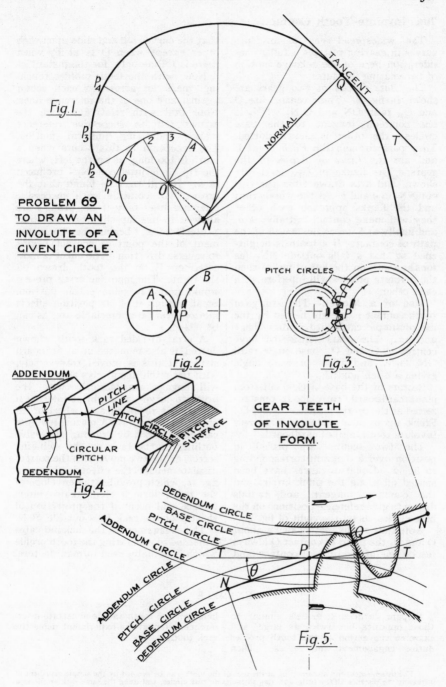

Fig.1.

**PROBLEM 69
TO DRAW AN
INVOLUTE OF A
GIVEN CIRCLE.**

TANGENT

NORMAL

P_4 P_3 P_2 P_1

0 1 2 3 4

O

N

Q

T

PITCH CIRCLES

A B P

Fig.2.

Fig.3.

**GEAR TEETH
OF INVOLUTE
FORM.**

ADDENDUM

PITCH
LINE

PITCH

CIRCULAR
PITCH

DEDENDUM   Fig.4.

PITCH CIRCLE   PITCH SURFACE

DEDENDUM CIRCLE
BASE CIRCLE
PITCH CIRCLE
ADDENDUM CIRCLE

ADDENDUM CIRCLE
PITCH CIRCLE
BASE CIRCLE
DEDENDUM CIRCLE

T   P θ Q   T   N

N

Fig.5.

70a. Involute Tooth Contact.

The widespread use of involute curves in gearing justifies a fuller consideration here of the relative motion of two mating involutes.

The *Base Circles* of two gears are shown in fig. 2. Their centres are O and O_1, radii ON and O_1N_1. NN_1 is one common tangent to the base circles; the other is shown dotted. They pass through the *Pitch Point* P and are the *Lines of Action* of the gears. The flanks of two teeth are shown, and arcs drawn from the tips cut NN_1 in *a* and *b*. As the wheels turn and the flanks approach each other they will make contact initially at *a*, and finally at *b*; *ab* is the length of the path of contact. If a tooth is heightened so that *a* falls outside NN_1 (as for the tooth on the left of fig. 1) then the involute profiles will interfere with each other.

Use of a Model.—The study of tooth contact is greatly helped by the use of simple cardboard models, figs. 1 and 1*a*. The gears considered have centres at O and O_1, base circle radii ON, O_1N_1, and a 20° pressure angle giving a pitch point P.

Sectors of the base circles, cut from stout cardboard, turn about pins inserted at the centre positions O and O_1. Stuck upon each sector is a correct involute tooth, also of cardboard.

The two sectors are placed in position over a diagram corresponding to fig. 2. Equal distances have been spaced off along the pitch circles and the common tangent, and radials drawn to give reference positions on the base circles, as on the right of fig. 1.

Note that as the sectors swing about O and O_1 the point of contact Q always lies on one of the lines of contact, and

that the curves roll and slide upon each other except when Q is at P, when there is rolling only, for the instant.

Now, with the teeth profiles touching, mark an arrow on each sector against any one of the reference lines. Note that with relative motion the arrows fall on successive reference marks, indicating uniform motion. The exception to this occurs when a tooth is too long, as on the left, where the tip circle intersects N_1N produced in *a*. It will then be found that the arrows will coincide with successive reference lines only if the profiles are allowed to overlap slightly; and that "interference"[*] begins when the movement of the point of contact begins to reverse direction. The effect is indicated on E on the tooth drawn on the right. The manufacturing process would normally remove this portion. Small though it is, its position effects motion over an appreciable arc, as can be tested.

A straight-sided rack tooth, shown dotted, is also mounted on a cardboard strip. If this is moved transversely, with the pitch line always along TT, it will engage satisfactorily with the involute curve—as might be inferred if the rack is regarded as a wheel of infinitely large radius. A mating involute can be obtained by drawing successive outlines of the rack on the swinging sector, as the two move with the correct displacements; the effect is shown in fig. 1*b*, which provides the envelope of the tooth form. Here again, interference will occur if the projection of the tip of the rack falls outside of N.

The engagement of the rack indicates a method of generating the tooth profile which is actually used in manufacture.

EXAMPLE

Prepare cardboard models similar to those opposite but twice as large, and **examine** the action of the tooth profiles **during** engagement and recess. Then heighten one tooth and demonstrate interference. Repeat the experiment using the rack tooth.

[*] The interference, and undercutting at the root of the tooth, can be avoided by the simple expedient of increasing the distance OO_1 a little, adhering to the same base circles, and using the same rack profile as a generator for the teeth.

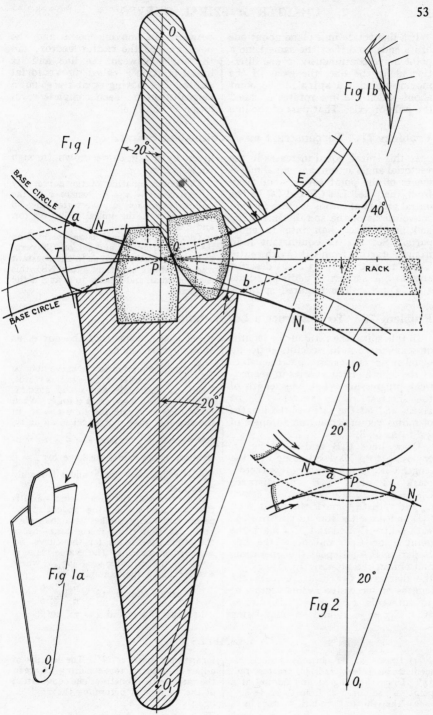

Fig 1

Fig 1b

BASE CIRCLE

BASE CIRCLE

20°

20°

40°

RACK

Fig 1a

Fig 2

20°

20°

If a line rotate in a plane about one of its ends, and if at the same time a point move continuously in one direction along the line, the locus of the moving point is a **spiral.** The point about which the line rotates is called the **pole** or axis. That part of the line between the moving point and the pole is called the **radius vector,** and the angle between the line and its initial position is called the **vectorial angle.** The moving point traces out a **convolution** for each complete revolution of the line.

Problem 71. To construct an Archimedian Spiral.

In this spiral, equal increases in the vectorial angle accompany equal movements of the point towards, or away from, the pole. Let CP and CQ be the initial and final radius vectors for two convolutions of the spiral. Bisect PQ and divide each half into 12 equal parts. Set out 12 equidistant radii from C, starting from CP. With centre C and radii C1, C2, C3 . . . , describe arcs to cut successive radii in points 1, 2, 3 . . . . The required spiral is given by a fair curve drawn through these points.

The polar equation of the curve may be written $r = a + c\theta$, where r is the radius vector for any vectorial angle θ (radians), a is the initial radius vector, and c is a constant.

Tangent and Normal at any point S *on the curve.*—Join CS and set off CN perp. to CS and equal in length to the constant c in the above equation. Join NS: this is the normal, and a perp. to it at S is the tangent.

Problem 72. To construct a Logarithmic Spiral.

In this spiral the ratio of the lengths of consecutive radius vectors enclosing equal angles is constant ; i.e. the values of the lengths form a series in geometrical progression. Let the length of the shortest radius vector, CP, be given, and let the ratio of the lengths of radius vectors enclosing an angle of 30° be as 9 : 8.

Draw lines C_1B and C_1D, fig. 2, enclosing 30°. Mark off along C_1D the length C_1P_1 of the given radius vector; mark off along C_1B the distance $C_1A = \frac{8}{8}C_1P_1$. Join AP_1. With centre C_1 and rad. C_1A mark the point 1 on C_1D and draw the line $1a$ parl to P_1A; with centre C_1 and rad. C_1a mark the point 2 on C_1D and draw the line $2b$ parl to P_1A. Repeat this operation, and obtain the points 1 to 12 on C_1D: the distances $C_1 1$, $C_1 2$ . . . are the lengths of successive radius vectors at 30° intervals, for $C_1 1 = \frac{8}{8}C_1P_1$, $C_1 2 = \frac{8}{8}C_1 1$, $C_1 3 = \frac{8}{8}C_1 2$, and so on. Using these lengths construct the curve, as shown clearly in fig. 3.

The polar equation to the curve may be written $r = a^\theta$, where r is the radius vector, θ the vectorial angle, and a is a constant. Hence $\log r = \theta \log a$, i.e. $\theta \propto \log r$. When $\theta = 0$, $\log r = 0$, and therefore $r = 1$. In the example given CP is taken as unity, and when $\theta = \frac{\pi}{6}$ radians, $r = \frac{9}{8} \times 1$. Substituting these values we have $\log \frac{9}{8} = \frac{\pi}{6}$ $\log a$, i.e. $\log a = \frac{6}{\pi} \log \frac{9}{8}$, and a is readily calculated.

A property of the logarithmic spiral is that the angle α which the tangent at any point on the curve makes with the radius vector at that point is constant—hence the alternative name for the curve, *the equiangular spiral.* The value $\tan \alpha$ is given by the ratio $\log_{10} e \div \log_{10} a$, where $e = 2.718$; in the example,

$$\tan \alpha = \log_{10} 2.718 \div \frac{6}{\pi} \log_{10} \frac{9}{8},$$

i.e. $\tan \alpha = 4.45$ and $\alpha = 77° \; 20'$.

EXAMPLES

(1) Draw two convolutions of an Archimedian spiral, least rad. $\frac{1}{2}''$, greatest rad. $3\frac{1}{2}''$. Draw the tangent and normal at a point on the curve 2″ from the pole.

(2) The shortest radius vector in a logarithmic spiral is $1\frac{1}{2}''$. The lengths of adjacent radius vectors enclosing 30° are in the ratio 9 : 10. Construct one convolution of the curve and determine the angle α, fig. 3.

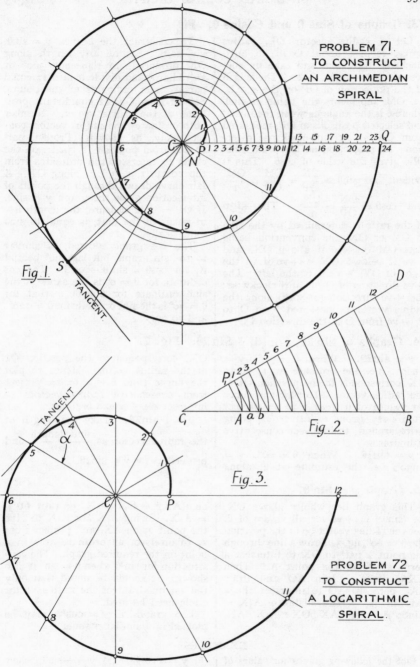

PROBLEM 71.
TO CONSTRUCT
AN ARCHIMEDIAN
SPIRAL

Fig. 1.

TANGENT.

Fig. 2.

Fig. 3.

PROBLEM 72
TO CONSTRUCT
A LOGARITHMIC
SPIRAL.

TANGENT

73. Graphs of Sine θ and Cosine θ. Fig. 1.

Let a radius vector OP, radius unity, revolve about O in an anti-clockwise direction, and take up successive positions O1, O2 .... The length of the projection of OP on the y axis, i.e. OM, represents the value of $\sin\theta$, where θ is the angle between the initial and selected positions of OP, measured from OX in an anti-clockwise direction; its projection on the x axis, i.e. ON, gives the value of $\cos\theta$. This is evident, for $\sin\theta = \dfrac{PN}{OP} = \dfrac{OM}{OP} = \dfrac{OM}{1}$;

and $\cos\theta = \dfrac{ON}{OP} = \dfrac{ON}{1}$. The **sign** of the ratio is determined by the sign of OM or ON, the convention being that OM is +ve if above OX, and —ve if below; ON is +ve if to the right of OY, —ve if to the left. The +ve direction of OP is anti-clockwise; the +ve direction (or way) along the radius vector is measured from O to P, —ve from O in the direction PO.

To construct the curves $y = \sin\theta$, $y = \cos\theta$. Set off any length along OX produced, to represent 360° or 2π radians, and divide it into 12 equal parts; erect ordinates at the points. Set off from OX 12 equidistant positions of the radius vector. Number both ordinates and radius vectors consecutively, as shown. Project par[l] to OX from points 1, 2, 3 . . . on the circle to intersect the ordinates from the points 1, 2, 3 . . . along OX. A fair curve drawn through the points of intersection gives the graph $y = \sin\theta$. If OP = unit measure, the length np, in terms of the unit, is equal to $\sin\theta$ numerically.

The cos graph, dotted, is similar to the sin graph, but lags 90° behind it, for $\cos\theta = \sin(\theta + 90°)$. The cos ordinate for θ = 0° is the same as the sine ordinate for θ = 90°; that for θ = 30° is the sine ordinate for θ = 120°; and so on.

74. Graphs of Sin 2θ, Sin ½θ, ½ Sin 2θ. Fig. 2.

$y = \sin 2\theta$. When θ = 30°, $y = \sin 60°$; i.e. the ordinate at 1, along OX, corresponds to the position O2 of the radius vector. To plot the curve for values of θ from 0 to 2π project from *every second radius vector*, going twice around, to intersect consecutive ordinates.

$y = \sin\frac{1}{2}\theta$. When θ = 60°, $y = \sin 30°$, i.e. the ordinate at 2, along OX, corresponds to the position O1 of the radius vector. Hence to plot the curve from θ = 0 to 2π project from consecutive radius vectors to intersect *every second ordinate*.

$y = \frac{1}{2}\sin 2\theta$. Take the length of the radius vector as $\dfrac{\text{unit length}}{2}$ and proceed as for $y = \sin 2\theta$.

75. Graph $y = \text{Sin}^2\theta$.

This graph lies wholly above OX, for $(\sin\theta)^2$ is +ve for all values of θ. Let the radius vector be in the position O2, θ = 60°, fig. 3. Draw a line through the point 2 par[l] to OX to intersect a perp. at X in the point A. Then AX = $\sin 60°$. Join AO and draw AA₁, perp. to AO to intersect the x axis in A₁. The distance $A_1X = (\sin 60°)^2$; for AX : OX :: A_1X : AX,

i.e. $(AX)^2 = OX \cdot A_1X$, or $(\sin 60°) = A_1X$. Swing A_1X about X to give the point 2₁ on AX, and project from 2₁ to intersect an ordinate at 2 in a point on the required graph. The construction for $\sin^2\theta$ when θ = 30° is also shown. It should be noted that only the 1st quadrant of the radius vector circle need be used.

The graph of $y = \text{cos}^2\theta$ may be plotted in a similar manner.

EXAMPLES

Plot the following graphs for values of θ from 0 to 2π taking $1''$ as the unit.

(1) $y = \sin\theta$; (2) $y = \cos\theta$; (3) $y = 2\sin 2\theta$; (4) $y = 2\cos 2\theta$; (5) $y = 2\sin\frac{1}{2}\theta$; (6) $y = 2\cos\frac{1}{2}\theta$; (7) $y = -\sin\theta$; show that this is the same as the graph $y = \sin(\theta + 180°)$; (8) $y = \sin^2\theta$; (9) $y = \cos^2\theta$.

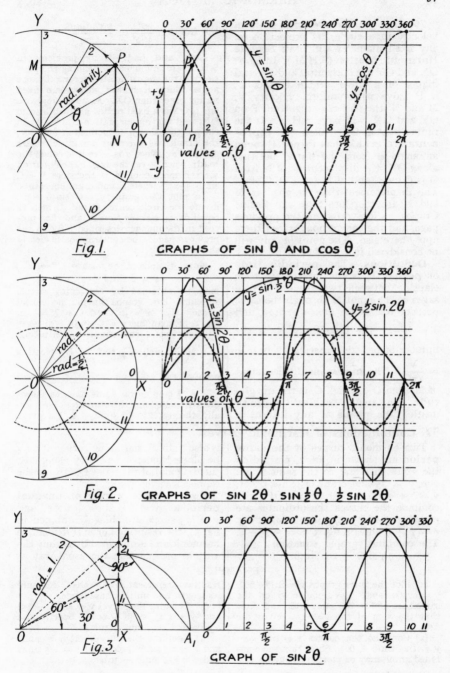

Fig.1.   CRAPHS OF SIN θ AND COS θ.

Fig.2.   GRAPHS OF SIN 2θ, SIN ½θ, ½ SIN 2θ.

Fig.3.   GRAPH OF SIN² θ.

76. If a point move with uniform velocity in a circle, its projection on any fixed diameter will have **Simple Harmonic Motion** (S.H.M.). In fig. 1, OX and OY are horizontal and vertical diams. of a circle, centre O rad. C; if P move with uniform velocity V, M and N, the feet of perps. from P on OY and OX, will have S.H.M. If the radius vector OP, called the **representative crank,** has turned through an angle φ from the initial position along OX, the displacement of N from O is C cosφ, and of M from O, C sinφ. The displacement-time graphs for M and N are therefore Sine and Cosine Graphs, as illustrated on the previous page, and the conventions stated there apply here also. The point M only will be considered in what follows.

Definitions.—The **periodic time** or **period** T, is the interval of time elapsing between two successive passages of the point M through the same position *and in the same direction*, and $T = \dfrac{2\pi C}{V}$. The **frequency** N, is the number of periods per second; $N = \dfrac{1}{T} = \dfrac{V}{2\pi C}$. The maximum displacement of M from O, i.e. rad. C, is the **amplitude.** The angle φ is called the **phase angle:** the phase at any instant is

the fraction $\dfrac{\varphi^{o}}{360^{\circ}}$ or $\dfrac{\varphi \text{ radians}}{2\pi}$

Lead and Lag.—Many problems in S.H.M. are simplified by defining the position of the representative crank by the angle which it makes with some fixed radius other than OX. As an example, in fig. 1, OA is a fixed radius inclined at α to OX, and the representative crank OP, in the position shown, is inclined at θ to OA; the crank has therefore an initial phase angle of α. When α is +ve it is called the **lead;** when −ve, the **lag.** The equation for the motion of M is therefore $y = \text{C} \sin(\theta \pm \alpha)$,* where C and α are constants.

To plot the graph $y = \text{C} \sin(\theta + \alpha)$, divide the circle centre O rad. C into 12 equal parts, *starting from* A, and then proceed in the manner described fully on the previous page: the construction is clearly shown in fig. 1.

Note: Graphs of the form
$$y = \text{C} \sin(\theta \pm \alpha)$$
may be obtained by combining simple sine and cos. graphs having no initial phase. Let $\alpha = 30^\circ$ and $\text{C} = \frac{3}{4}''$ as in fig. 1. Expanding $y = \text{C} \sin(\theta + \alpha)$ gives C $(\sin\theta \, \cos\alpha + \cos\theta \, \sin\alpha)$; substituting, $y = \frac{3}{4}\left(\dfrac{\sqrt{3}}{2} \sin\theta + \frac{1}{2} \cos\theta\right)$. By plotting each of the graphs $y = \frac{3}{4} \cdot \dfrac{\sqrt{3}}{2} \sin\theta$ and $y = \frac{3}{4} \cdot \frac{1}{2} \cos\theta$ on the same base, as shown dotted, and adding the ordinates algebraically, the graph $y = \frac{3}{4} \sin(\theta + 30^\circ)$ is given.

77. Combinations of Harmonic Curves.

Two harmonic curves **of the same period** are shown in fig. 2, plotted on the same base and to the same scale: one, $y = \cdot 75 \sin\theta$, crank OQ; the other, $y = 1 \cdot 1 \sin(\theta + 60^\circ)$, crank OP. To combine the curves the ordinates are added, regard being given to sign: e.g. $r = p + q$ and (in fig. 3) $r = -p + q$. The combined graph, equation
$$y = \cdot 75 \sin\theta + 1 \cdot 1 \sin(\theta + 60^\circ),$$

is also a simple harmonic curve: it may be plotted independently by using OR, the diagonal of the par^m POQR, as the representative crank.

Two harmonic curves **of unequal periods,** $y = 1 \cdot 1 \sin(\theta + 60^\circ)$ and $y = \cdot 75 \sin 2\theta$, are shown combined in fig. 3: the resulting curve is *not* simple harmonic and is obtained by adding the various ordinates in the usual way.

EXAMPLES

(1) Plot the three graphs $y = \sin(\theta + 60)$, $y = \cos 60 \cdot \sin\theta$, $y = \sin 60 \cdot \cos\theta$ on the same base and to the same scale and show that the sum of the last two gives the first.

(2) Combine the curves $y = \sin\theta$ and $y = 1 \cdot 25 \sin(\theta + 60)$. Show that the combined graph may be plotted by means of a

resultant representative crank. Measure its length and initial phase angle.

(3) Combine the curves $y = \sin(\theta - 45)$ and $y = 1 \cdot 25 \sin(\theta + 60)$. Specify the equivalent resultant crank.

(4) Combine the curves (a) $y = \sin 2\theta$ and $y = 1 \cdot 5 \sin(\theta + 60)$; (b) $y = 1 \cdot 5 \sin 2\theta$ and $y = \cdot 75 \sin(\theta - 30)$.

---

* If the angular velocity of P is ω and t is the time taken, the equation may be written:
$$\text{Displacement} = \text{C}(\sin \omega t + a) = \text{C}\left(\sin \dfrac{2\pi t}{T} + a\right).$$

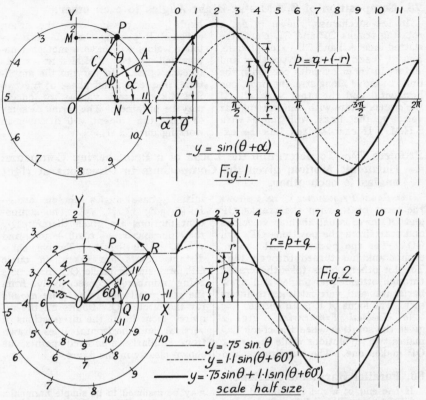

$-p = q + (-r)$

$y = \sin(\theta + \alpha)$

Fig. 1.

$r = p + q$

Fig. 2.

$y = \cdot 75 \sin \theta$

$y = 1 \cdot 1 \sin(\theta + 60°)$

$y = \cdot 75 \sin \theta + 1 \cdot 1 \sin(\theta + 60°)$
scale half size.

COMBINATION OF HARMONIC CURVES OF EQUAL PERIODS.

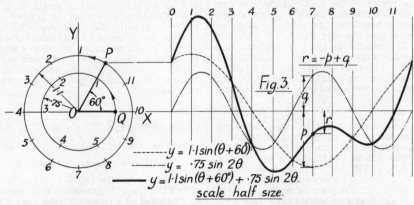

$r = -p + q$

Fig. 3.

$y = 1 \cdot 1 \sin(\theta + 60)$

$y = \cdot 75 \sin 2\theta$

$y = 1 \cdot 1 \sin(\theta + 60°) + \cdot 75 \sin 2\theta.$
scale half size.

COMBINATION OF HARMONIC CURVES OF UNEQUAL PERIODS.

78. Composition of S.H.M.s at Right Angles to each other.

In the mechanism shown in fig. 1 rotating cranks OP and OQ cause the slotted links AB and CD to reciprocate, pins at P and Q entering the slots and sliding in them. Guides constrain the links to move along straight lines, one horizontal and the other vertical. If the cranks rotate with uniform angular velocity, the links will move with S.H.M. If each slotted link be extended, one passing over the other, a pencil R passing through both extended slots will have a motion compounded of two S.H.M.s at rt. ∠s to each other. By varying the amplitude, frequency, and phase of the two S.H.M.s, a great variety of curves may be obtained. The forms of some of these the student will discover in working out Ex. 1.

Problem 79. To determine the Locus of a Point having Compound Harmonic Motion, given the Components in Directions at right angles to each other.

(a) *Equal Frequencies.* Fig. 1 shows the locus of R when the cranks OP and OQ have equal angular velocities and start from the positions OP_0 and OQ_0. For the position of R shown, each crank has turned through 180°. To plot other points take the cranks through other equal angles and draw horizontal and vertical lines to intersect. The locus of R is an ellipse.

(b) *Unequal Frequencies.* Fig. 2 gives the solution when the crank OQ makes two revolutions while the crank OP makes one. The amplitudes and initial phase angles chosen are:— P, $\frac{7}{8}''$, 30°; Q, $\frac{3}{4}''$, 45°. The component motions are given therefore by the equations $x = \frac{7}{8} \cos(\theta + 30°)$, and $y = \frac{3}{4} \sin(2\theta + 45°)$. Twelve equidistant points are taken on circle OP, and six on circle OQ, both sets being numbered consecutively from the zero positions, *going twice around* for the circle OQ. Points on the required locus lie at the intersections of vertical and horizontal lines drawn from similarly numbered points, as shown clearly in the figure.

80. Parallel Harmonic Motions.

If one end of a rod is attached to a crank rotating with uniform velocity, and the other end to a part which is free to move in a straight line passing through the centre of the crank circle, the part will have a motion which is approximately simple harmonic; an infinitely long rod would give the part true S.H.M.

The rod EF in fig. 3 is displaced in a horizontal direction by an oscillating link BD. The ends of BD are actuated by rods from the cranks OP and OQ, and the horizontal components of their motions may be assumed to be simple harmonic. *Any point in BD, or BD produced, has S.H.M.*, and the equivalent crank for the point E may be obtained as in fig. 4. *op* and *oq* are proportional in length to OP and OQ and par[1] to them; *r* divides *pq* in the same ratio that E divides BD, and *or* gives the phase of an equivalent crank and its length to scale. In other words, the motion given to EF by a crank OR, dotted, is very nearly the same as the resultant displacement given to it by the cranks OP and OQ.

EXAMPLES

(1) Compound the following harmonic motions at rt. ∠s as in figs. 1 and 2. In each take the amplitude of the horizontal S.H.M. = $1\frac{3}{4}''$ and of the vertical S.H.M. = $1\frac{1}{2}''$. (a) Frequencies equal, α = 90°, β = 0. (b) Frequencies equal, α and β = 0. (c) Frequencies equal, α = 30°, β = 45°. (d) Frequencies OQ : OP :: 2 : 1, α = 30°, β = 45°. (e) Frequencies OQ : OP :: 2 : 3, α = 30°, β = 90°.

(2) A point has a compound harmonic motion and the components in the directions OX and OY are given by the equations $x = 1.5 \cos\left(\theta + \frac{\pi}{3}\right)$ in. and $y = .9 \sin\left(2\theta + \frac{\pi}{2}\right)$ in. Plot its locus.

(3) In fig. 3 take OQ = 3″, β = 60°, BE : ED :: 2 : 3. Determine the length and position of a crank OP such that EF will have a travel of $4\frac{1}{2}''$ and be in the extreme outward position when OQ is in the position shown. Then take β = 30° and β = 15° and repeat the construction.

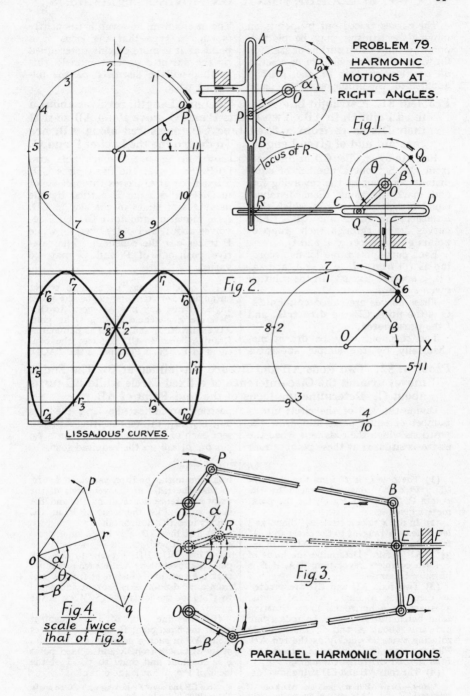

PROBLEM 79.

HARMONIC
MOTIONS AT
RIGHT ANGLES.

Fig. 1.

locus of R

Fig. 2.

LISSAJOUS' CURVES.

Fig. 4.
scale twice
that of Fig. 3.

Fig. 3.

PARALLEL HARMONIC MOTIONS.

The curves traced out by points on moving mechanisms may be plotted, point by point, from outline or skeleton drawings of the mechanism in a series of positions. This is the method adopted for the problems on this page.

The mechanism, however, is frequently of such a type that the locus of a point on it is more quickly determined by the aid of a paper trammel: this method will be discussed on the following page.

Problem 81. A straight Line AB, of indefinite Length, revolves about a fixed Point O, in AB. Two Points P and Q move along AB so that their Distances from a fixed Line CD, measured along AB, are constant and of given Length l. To determine the Loci of P and Q.

Refer to fig. 1. Let CD be distant d from O. Set out a number of equidistant positions of the revolving line and mark off along them distances equal to l, from points in which they intersect the fixed line CD. Fair curves drawn through each group of points give the loci of P and Q.

Each curve has three forms according as (1) l is $> d$, (2) $l = d$, and (3) l is $< d$. (See Ex. 1.) In the solution given l is $> d$.

These curves are called **conchoids**; O is the **pole**, CD the **directrix**, and l the **parameter**.

The conchoid may be drawn mechanically by the simple apparatus shown in fig. 3. The movable arm OEP slides over the fixed piece CD; a fixed pin at O passes through a slot in OEP, and a pin E carried by the moving arm slides in the slot in CD. As E moves up and down CD, OEP revolves about, and slides over, O, while P traces out the conchoid. The relative positions of P and O may be varied.

The conchoid may be used to trisect an angle. Let AOD, fig. 2, be the given angle. Draw any line AD perp. to OD. Describe a conchoid having O as pole, AD as directrix, and 2 . AO as parameter. Draw AP par¹ to OD to meet the conchoid in P. Then PO trisects the ∠AOD.

Problem 82. Two Rods AB and BC are pin-jointed at B. The End A moves around the Circumference of a fixed Circle while BC turns about C. Determine the locus of the mid-Point of AB.

Divide the ◯ᶜᵉ of the circle into a number of equal parts and draw the positions which the rods take when the end A is situated at these points: four positions are indicated in fig. 4, the lines 00_1, 11_1, 22_1, 33_1, representing AB. Bisect each of these lines: the points p_0, p_1, p_2, p_3 are on the required locus.

EXAMPLES

(1) The pole O is 1″ from the directrix CD. Plot the conchoids traced by the points P, Q, P_1, Q_1, P_2, Q_2 for the parameters given.

(2) In fig. 4 take a circle $2\frac{3}{4}$″ diam. and let C lie on a tangent to the circle and $5\frac{1}{4}$″ from the point of contact. AB = BC = $3\frac{1}{8}$″. AP = BP. Determine the locus of P for a complete revolution of A, if B is initially below AC.

(3) Two rods AB and CD are pivoted at their ends A and C to fixed points, 3″ apart on a horizontal line. Starting from horizontal positions the rods swing outwards about A and C, the rod CD rotating twice as quickly as the rod AB. Plot the locus of the point of intersection P as the rod CD turns through 90°.

(4) The rods AB and CD swing inwards, from the initial position shown in figure, so that the ratio of the velocities of the point of intersection P along DC and BA is as 3 : 2. Plot the locus of P, as DC swings into the horizontal.

(5) The line AB, 5·2″ long, moves par¹ to itself and with its end B always in the fixed line BD; BD is perp. to AB. A point P moves along AB in such a manner that the sum of its distances from A and from a fixed point F is contant and equal to 5″. Plot the locus of P.* What is the curve?

(6) The line AB, 6″ long, revolves about the fixed point B. A point P moves along AB in such a manner that the sum of its distances from A and a fixed point F is constant and equal to 5″. Plot the locus of P.* What is the curve?

* *Hint.*—Draw AB in any position. Mark off AC = 5″. Join CF and draw FP making with FC an angle equal to FCP. P is the position of the moving point for the position of the line chosen.

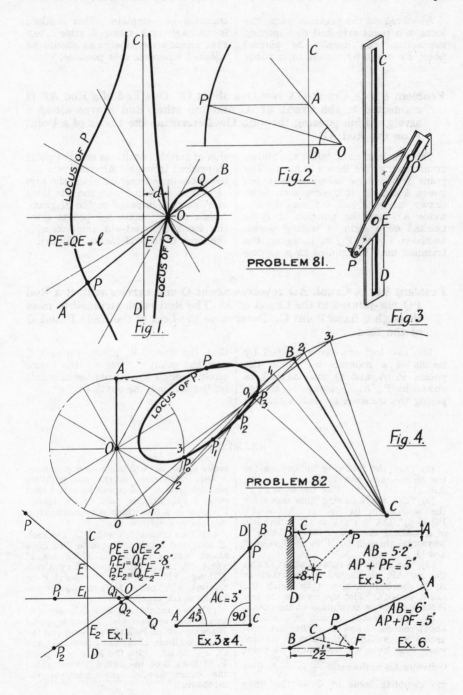

LOCUS OF P

LOCUS OF Q

$PE = QE = l$

d

C

B

Q

O

E

D

P

A

Fig. 1.

Fig. 2.

C

P

A

D

O

PROBLEM 81.

C

O

E

P

D

Fig. 3.

A

P

B

2

3

1

LOCUS OF P

O

O_1

P_3

P_2

P_1

P_0

3

2

1

C

Fig. 4.

PROBLEM 82

P

C

$PE = QE = 2''$
$PE_1 = QE_1 = .8''$
$P_2E_2 = Q_2E_2 = 1''$

E

P_1

E_1

Q_1

O

Q_2

Q

E_2

D

P_2

Ex. 1.

D

B

P

$AC = 3''$

A

45°

90°

C

Ex. 3 & 4.

B

C

P

A

$AB = 5.2''$
$AP + PF = 5''$

.8″

F

D

Ex. 5.

A

C

P

B

$AB = 6''$
$AP + PF = 5''$

$2\frac{1}{2}''$

F

Ex. 6.

As stated on the previous page, the locus of a point attached to a moving mechanism can usually be plotted point by point by means of a paper trammel or template. This method is quicker and more flexible than that already described and should be adopted wherever it is possible.

Problem 83. A Crank OA revolves about O. One End of a Rod AB is connected to the Crank at A, and the other End moves along a straight Line passing through O. Determine the Locus of a Point P on the Rod AB.

This mechanism is called the slider-crank pair, and is shown in fig. 1. The point P has been taken midway between A and B. If the crank circle be drawn, centre O, rad. OA, and the distance AB and the position of P be marked on a strip of tracing paper, as shown by A_1, P_1, B_1 in figure, the trammel may be applied in a succession of suitable positions and the point P_1 pricked through. Alternatively the distances may be set off along the *edge* of a strip of paper and the positions of P_1 marked in pencil on the diagram. A fair curve joining the points gives the locus required—in this example the locus is an oval (not an ellipse).

Problem 84. A Crank AO revolves about O and carries with it a Rod PQ pin-jointed to the Crank at A. The Rod is constrained to pass through a fixed Point C. Determine the Loci of the Ends P and Q of the Rod.

The two loci are easily plotted by means of a trammel on which the points P, A and Q are marked as shown by P_1, A_1, Q_1 in fig. 2. By applying the trammel in such a manner that the line Q_1P_1 passes through C and the point A_1 lies on the crank circle through A, any number of points on the loci may be obtained.

EXAMPLES

(1) Take the following dimensions for the slider-crank pair in fig. 1: OA, 2″, AB, $4\frac{1}{2}$″, AP = PB. Plot the locus of P.

(2) Take the following dimensions for the mechanism in fig. 2: AO = $1\frac{1}{2}$″, PQ = $5\frac{1}{4}$″, QA = 4″; C is $2\frac{1}{4}$″ to the left of a vertical through O and $\frac{3}{4}$″ below a horizontal through O. Plot the loci of P and Q.

(3) The crank OA rotates about O and the rod AP, pin-jointed to the crank at A, is constrained always to touch the circle, centre C. Plot the locus of the end P for a complete revolution of the crank.

(4) The two rods AO and BO_1 oscillate about centres O and O_1. They are connected by the link AB, which is vertical when both links are horizontal. A point C divides AB in the ratio $\dfrac{AC}{CB} = \dfrac{BO_1}{AO}$. Plot the complete locus of C as the links move as far as possible in each direction.

This mechanism is the simple Watt parallel motion, and it will be found that over a considerable part of its travel, C moves in a line which is approximately straight and vertical.

(5) Two cranks AO and BO_1 are each 1″ long and rotate in opposite directions about centres O and O_1 which are 4″ apart. A and B are connected by a rod, 4″ long. Plot the locus of a point P on the link, $1\frac{1}{4}$″ from A, for a complete revolution of the cranks.

(6) Two cranks AO and BO_1 are 1·3″ and 2″ long respectively. The ends A and B are connected by a rod $3\frac{1}{2}$″ long and the cranks oscillate about fixed centres O and O_1 2·1″ apart. Plot the locus of a point P, 1″ from B on the link AB, throughout the entire possible movement of the mechanism.

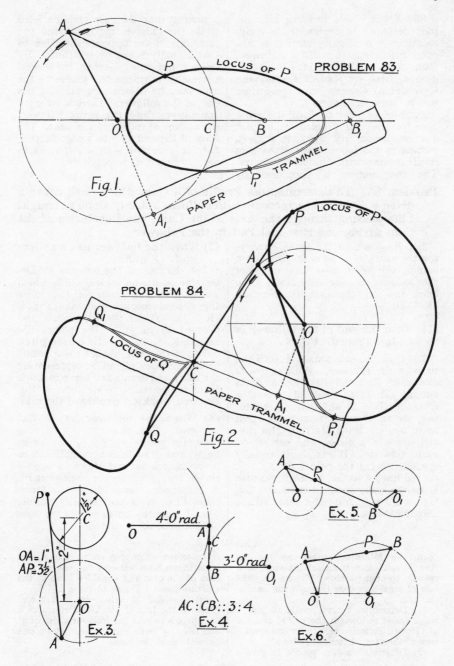

PROBLEM 83.

LOCUS OF P

Fig.1.

PAPER TRAMMEL

PROBLEM 84.

LOCUS OF P

LOCUS OF Q

PAPER TRAMMEL.

Fig.2

Ex.3.
OA=1"
AP=3½"

Ex.4.
4'-0"rad.
3'-0"rad.
AC:CB::3:4.

Ex.5.

Ex.6.

85. A cam is a rotating machine part designed to communicate reciprocating or oscillating motion to another machine part called a **follower.** Cams may be divided into two main groups, **Disc or Radial Cams,** and **Cylindrical Cams;** only the former will be considered here.

Profiles of Disc or Radial Cams. The fundamental problem is to plot a curve which will transmit a given motion to the follower when the cam itself rotates with uniform velocity. The construction is simplified by assuming that the cam remains fixed while the follower moves around the cam axis in the opposite direction to that actually taken by the came. By setting out the follower in several successive positions the profile of the cam may be drawn tangential to the end of the follower. There is no need actually to draw the entire follower: the outline of its end is sufficient. The form of the end may be wedge-shaped, flat, or provided with a roller, as in figs. 1, 3, and 2, respectively.

Problem 86. To determine the Profile of a Cam which shall cause a given Follower to reciprocate with uniform Velocity along a straight Line passing through the Axis of the Cam, one Revolution of the Cam giving one Rise and Fall to the Follower.

Let R = rad. of the spindle carrying the cam; D = least distance from spindle to profile; L = total travel of the follower in one direction. The least rad. of the cam is therefore $(R + D)$.

(1) When the end of the follower is wedge-shaped. Fig. 4.

Let O be the cam axis and AOB the path of the follower. With centre O and rad. $(R + D)$ describe a circle cutting OB in C. Set off from C a distance CB = L, and divide CB into, say, six equal parts, numbering them as in figure. With centre O and rad. OB describe a circle, and set off on each side of AB six radii equally spaced. Swing the points 1, 2, 3 . . . on the line of stroke, about the centre O to the corresponding radii O1, O2, O3 . . . , and draw the cam outline through the points p_1, p_2, p_3 . . . .

(2) When the follower has a roller-end. Fig. 5.

Let the rad. of the roller = r. Determine points on a cam profile which would give the *centre* of the roller the desired motion, i.e. take a rad. $(R + D + r)$ for the base circle and proceed as in (1). The points obtained, c_1, c_2, c_3 . . . , lie on the **pitch curve** of the cam. About these points describe circles of rad. r, representing the roller, and draw a fair curve to touch them tangentially. This curve is the required **working profile** of the cam.

(3) When the follower has a flat end. Fig. 6.

Obtain points p_1, p_2, p_3 . . . as in fig. 4, and draw lines through them to represent the flat end of the follower—shown here perp. to the various radii. A continuous curve drawn to touch these lines gives the required profile. The differences between the various outlines should be noted.

EXAMPLES

In the following, take the rad. of spindle = 1" and the least distance from spindle to cam profile = ¾". Assume the line of stroke to pass through the axis of the cam.

(1) Determine a cam profile to give a roller-ended follower a rise of 2½" and fall of 2½" with uniform velocity for one revolution of the cam. Diam. of roller 1".

(2) Determine a cam profile to give a roller-ended follower a continuous rise of 3" with uniform velocity for one revolution of the cam and a sudden drop to the starting-point. Diam. of roller 1¼".*

(3) The end of a follower is flat and perp. to the line of stroke. Determine a cam profile to give the follower a rise of 3" and fall of 3" with uniform velocity for one revolution of the cam.

---

* *Note.*—A strict adherence to the construction of fig. 5 is not permissible if the full travel of 3" is required. It will be found that the roller circles *interfere* with one another where the sudden change in motion occurs, and that to give the roller its full lift the cam profile must be continued.

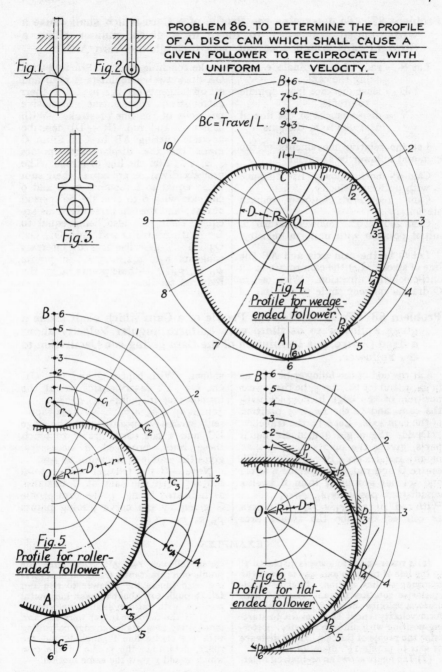

Fig. 1.

Fig. 2.

Fig. 3.

PROBLEM 86. TO DETERMINE THE PROFILE OF A DISC CAM WHICH SHALL CAUSE A GIVEN FOLLOWER TO RECIPROCATE WITH UNIFORM VELOCITY.

BC = Travel L.

Fig. 4.
Profile for wedge-ended follower.

Fig. 5.
Profile for roller-ended follower.

Fig. 6.
Profile for flat-ended follower.

Problem 87. To determine the Profile of a Cam which shall cause a wedge-ended Follower * to move in a given manner along a Straight Line which does not pass through the Cam Axis.

Let R = rad. of the spindle carrying the cam.

D = least distance from spindle to profile.

Y = displacement of the line of stroke from the axis.

Let the relative movements of the cam and follower be as under:

Cam 0°–180°: follower, rise distance L with uniform velocity.

Cam 180°–270°: follower, remain stationary.

Cam 270°–360°: follower, return to initial position with uniform velocity.

Let O be the cam axis and AB the line of stroke. Let the cam rotate in an anti-clockwise direction. With centre O draw a tangent circle to the line of stroke touching it at A. Starting with OA draw twelve equidistant radii and set off tangents $a_1 1$, $a_2 2$, $a_3 3$ . . . at their extremities, representing successive positions of the line of stroke. With centre O and rad. (R + D) describe a circle cutting AB in C. From C mark off the displacement points 1, 2, 3 . . . on the line of stroke. The first six divisions are equal, their sum being equal to L; points 7, 8, and 9 coincide with 6 to give the 90° period of rest; and the last three divisions are equal, their sum also being equal to L. With centre O and radii O1, O2, O3 . . . , describe arcs to intersect tangents $a_1 1$, $a_2 2$, $a_3 3$ . . . in points p_1, p_2, p_3 . . .—these points lie on the required profile.

Problem 88. To determine the Profile of a Cam which shall cause a given Follower to oscillate with uniform angular Velocity about a fixed Centre, one Revolution of the Cam giving one Oscillation to the Follower.

Let the end of the follower be flat, as in fig. 4, and let EC, fig. 5, be the lowest position of the edge in contact with the cam, and O the relative position of the cam axis. Let θ be the travel.

Divide the angle θ into six equal parts, giving the points 1, 2, 3 . . . on *any suitable arc* CB, rad. EC. With centre O and rad. OE describe a circle and set out around it from E twelve equidistant positions e_1, e_2, e_3 . . . . With rad. EC and e_1, e_2, e_3 . . . in turn as centres, describe the twelve arcs shown. With centre O and rad. O1, O2, O3 . . . describe arcs to cut the former arcs in points p_1, p_2, p_3 . . . . Join $e_1 p_1$, $e_2 p_2$, $e_3 p_3$ . . .—these represent successive positions of the edge EC, and a fair curve drawn to touch these lines, produced if necessary, gives the profile of the required cam.

Note.—If the follower makes *point* contact with the cam at C initially, as indicated in fig. 3, the cam profile is given by the curve joining points p_1, p_2, p_3 . . . .

EXAMPLES

(1) A roller-ended follower is displaced 1″ to the left of the cam axis and is given the following st. line motion. First 90° anti-clockwise rotation of cam, rise 1½″ with uniform velocity; next 90°, rise ¾″ with uniform velocity; next 180°, return to starting-position with uniform velocity. Determine the profile of the cam. Least distance of axis to profile 1¼″, diam. of roller 1″.

(2) The figure shows an oscillating roller-ended follower, rad. 5″. Determine the profile of a cam, centre O, which in revolving once causes the follower to rise and fall through 30°, about a mean horizontal position, with uniform angular velocity.

(3) If the follower in (2) were flat and indefinitely long, its lower edge coinciding with the centre line through E and the roller, determine the profile of the cam which would give it the same motion.

* In practice either a roller or flat end would be used in preference to a wedge; the modification involved to the cam profile may be carried out as shown on the previous page.

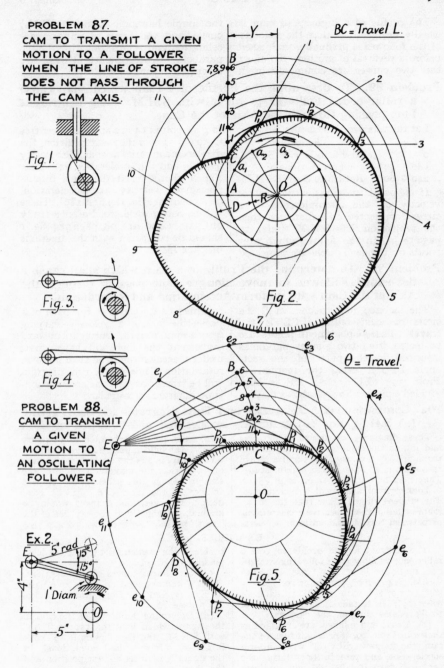

PROBLEM 87.
CAM TO TRANSMIT A GIVEN
MOTION TO A FOLLOWER
WHEN THE LINE OF STROKE
DOES NOT PASS THROUGH
THE CAM AXIS.

Fig. 1.

BC = Travel L.

Fig 2.

Fig. 3.

Fig. 4.

PROBLEM 88.
CAM TO TRANSMIT
A GIVEN
MOTION TO
AN OSCILLATING
FOLLOWER.

θ = Travel.

Ex. 2.
E
5" rad
15°
15°
1' Diam.
4"
O
5"

Fig. 5.

The profile of a high speed cam is usually designed so that the velocity of the follower is gradually diminished before a reversal of motion. To effect this two curves are commonly used, the simple harmonic motion (S.H.M.) curve, and the curve of uniform acceleration and retardation (the double parabola). These important cases are discussed below.

Problem 89. To determine the Profile of a Cam which shall cause a roller-ended Follower to move with S.H.M. along a straight Line passing through the Axis of the Cam.

Let the relative movements be:

Cam	Follower
$0°$–$180°$.	Rise C to B with S.H.M.
$180°$–$240°$.	Remain stationary.
$240°$–$360°$.	Return to C with S.H.M.

If a point moving with uniform velocity on the circumference of a circle is projected on a diameter, then the projection has S.H.M. (refer to page 58). In the figure, equidistant points on the semi-circle are projected to give points 1 to 12 and hence centres C_1 to C_{12}. From the mean position, the displacement varies as $\sin\theta$, the velocity as $\cos\theta$, and the acceleration as $-\sin\theta$. Fig. 1 shows linear diagrams of displacement, velocity, and acceleration, on a time base, for the first $180°$. These will be better understood after a study of Chap. 7 and Art. 98, page 78 ; fig. 1 should be compared with the diagrams on page 75.

Problem 90. To determine the Profile of a Cam which shall cause a flat-ended Follower to move along a Line passing through the Axis of the Cam with uniform Acceleration and Retardation.

The inclined foot touches at C the circle of least radius; CB is the travel. The displacement points are projected from two similar parabolas inscribed in the parts of the rectangle EFGH, using the method of Prob. 39. FG = CB; EF is subdivided equally. The parabolas may be regarded as the parts of the path of a projectile having a velocity varying from a max. to zero. During its upward movement the foot will have, first, uniform acceleration, and then uniform retardation. The velocity-time diagram will be triangular, and the acceleration-time diagram rectangular.

90a. Comparison of Acceleration and Time diagrams for Cams giving (a) S.H.M. and (b) uniformly accelerated and retarded motion.

These diagrams are shown in fig. 2(a) and (b), to the same scale, for equal lifts through $180°$, a pause of $60°$, and a return through $120°$. They repay careful study. They may be obtained as on page 75 or, of course, from analytical considerations. The student should show that the maximum values of the respective accelerations, as between S.H.M. and uniform accelera- tion, are in the ratio π^2 : 8, i.e. as 5 : 4 very nearly. It would appear, however, that the S.H.M. profile has the advantage over the other in that the maximum value of the abrupt *change* of acceleration is the smaller, in the ratio 5 : 8. In any good design, however, the changes would be gradual, e.g. slightly at the expense of the " rest " period. (Refer to Exs. 3 & 4, p 78.)

EXAMPLES

(1) Construct a cam profile to give a roller-ended follower the following straight line motion: First $140°$, rise $2\frac{1}{2}''$ with S.H.M.; $140°$–$200°$, remain at rest; $200°$–$360°$, drop $2\frac{1}{2}''$ with S.H.M. Diam. of roller $1\frac{1}{2}''$; least rad. of cam $2''$; line of stroke passes through the cam axis.

(2) A cam and follower are arranged as shown in Prob. 90. For a half turn of the cam, the follower rises $1\cdot8''$ with uniform acceleration and retardation; during the next half turn it falls $1\cdot8''$ with uniform acceleration and retardation. Taking CO = $1\frac{1}{2}''$ and $x = 60°$ construct the profile of the cam and determine the least possible length of foot.

(3) Plot linear curves of displacement, velocity and acceleration, on a time base, for Exs. (1) and (2). Determine the scales, as on page 75

(4) Determine the profile of the cam in the figure which lifts the tappet T $1\frac{1}{2}''$ with S.H.M., and then lowers it $1\frac{1}{2}''$ with S.H.M. in successive half revolutions. In the figure T is in its highest position and the short arm CR of the double lever is horizontal.

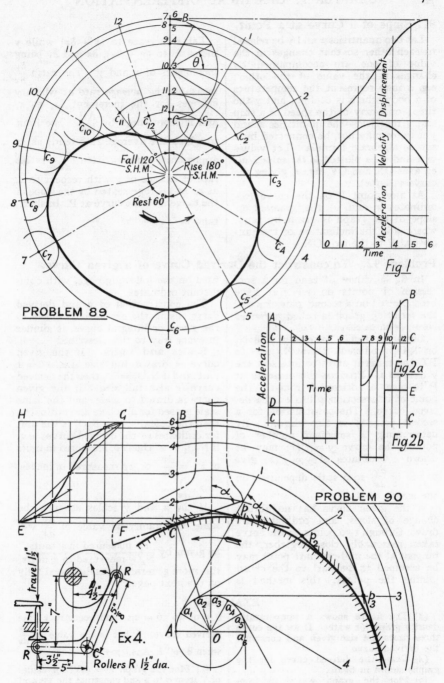

PROBLEM 89

Fall 120° S.H.M.
Rise 180° S.H.M.
Rest 60°

Fig 1

Displacement
Velocity
Acceleration
Time

Fig 2a
Acceleration
Time
Fig 2b

PROBLEM 90

Ex 4.
travel 1½"
7"
Rollers R 1½" dia.

91. Slope of a Curve at a Point.

Let two quantities x and y be related to each other so that changes in the value of one are accompanied by changes in the value of the other: e.g. x may represent the temperature of a body which is cooling, and y the time; or x may be the speed of a ship and y the H.P.—the quantities x and y may, or may not, be connected by a simple algebraic formula. Let values of x and y be plotted *to the same scale* on axes OX and OY, fig. 1, giving the curve $y = f(x)$.

At any point P on the curve (co-ordinates x and y) the gradient or slope of the graph is given by $\tan\psi$, where ψ is the inclination of the tangent at P to the axis OX.

Let x increase to $(x + \delta x)$ while y increases to $(y + \delta y)$, δx and δy being increments of x and y. The ratio $\frac{\delta y}{\delta x}$ measures the *average* rate at which y changes for the increment δx, at x, and is represented by $\frac{AB}{PA}$. The limit to which the average rate tends as δx approaches zero—written $\frac{dy}{dx}$—is *the* rate of change of y with respect to x, at x, and is represented by the slope of the tangent to the curve at P. In short,

$$\tan\psi = \frac{dy}{dx}.$$

Problem 92. To construct the Derived Curve of a given Curve.

In fig. 2 values of $\tan\psi$ for a succession of points on $y = f(x)$ have been plotted on a second pair of axes; the resulting graph is called a *derived curve, or slope curve.*

Construction. Take any pole N_1 on X_1O_1 produced. Draw $N_1 t$ par¹ to PT, the tangent at P, to intersect the y axis in t. Produce the ordinate at P to meet a horizontal from t. The point of intersection p lies on the derived curve. The construction for a second point q is also shown. By determining a sufficient number of points, the curve $y = f'(x)$ may be drawn. Ordinates of $y = f'(x)$ give values of $\frac{dy}{dx}$, to a scale dependent on the length of N_1O_1.

If the given curve has no abrupt change of slope, and ordinates are drawn through two points on the curve taken reasonably close together, the tangent at the mid-ordinate point may be assumed to be par¹ to the chord joining the points: this method is used on the following page, with equidistant ordinates.

The construction of a 2nd derived curve, using the 1st derived curve as the given or integral curve, is similar in every way to that described.

Scales and Units. If the given curve is drawn full size, i.e. $1'' = 1$ unit, and if $N_1O_1 = 1''$, then the derived curve is also full size. If the given curve is drawn to scale, and the same scale is used for abscissæ and ordinates, then the pole distance N_1O_1 is the unit for ordinates of the derived curve: e.g. if $N_1O_1 = 2''$ then cp, measured in units of $2''$, $= \frac{dy}{dx}$; or, cp measured in inches,

$$= 2 \cdot \frac{dy}{dx}.$$

If the x and y scales are not the same, the numerical value of $\frac{dy}{dx}$ will be given by $\dfrac{cp \text{ measured in } y \text{ units}}{N_1O_1 \text{ measured in } x \text{ units}}$: this more general case is discussed fully on the next page.

EXAMPLES

(1) The figure shows a temperature-density graph for water. Draw the curve three times the size given and construct the derived curve.

(2) Draw the derived curve for the graphs shown in figure.

(3) Plot the graph $y = \sin\frac{1}{2}\theta$ from $\theta = 0$ to $\theta = 2\pi$ and construct the derived curve. What is the value of $\frac{dy}{dx}$ when $\theta = \frac{\pi}{2}$? Ans. ·707.

(4) Plot the graph $F = \frac{1}{10}V^3$ for values of V from 0 to 4 and construct the 1st and 2nd derived curves.

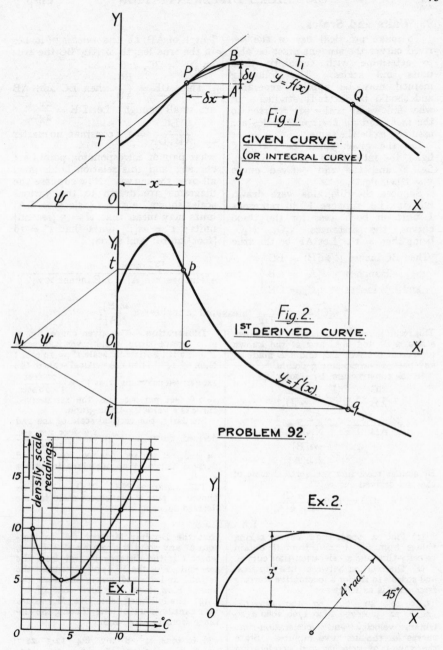

Fig. 1.

GIVEN CURVE.
(OR INTEGRAL CURVE)

$y = f(x)$

Fig. 2.
1ST. DERIVED CURVE.

$y = f'(x)$.

PROBLEM 92.

Ex. 1.

density scale readings

°C

Ex. 2.

3″

4″ rad.

45°

93. Units and Scales.

To make practical use of the derived curves the student must be able to determine with certainty their units and scales. The following method may be applied generally, and should be carefully studied. In what follows all scales are referred to the *inch*, and all *horizontal* scales are assumed to be the same.

Let the given curve be $y = f(x)$, fig. 1, the 1st derived curve $y = f'(x)$, fig. 2, and the 2nd derived curve $y = f''(x)$, fig. 3.

Suppose the diagrams were drawn full size, i.e. suppose **1″** to represent **1 unit** on both axes for the three curves, the distances N_1O_1, N_2O_2 being then $= 1''$. Let $\overline{AB}$ be the true

length of AB, i.e. the *number* of inches in the true length of AB, $\overline{BC}$ the true length of BC, and so on.

Then $\overline{DE} = \dfrac{\overline{BC}}{\overline{AB}}$ when $\overline{BC}$ and $\overline{AB}$ are small enough; i.e. $\overline{DE} = \dfrac{dy}{dx}$.

$\therefore \dfrac{\overline{BC}}{\overline{AB}.\overline{DE}} = 1$, a constant, no matter

what pair of corresponding points are chosen; and this relation holds good all over the curves. Now suppose the diagrams are drawn to the various scales shown: e.g. the scale $1'' = h[A]$ units may mean that $1'' = 5$ [second] units; $1'' = m[B]$ units that $1'' = 10$ [foot] units; and so on.

Then BC inches $\times m[B] = \overline{BC}$
AB inches $\times h[A] = \overline{AB}$
and DE inches $\times n[C] = \overline{DE}$

$\left. \right\} \therefore 1 = \dfrac{BC \text{ inches} \times m[B]}{AB \text{ inches} \times h[A] \times DE \text{ inches} \times n[C]}.$

$$\therefore n[C] = \frac{\overline{BC}}{\overline{AB}.\overline{DE}} \text{ (measured in inches) } \times \frac{m[B]}{h[A]}.$$

This result gives $n[C]$ in terms of the known scales $m[B]$ and $h[A]$ and of the known measurements AB, BC, and DE made at any pair of corresponding points.

In the construction, by similar $\triangle$ s,

$$\frac{BC}{AB} = \frac{O_1F}{N_1O_1} = \frac{DE}{N_1O_1};$$

i.e. $\dfrac{BC}{AB.DE} = \dfrac{1}{N_1O_1} = \dfrac{1}{d}.$

$$\therefore n[C] = \frac{m[B]}{d.h[A]}.$$

By similar reasoning the vertical scale of the 2nd derived curve,

i.e. $o[D]$ units, is $\dfrac{n[C]}{d_1.h[A]}$.

Illustration.—The given curve, fig. 1, is a **space-time** graph, vertical scale $1'' = 20$ ft., horizontal scale $1'' = 10$ secs.; $N_1O_1 = 1\frac{1}{4}''$. Then the vertical scale of 1st derived curve, fig. 2, is $1'' = \dfrac{20 \text{ feet}}{\frac{1}{4} \times 10 \text{ secs.}}$

$= 1\cdot6$ feet per sec., i.e. the 1st derived curve is a **velocity-time** graph.

Similarly, the vertical scale of the 2nd derived curve is $1'' = \dfrac{1\cdot6 \text{ feet per sec.}}{\frac{1}{4} \times 10 \text{ secs.}}$

$= \cdot128$ feet per sec. per sec., i.e. the 2nd derived curve is an **acceleration-time** graph.

The distances N_1O_1, N_2O_2 should be chosen to give a convenient scale for the derived curve.

EXAMPLES

(1) Plot a space-time curve taking values from fig. 1, and from it obtain velocity-time and acceleration-time curves.

(2) The relation between time t in secs. and space s in ft., for a locomotive starting from rest, is as follows:

t,	0	20	40	60	80	100	120	140
s,	0	72	320	790	1320	1760	2080	2320

Obtain velocity and acceleration-time curves for the first two minutes. State the values of velocity and acceleration after 2 min. Ans. 14 f.s., $-\cdot18$ f.s.s.

(3) A cantilever 6′ 0″ long is loaded so

that the Bending Moment (M) in foot-tons at any point is given by $M = l^2 \div 5$, where l is the distance in feet from the free end. Plot the Ml curve, construct the 1st and 2nd derived curves and state the units. How is the cantilever loaded? *

(4) The following table gives M, l values for a cantilever 12′ 0″ long; l being the distance from the free end.

M, ft. tons,	$\cdot1$	$\cdot8$	$2\cdot7$	$6\cdot4$	$12\cdot5$	$21\cdot7$
l, ft.,		2	4	6	8	10	12

Determine the shear and load curves.

Note.—An additional example (No. 5) is given on p. 86.

* $\dfrac{dM}{dl}$ gives the shear, S, and $\dfrac{dS}{dl}$ gives the load per foot.

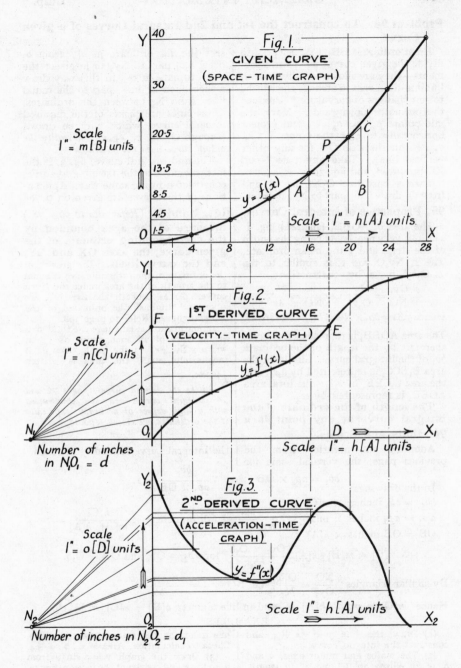

Fig. 1.
GIVEN CURVE
(SPACE-TIME GRAPH)

Scale
$1'' = m[B]$ units

$y = f(x)$

Scale $1'' = h[A]$ units.

Fig. 2.
1ST DERIVED CURVE
(VELOCITY-TIME GRAPH)

Scale
$1'' = n[C]$ units

$y = f'(x)$

Scale $1'' = h[A]$ units

Number of inches
in $N_1O_1 = d$

Fig. 3.
2ND DERIVED CURVE.
**(ACCELERATION-TIME
GRAPH)**

Scale
$1'' = o[D]$ units

$y = f''(x)$

Scale $1'' = h[A]$ units

Number of inches in $N_2O_2 = d_1$

Problem 94. To construct the 1st and 2nd Integral Curves of a given Curve.

Erect ordinates BB_1, CC_1, EE_1 . . . to divide the given curve, fig. 1, into segments—they are shown equally spaced, but it is often best to arrange the widths to suit changes of curvature.* Produce the ordinates downwards. Mark the mid-points p_1, p_2, p_3 . . . and project horizontally to give the points y_1, y_2, y_3 . . . on the axis OY (or any other vertical line). Take any pole N on XO produced, and join to y_1, y_2, y_3 . . . .

Take a base line O_1X_1, fig. 2, and from O_1 draw O_1b_1 par¹ to Ny_1 to inter-sect the 1st ordinate in b_1; from b_1 draw b_1c_1 par¹ to Ny_2 to intersect the 2nd ordinate in c_1; in this way draw a succession of lines, par¹ to the radial lines from N, between the ordinates. These lines are *chords* of the required integral curve, which may be drawn through O_1, b_1, c_1, e_1 . . . . *Usually the actual curve need not be drawn.*

The 2nd integral curve, fig. 3, is the integral curve of the 1st integral curve, constructed in the same way. Applications of the curves are given on p. 78

95. Properties of Integral Curves.

The area of $AOBB_1$ (shaded), fig. 1, $= y \cdot \delta x$, very nearly, δx being the width of the strip, and y the mean ordinate. The $\triangle Ny_1O$, fig. 1, is similar to the $\triangle O_1b_1b$, fig. 2, by construction;

$$\therefore \frac{Oy_1}{NO} = \frac{bb_1}{O_1b}; \text{ i.e. } \frac{y}{NO} = \frac{bb_1}{\delta x}.$$

Hence $y.\delta x = bb_1 \times NO = bb_1 \times constant.$

The area $AOBB_1$ is represented to scale therefore by the length of the ordinate bb_1 of the integral curve. Similarly the area B_1BCC_1 is represented by c_2c_1, and the area C_1CEE_1 by e_2e_1: the total area $AOEE_1$ is represented by ee_1.

The length of the ordinate of the integral curve at any point is a

Figs. 1 and 2. (*Refer also to page 78.*)

measure of the area bounded by the corresponding ordinate of the given curve, the axes OX and OY, and the curve itself. The growth of the ordinate of the integral curve is equal to the growth of the area under the given curve. Areas beneath the axis OX are −ve, and diminish the ordinates of the integral curve. Refer to page 368.

In mathematical language the total area is the sum of all such strips as $y \cdot \delta x$, written $\Sigma y \cdot \delta x$; in the limit, when δx is indefinitely small, the area is written $\int y \cdot dx$.

Unlike the derived curves, pages 72 and 74 integral curves may be constructed with a high degree of accuracy, especially if the ordinates are closely spaced.

96. Units and Scales.

Adopting the notation from the previous page, the vertical scale for the integral curve is readily deduced.

In the drawings: $\underline{bb_1} = \underline{ap_1} \times \underline{OB}.$ $\quad \therefore \frac{\underline{bb_1}}{\underline{ap_1} \times \underline{OB}} = 1.$

$$\left.\begin{array}{l} \underline{bb_1} = bb_1 \text{ inches} \times n[C] \\ \underline{ap_1} = ap_1 \text{ inches} \times m[B] \\ \underline{OB} = OB \text{ inches} \times h[A] \end{array}\right\} \therefore \frac{bb_1}{ap_1 \times OB} \text{ (all in inches)} \times \frac{n[C]}{m[B] \cdot h[A]} = 1.$$

$$\therefore n[C] = m[B] \cdot h[A] \cdot \frac{Oy_1 \times O_1b}{bb_1}; \text{ for } ap_1 = Oy_1 \text{ and } OB = O_1b.$$

By similar triangles $\frac{NO}{Oy_1} = \frac{O_1b}{bb_1}$; i.e. NO or $d = \frac{Oy_1 \times O_1b}{bb_1}$.

Hence $n[C] = m[B] \cdot h[A] \cdot d$, and in like manner $o[D] = n[C] \cdot h[A] \cdot d_1$.

EXAMPLES

(1) Draw the line $y = \cdot 5x + 4$ and construct the integral curve.

(2) The major and minor axes, a and b, of an ellipse are 6″ and 4″ in length. Determine its area graphically by plotting one quadrant and integrating the curve. Check by calculation. Area $= \pi \cdot a \cdot b \div 4$.

(3) Draw the graph $y = 2 \sin\frac{1}{2}\theta$ from $\theta = 0$ to 2π and construct the integral curve. What is the mean value of y? Ans. $1\cdot273$

* Refer to figs. 4 and 5, p.85

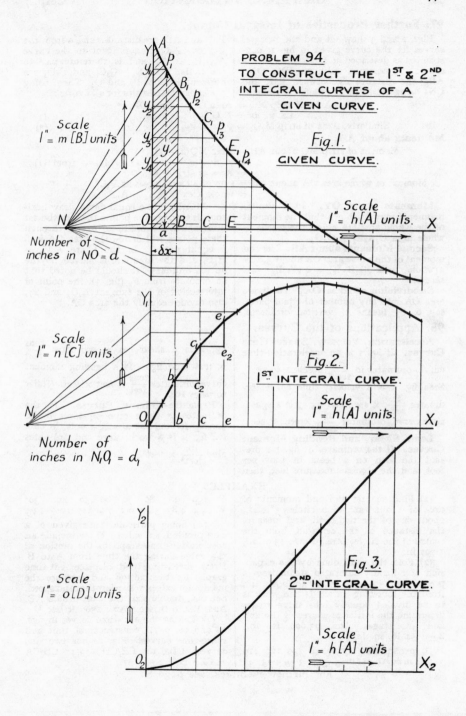

PROBLEM 94.
TO CONSTRUCT THE 1ST & 2ND
INTEGRAL CURVES OF A
GIVEN CURVE.

Fig. I.

GIVEN CURVE.

Scale
$1'' = m [B]$ units

Scale
$1'' = h [A]$ units.

Number of
inches in NO = d

δx

Fig. 2.

1ST INTEGRAL CURVE.

Scale
$1'' = n [C]$ units

Scale
$1'' = h [A]$ units.

Number of
inches in $N_1 O_1 = d_1$

Fig. 3.

2ND INTEGRAL CURVE.

Scale
$1'' = o [D]$ units

Scale
$1'' = h [A]$ units

97. Further Properties of Integral Curves.

Figs. 2 and 3 show 1st and 2nd integral curves for the curve given in fig. 1, constructed as described on p. 76 The curves are cut by common ordinates PM_2, QN_2, AB_2, the distance dx between PM_2 and QN_2 being regarded as indefinitely small; PM_2 and AB_2 are distant x and a from the y axis. A_2G_2 is a tangent to the curve $A_2Q_2P_2$ at A_2—and is therefore par[l] to the radial from E_1. Q_1C_1 and P_1D_1 are horizontals from P_1 and Q_1 Then, using the appropriate scales for all figures:

$$\text{Area of strip } MQ = \text{Area } OQ - \text{Area } OP;$$
$$\text{i.e. } y \cdot dx = N_1Q_1 - M_1P_1 = dy_1. \quad \ldots \quad (1)$$
$$\text{Similarly, Area of strip } M_1Q_1 = y_1dx = dy_2. \quad \ldots \ldots \quad (2)$$

Moments about AB.

$$\text{Moment of strip } MQ \text{ about } AB = \text{Area } MQ(a - x)$$
$$= y \cdot dx \cdot (a - x) = (a - x)dy_1 \ldots \text{ from (1)}$$
$$= \text{Area of strip } P_1C_1.$$

$\therefore$ Moment of whole area OA about AB = sum of all such areas as P_1C_1
$$= \text{whole area } O_1B_1A_1 = A_2B_2.$$

Moments about OY. In a similar manner it may be shown that the moment of the whole area OA about the y axis = whole area $O_1A_1E_1 = O_2G_2$.

Second Moments about AB. The 2nd moment of the whole area OA about AB = 2(whole area $O_2B_2A_2$) = $2 \cdot A_3B_3$ (not shown).

Subdivision of Area OA. To divide area OA into any number of equal parts, say 6, by means of vertical ordinates, divide A_1B_1 into 6 equal parts, draw horizontals through the points to meet the 1st integral curve; ordinates drawn through these points will divide the area OA into 6 equal parts. In the figure $A_1H_1 = H_1B_1$ and the area OA is halved by the dotted ordinate. It should be noted that an ordinate from F_2 (fig. 3), the point of intersection of the tangent A_2G_2 and xy, also divides equally the area OA.

98. Applications of the Curves.

Acceleration, Velocity, Space-Time Curves. If fig. 1 is an acceleration-time curve, ordinates in $\frac{\text{ft.}}{\text{sec.}^2}$, abscissæ in secs., then fig. 2 is a velocity-time curve, ordinates $\frac{\text{ft.}}{\text{sec.}^2} \times \text{sec.}$, and fig. 3 is a space-time curve, ordinates $\frac{\text{ft.}}{\text{sec.}^2} \times \text{sec.} \times \text{sec.}$

Load, Shear, and Bending Moment Curves. If the ordinates of fig. 1 represent the load on a beam in tons per foot, and the abscissæ measure feet, then fig. 2 is a shear curve, ordinates $\frac{\text{tons}}{\text{feet}} \times \text{feet}$, and fig. 3 is a bending moment curve, ordinates $\frac{\text{tons}}{\text{feet}} \times \text{feet} \times \text{feet}$. (Refer to page 104

Pressure-Volume Curves. If the ordinates of fig. 1 give pressures in lb. per sq. ft. and the abscissæ vols. in cub. ft., fig. 2 is a work curve and ordinates give $\frac{\text{lb.}}{\text{feet}^2} \times \text{feet}^3$.

EXAMPLES

(1) Find the 1st and 2nd moments of area of a quadrant of a circle, 3″ rad., about one of the two radii, and measure the distance of the centroid from the centre of the circle. Ans. 9 in.[3], 15.9 in.[4], 1.797 in.

(2) Find the work done by the expansion of 1 lb. dry saturated steam from a pressure of 150 lb./sq. in. to 30 lb./sq. in., from the following data. If the work is to be divided equally into three stages determine the initial pressures of the 2nd and 3rd stages. Ans. 93,600 ft. lb.; 89.5, 52 lb./sq. in.

P. lb. per sq. inch. 150 140 130 120 110
V. vol. in cu. ft. per lb. 3 3.22 3.45 3.73 4

P. 100 90 80 70 60 50 40 30
V. 4.4 4.89 5.47 6.2 7.2 8.5 10.5 13.7

(3) Some particulars are given of a cam profile for a valve. These include an acceleration-time graph of the motion of the valve during the time from A to B. Draw speed-time and displacement-time graphs for the motion and measure the maximum values of acceleration and speed. Set out, double size, the cam profile AB. Ans. Vel. 6 ft./sec.; Acc. 1500 ft./sec.[2].

(4) Taking the acc.-time curves in figs. 2a and 2b, p. 71, construct vel.-time and space-time curves. Fig. 2a has two sine curves: CB/CA = CE/CD = 9/4; CD/CA = CE/CB = 4/5.

For further examples, see page 86

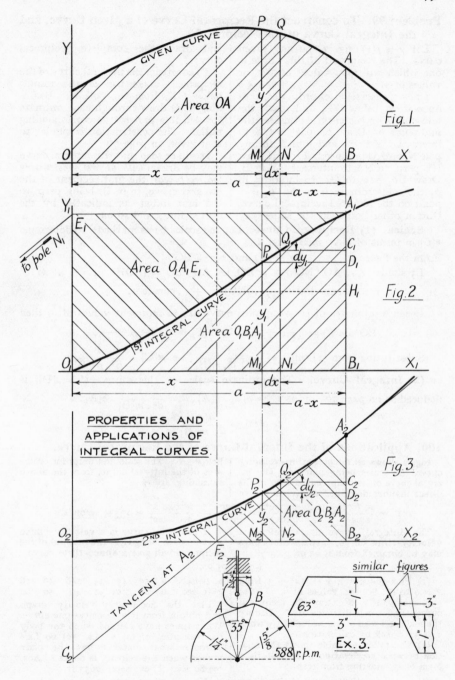

Area OA

Fig.1

X

Y

O

M   N   B

x

dx

a

a−x

P   Q

y

A

GIVEN CURVE

Y₁

E₁

To pole N₁

Area O₁A₁E₁

Area O₁B₁A₁

1ˢᵗ INTEGRAL CURVE

Fig.2

O₁

M₁   N₁   B₁   X₁

x

dx

a

a−x

y₁

P₁   Q₁

C₁

D₁

H₁

A₁

dy₁

PROPERTIES AND
APPLICATIONS OF
INTEGRAL CURVES.

Fig.3

2ⁿᵈ INTEGRAL CURVE

Area O₂B₂A₂

O₂

M₂   N₂   B₂   X₂

y₂

P₂   Q₂

C₂

D₂

dy₂

A₂

TANGENT AT A₂

F₂

C₂

similar figures

63°

3″

3″

Ex.3.

35°

¼″

⅝″

¾″

A   B

588 r.p.m.

Problem 99. To construct the Reciprocal Curve of a given Curve, and the Integral Curve of the Reciprocal.

Let $y = f(x)$, fig. 1, be any given curve. The reciprocal of this curve is one which will give values of $1/y$ for values of x.

Draw a succession of ordinates, produce them below OX, and mark the mid-points on the curve. Consider the mid-point p. Draw pA par[1] to OX to intersect OY in A, and join A to any pole N, on XO produced. From N draw NB perp. to NA to intersect OY in B. Draw Bp_1 par[1] to OX to intersect in p_1 the mid-ordinate from p: p_1 is a point on the required reciprocal curve. Obtain other points in the same way

and draw the complete reciprocal curve.

To construct the integral curve of the reciprocal, make use of the radials below OX, from the pole N. Beginning at O draw through each ordinate space a line par[1] to the corresponding radial. The construction is similar to that given on page 76

It will be seen that this integral curve may be drawn *without first constructing the reciprocal*, the first segment of the integral curve, from O, being perp. to the first radial, as indicated by the dotted line; and so on.

Scales. (1) Reciprocal Curve. Let the scales be as marked: to determine n[C] in terms of h[A], m[B], and d.

From the figure, $\qquad\qquad po = $ AO and $p_1o = $ BO.

By similar $\triangle$s, AO : NO :: NO : BO, i.e. AO . BO = (NO)^2 . . . (1)

$$\text{i.e. BO} = \frac{1}{\text{AO}} \cdot (\text{NO})^2 = \frac{1}{\text{AO}} \cdot \text{constant.}$$

Because ordinates of the reciprocal curve are to represent values of $\dfrac{1}{y}$, then

$$\text{BO}'' \cdot n[\text{C}] = \frac{1}{\text{AO}'' \cdot m[\text{B}]}; \text{ i.e. } n[\text{C}] = \frac{1}{\text{AO}'' \cdot \text{BO}'' \cdot m[\text{B}]}.$$

Substituting from (1), $n[\text{C}] = \dfrac{1}{(\text{NO}'')^2 \cdot m[\text{B}]}$; i.e. $\boldsymbol{n[\text{C}] = \dfrac{1}{d^2 \cdot m[\text{B}]}}$

(2) Integral Curve. The ordinate scale for this curve, $1'' = o[\text{D}]$, is deduced as on page 76 and is $1'' = n[\text{C}] \cdot h[\text{A}] \cdot d = \dfrac{1}{d^2 \cdot m[\text{B}]} \cdot h[\text{A}] \cdot d$.

$$\therefore \ o[\text{D}] = \frac{h[\text{A}]}{d \cdot m[\text{B}]}.$$

100. Applications of the Integral Curve of the Reciprocal Curve.

Fig. 2 shows an **acceleration-velocity** curve, the scales being marked. The integral curve of its reciprocal, drawn in the direct manner described above, is shown

below OX. The scale and units for ordinates of the integral curve, from the above reasoning, are:

$$1'' = \frac{20 \text{ feet}}{\text{secs.}} \div \text{NO} \cdot \frac{\cdot 02 \text{ feet}}{(\text{secs.})^2} = \frac{20}{\cdot 75 \times \cdot 02} \cdot \frac{\text{feet secs.}^2}{\text{secs. feet}} = 1333\tfrac{1}{3} \text{ seconds.}$$

The integral curve is therefore a **velocity-time** curve; a space-time curve may be obtained from it as on page 78.

If the given curve is a **velocity-space** curve (space along OX) the integral of the reciprocal will give a **space-time** curve.

EXAMPLES

(1) Plot the graph $y = \cdot 1 x^2 + 3$ from $x = 0$ to 6. Scales: values of x, $1'' = 1$; values of y, $1'' = 2$. Draw the reciprocal curve using a pole distance of $2''$. Deduce the scale and read off the value of $1/y$ when $x = 3$. Check by calculation.

(2) The following table shows the relation between acceleration x and velocity v for a body, starting from rest.

α, ft./sec.^{2}	·6	·55	·45	·34	·246	·19	·16
v, ft./sec.	o	10	20	30	40	50	60

Plot the acceleration-velocity graph and obtain from it a velocity-time curve. What times have elapsed when the body has velocities of (a) 35 f.s., (b) 60 f.s.? Through what distance has the body moved when the velocity is 60 f.s.? Ans. (a) 80 sec., (b) 205 sec.; 7655 ft.

Note.—An additional example (No. 3) is given on page 86

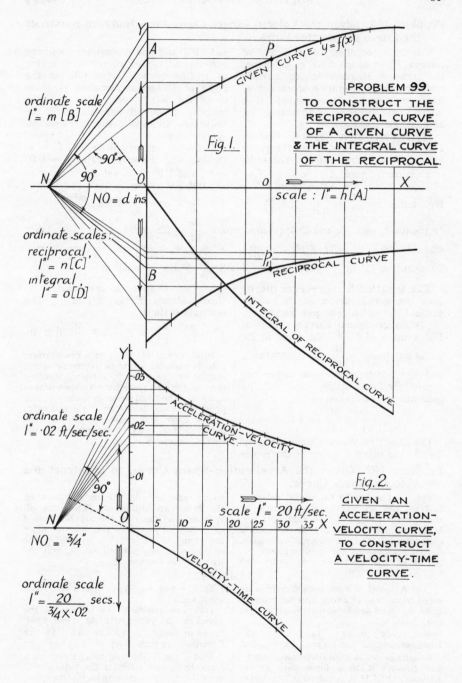

ordinate scale
1" = m [B]

90°
90°
N   O
NO = d ins

ordinate scales:
reciprocal,
1" = n [C],
integral
1" = o [D]

A   P   GIVEN CURVE y = f(x)

PROBLEM 99.
TO CONSTRUCT THE
RECIPROCAL CURVE
OF A GIVEN CURVE
& THE INTEGRAL CURVE
OF THE RECIPROCAL.

Fig. 1.

O   scale : 1" = h [A]   X

B   P₁   RECIPROCAL CURVE

INTEGRAL OF RECIPROCAL CURVE

ordinate scale
1" = ·02 ft./sec./sec.

·03
·02
·01

ACCELERATION-VELOCITY
CURVE.

90°

N   O
NO = ¾"

5   10   15   20   25   30   35   X

scale 1" = 20 ft./sec.

Fig. 2.
GIVEN AN
ACCELERATION-
VELOCITY CURVE,
TO CONSTRUCT
A VELOCITY-TIME
CURVE.

VELOCITY-TIME CURVE

ordinate scale
1" = $\frac{20}{¾ \times ·02}$ secs.

Problem 101. Given the Velocity-Space Curve for a Body, to construct the Curve of Effective Work.

The curve of effective work is the integral curve of the force-space curve: the latter is represented (to another scale) by the acceleration-space curve, for force = mass × acceleration. The initial step, therefore, is to obtain the acceleration-space curve.

(1) Acceleration - Space Curve. Fig. 1. Let P be a point on the given velocity-space curve. Draw the ordi- nate PR and the normal PS; RS, the sub-normal, represents the acceleration at P (see below). If S falls to the right of R the acceleration is $+$ve; if to the left, $-$ve. Mark off the lengths of the sub-normals along the ordinates and draw the acceleration-space curve.

Let the velocity-scale be $1'' = m$ ft./sec., and the space scale $1'' = h$ ft.

Then from figure, $AB \cdot m = \delta v$, $PA \cdot h = \delta s$, and $RP \cdot m = $ vel. v;

i.e. $AB = \delta v/m$, $PA = \delta s/h$, and $RP = v/m$.

By similar $\triangle$s $\dfrac{AB}{PA} = \dfrac{RS}{RP}$; i.e. $\dfrac{\delta v}{\delta s} \cdot \dfrac{h}{m} = RS \cdot \dfrac{m}{v}$; or $v \cdot \dfrac{\delta v}{\delta s} = RS \cdot \dfrac{m^2}{h}$.

In the limit, when δv and δs approach zero, $v \cdot \dfrac{dv}{ds} = RS \cdot m^2/h$. . . . (1)

But $v \cdot dv/ds = v \cdot dv/dt \cdot dt/ds = dv/dt$, $\therefore$ $v \cdot dv/ds = dv/dt = $ acceleration. (2)

Substituting (2) in (1), Acceleration $= RS \cdot \dfrac{m^2}{h} = RS \times $ constant.

The length RS represents therefore the acceleration at P, to the scale $1'' = m^2/h$ feet per sec.².

(2) Force-Space Curve. Fig. 2. For a mass of M lb. the force in lb. wt. at any point $= \dfrac{M}{g} \times $ (acceleration), and the acceleration-space curve becomes the force-space curve if the ordinate scale is taken as

$$1'' = \left(\frac{m^2}{h} \cdot \frac{M}{g}\right) \text{ lb. wt.}$$

(3) Effective Work Curve. Fig. 3. This is the integral curve of the force- space curve and the construction is clearly shown in figs. 2 and 3. The ordinate scale is

$$1'' = \left(\frac{m^2}{h} \cdot \frac{M}{g}\right) \cdot h \cdot d = \left(m^2 \cdot \frac{M}{g} \cdot d\right) \text{ ft. lb.}$$

In all practical examples a **resistance**, usually variable, has to be overcome before motion can be given to a body. A resistance curve on a space base is shown dotted in fig. 2. The **total force** necessary at any point is the algebraic sum of the ordinates of the force and resistance curves; the curve of total work is the integral of the total force curve.

Problem 102. Given the Acceleration-Space Curve, to construct the Velocity-Space Curve.

Fig. 4. Let OA be the initial velocity, and let the acceleration-space curve be that shown and similar to that obtained in fig. 1. Erect ordinates at reasonably close intervals, and draw the mid-ordinates, dotted. Consider the mid-ordinate ab. Set off $ac = ab$, and with c as centre and cA as rad., describe the arc AB between the first two ordinates. Then set off $df = de$, and describe the arc BC from f as centre. Proceed in this way to draw the complete curve. The process is the reverse of that described above, and the scales are as marked.

EXAMPLES

(1*a*) A vessel of 6000 tons increases its speed from 10 to 18 knots, and the relation between speed and total distance travelled is as follows:

Speed, knots	10	12	14	16	18
Distance, feet	0	500	1170	2170	3800

Construct an acceleration-space curve and convert it to a force-space curve. Estimate the I.H.P. required to accelerate the vessel at each of the given speeds.

1 knot $= 1 \cdot 69$ f.s. Ans. 1580, 1820, 1850, 1580, 1060.

(1*b*) The **resistance** to motion of the vessel in 1 (*a*) varies with the speed, thus:

Speed in knots	10	12	14·1	16	18
Resistance in tons	6·93	10	14·1	19·5	27

Plot a curve showing total force, on a space base, and estimate the total useful work done, in foot tons, in increasing speed from 10 to 18 knots. Ans. 127,300.

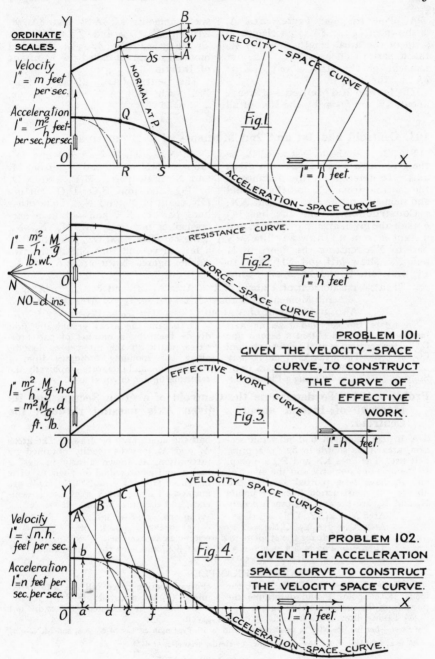

ORDINATE SCALES.

Velocity
$l'' = m$ feet per sec.

Acceleration
$l'' = \frac{m^2}{h}$ feet per sec. per sec.

Fig.1.

VELOCITY-SPACE CURVE

$l' = h$ feet.

ACCELERATION-SPACE CURVE

NORMAL AT P

$l'' = \frac{m^2}{h} \cdot \frac{M}{g}$ lb. wt.

$NO = d$ ins.

RESISTANCE CURVE.

Fig.2.

$l' = h$ feet

FORCE-SPACE CURVE

PROBLEM 101.

GIVEN THE VELOCITY-SPACE CURVE, TO CONSTRUCT THE CURVE OF EFFECTIVE WORK.

$l'' = \frac{m^2}{h} \cdot \frac{M}{g} \cdot h \cdot d$
$= m^2 \cdot \frac{M}{g} \cdot d$ ft. lb.

EFFECTIVE WORK CURVE

Fig.3.

$l'' = h$ feet.

Velocity
$l'' = \sqrt{n.h.}$ feet per sec.

Acceleration
$l'' = n$ feet per sec. per sec.

VELOCITY SPACE CURVE.

Fig.4.

PROBLEM 102.

GIVEN THE ACCELERATION SPACE CURVE TO CONSTRUCT THE VELOCITY SPACE CURVE.

$l' = h$ feet.

ACCELERATION-SPACE CURVE.

A plane irregular figure, area A, is shown in fig. 1; δA is an elemental strip of the area, length z, width δy, taken par^l to an axis XX. The 1st moment of the strip about XX is $\delta A . y$; the 2nd, $\delta A . y^2$.

The **1st and 2nd Moments** of area A about XX are given by the sum of all such moments as $\delta A . y$ and $\delta A . y^2$, written $\Sigma \delta A . y$ and $\Sigma \delta A . y^2$; i.e. $\Sigma z . \delta y . y$ and $\Sigma z . \delta y . y^2$. The 2nd moment of area is called the **Moment of Inertia.**

The **Centroid** C is distant $\bar{y}$ from XX such that A $. \bar{y} = \Sigma \delta A . y$, i.e.

$$\bar{y} = \frac{\Sigma \delta A . y}{A}.$$

103. Centroid and 1st and 2nd Moments of a given Area.

Fig. 2. Let the largest area, A, be the given area, and let XX be a given axis. To determine $\bar{y}$ the distance of the centroid from XX and the 1st and 2nd moments of the area A about XX.

Construction. Take any line PQ outside the given area, par^l to XX, and distant d from it. Draw any line MN par^l to XX, cutting the figure in M and N. Draw MB and ND perp. to PQ and join B and D to any point O on XX. These lines BO and DO intersect MN in M_1 and N_1; from M_1 and N_1 draw perps. M_1B_1 and N_1D_1 to PQ, and join B_1O, D_1O, cutting MN again in M_2 and N_2. Take other lines par^l to XX and obtain a succession of points as M_1, M_2, N_1, N_2, *using the same point* O. Join them by a fair curve and thus obtain the two smaller figures shown, areas A_1 and A_2.

Then*:—1st Moment of A about XX = Area $A_1 . d$; and $\bar{y} = A_1 . d/A$.
2nd Moment of A about XX, written I_{xx} = Area $A_2 . d^2$.
Also, 2nd Moment of A about CG, written $I_{CG} = I_{xx} - A(\bar{y})^2$.

To locate the centroid take an axis inclined to XX and obtain a second line C_1G_1: the centroid is at the intersection of CG and C_1G_1. The areas A, A_1 and A_2 may be obtained by graphical integration, or by Simpson's Rules, or by using a Planimeter. To obtain the areas graphically first divide them by a number of par^l ordinates, mark off the intercepted lengths along corresponding ordinates from a straight line, and integrate graphically the resulting figures of equal area.

Problem 104. To determine the Centroid of a given Section and its Moment of Inertia about a given Axis passing through the Centroid.

A British Standard bull-head rail section, area A, is shown in fig. 3, approx. half size. The lines XX and PQ are perp. to the axis of symmetry and the areas A_1 and A_2 have been plotted as described above; the construction lines for points N_1 and N_2 are shown. In practice only half the figure need be drawn. The areas A_1 and A_2 may be readily obtained by integration, as shown clearly in figs. 4 and 5, and as described on page 76. The distances ON ($2\frac{1}{2}''$) and O_1N_1 ($1\frac{1}{2}''$) are marked off along the Y axes for convenience, and the areas are as follows:

Area A = $2(AY . ON) = 2(1.96 \times 2.5)$ sq. ins. = 9.8 sq. ins.
Area $A_1 = 2(A_1Y_1 . O_1N_1) = 2(1.64 \times 1.5)$ sq. ins. = 4.92 sq. ins.
Area A_2 is left for the student to determine as an exercise.
Then $\bar{y} = 4.92 \times 6.5 \div 9.8 = 3.26''$ from XX, and is on the centre line.
$I_{xx} = A_2(6.5)^2$; $I_{CG} = A_2(6.5)^2 - 9.8(3.26)^2$.

EXAMPLES

(1) Refer to page 64, Ex. 2; plot the locus of the point P and determine the position of the centroid of the figure.

(2) Taking dimensions from fig. 3 draw the rail section twice full size and determine: (a) the centroid; (b) I_{xx}; (c) I_{CG}. Use your own judgment for radii not given.

* *Proof.*—Let the line MN be an elemental strip, width δy, length z. Let $M_1N_1 = z_1$ and $M_2N_2 = z_2$,

$\delta A = z . \delta y$; $\delta A_1 = z_1 \delta y$; $\delta A_2 = z_2 \delta y$, and, by similar triangles, $\dfrac{z_1}{z} = \dfrac{z_2}{z_1} = \dfrac{y}{d}$; i.e. $\dfrac{\delta A_1}{\delta A} = \dfrac{\delta A_2}{\delta A_1} = \dfrac{y}{d}$.

$\therefore A_1$, i.e. $\Sigma \delta A_1 = \Sigma y/d . \delta A = \dfrac{1}{d} \delta A . y = \dfrac{\text{1st moment of A about XX}}{d}$. Similarly $A_2 = \dfrac{1}{d^2} . \delta A . y^2 = I_{xx}/d^2$.

Further because $\bar{y} = \dfrac{\delta A . y}{A}$ by definition, it follows that $\bar{y} = \dfrac{A_1 . d}{A}$.

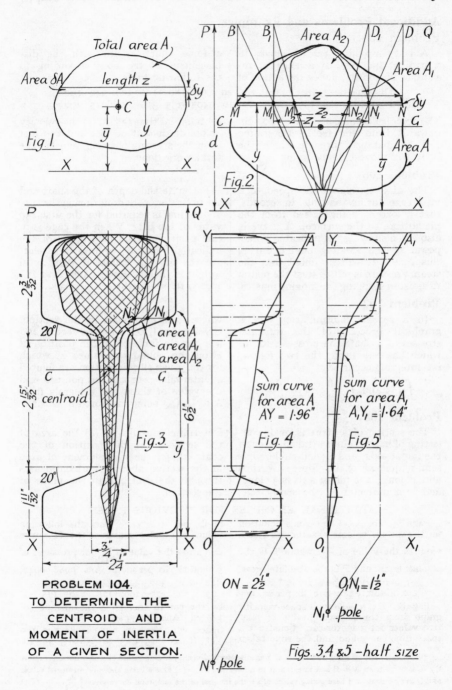

Total area A

Area δA   length z   δy

C

ȳ   y

Fig. 1.

X   X

Fig. 2.

P | B   B₁   Area A₂   D₁   D | Q

C   M   M₁   M₂   z₂   N₂   N₁   N   G

z   z₁   δy

Area A₁

Area A

d

y   ȳ

X   O   X

P   Q

2 3/32

20°   N₂   N₁

area A
area A₁
area A₂

C   centroid   G

2 15/32   6 1/2"

Fig. 3.   ȳ

20°

1 11/32"

X   3/4"   X
2 1/4"

PROBLEM 104.
TO DETERMINE THE
CENTROID AND
MOMENT OF INERTIA
OF A GIVEN SECTION.

Y   A

sum curve
for area A
AY = 1·96"

Fig. 4.

O   X

ON = 2 1/2"

N° pole

Y₁   A₁

sum curve
for area A₁
A₁Y₁ = 1·64"

Fig. 5.

O₁   X₁

O₁N₁ = 1 1/2"

N₁° pole

Figs. 3,4 &5 – half size

Additional Problems and Examples.

Problem 105.

A body weighing 322 lb. moves in a straight line against a constant resistance of 300 lb., under the action of a force which varies with the displacement of the body according to the following table.

Displacement, feet	0	20	40	60	80	100	120	140	160
Force, lb. wt.	..	600	520	460	405	355	300	255	215	200

The force remains constant at 200 lb. onwards from 160 feet. Determine (a) the distance through which the body will move before coming to rest, (b) the total time taken, (c) the velocity at points distant 60 feet and 120 feet from the starting position, and (d) the total work done.

Problem 106

The diagram shows the speed of a mine cage during the time, in seconds, that a load is being raised from the pit bottom to the surface. A curve is also given showing the indicated horse-power of the winding engine during this time. It will be noted that the steam pressure is put against the piston to assist in stopping the moving masses.

Estimate the depth of the shaft and the indicated work done during the run. What time is required for the first 150 yards of the lift? When the cage is in this position, what percentage of the indicated power is required to accelerate the moving masses if their total weight is then 30 tons? Ans. 2112 ft.; 12,200 ft. tons; 18¼ sec.; 16·3%.

Problem 107

In a recent determination, by a graphical process, of the torsional stresses in a shaft the cross-section of which has any form, the two following integrals occurred:

$$\int u \, d\theta \text{ and } \int \frac{\sin\theta}{OP} \, . \, du \text{ or } \int \frac{du}{OQ},$$

where O is any point in the section, ON a normal, OP a radius vector, and u a certain function of the position of P on the boundary, values of which for the given elliptic section are figured at intervals. For a given point O find the value of these two integrals from A to B, OQ being measured in cms.*

Problem 108.

The section of a strut is oval, consisting of half an ellipse (taken through the minor axis) and a semi-circle. The semi-major axis of the ellipse is vertical and 4″ long; the minor axis is 3″ long, and is a diameter of the semi-circle.

Determine graphically (a) the area of the section, (b) the position of the centroid, (c) the 2nd moment of area of the section about a horizontal axis through the centroid, in the plane of the section.

ADDITIONAL EXAMPLES FOR PREVIOUS PAGES:

Page 74 Ex. 5.—Clapeyron's equation for the specific volume of saturated steam requires the value of $\frac{dT}{dP}$, where T is the absolute temp. and P is the absolute press.

t, ordinary temperature in ° F.

p, absolute pressure, lb. per sq. inch

Page 80. Ex. 3.—Draw a space-velocity graph from the following table of relative values for a motor-car. Construct a space-time curve and find the time taken in lb. per sq. foot. From the following values of t and p construct the TP curve and state the value of $\frac{dT}{dP}$ for pressures of 50 and 125 lb. per sq. in. Ans. ·0088, ·0042.

240	281	307	328	344	358	371
25	50	75	100	125	150	175

by the car in traversing the distance of 120 yd. Ans. 12·53 sec.

Space, yd.	0	20	40	60	80	100	120
Vel., ft./sec.	15	19·5	26	33	39·5	45	48.

* Hint.—(1) Set off measured values of θ as abscissæ and corresponding values of u as ordinates. Draw the θ, u curve from A to B and integrate it to obtain $\int u \, . \, d\theta$. (2) Plot a curve showing measured values of OQ as ordinates on a base giving values of u; the integral of the reciprocal curve gives $\int \frac{du}{OQ}$.

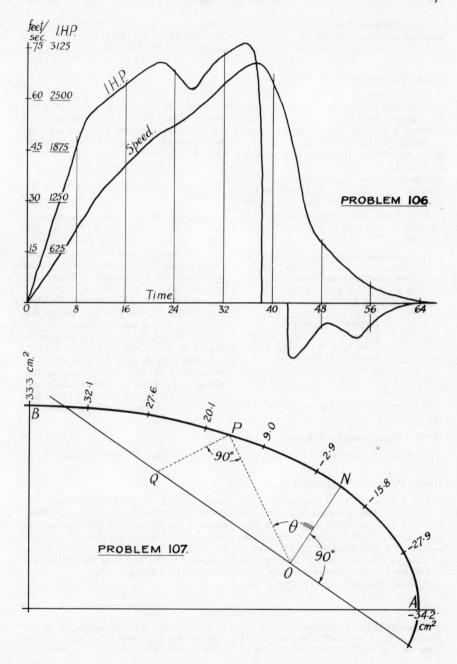

feet/sec. I.H.P.

75 — 3125
60 — 2500
45 — 1875
30 — 1250
15 — 625

I.H.P.

Speed.

Time

0   8   16   24   32   40   48   56   64

PROBLEM 106.

33·3 cm²

B

32·1

27·6

20·1

P

9·0

Q

90°

−2·9

N

−15·8

θ

90°

−27·9

PROBLEM 107.

O

A

−34·2
cm²

Some consideration is given here to the principles underlying the construction of instruments for the measurement of area and moments of area. The fundamental principles are important and readily comprehended.

108a. Planimeters *—Polar.

The mechanism in fig. 1 shows a link AP free to turn about a fixed pole P. Hinged to AP at A is a link AB carrying a tracing point B and a small recording wheel W having its axis along AB. When the point B is moved around an area, the wheel has a motion which is in part rotary and in part sliding. The extent of the rotary motion resulting from the movement of B once around the given area, is a measure of the area. One proof is as follows.

Two positions of the line AB and A_1B_1 are shown in fig. 2. They intersect in O and enclose the small angle $\delta\theta$. Their mid-points are C and C_1; CM is an arc about O.

The area ABB_1A_1
$$= \triangle OBB_1 - \triangle OAA_1$$
$$= \tfrac{1}{2}OB^2\delta\theta - \tfrac{1}{2}OA^2\delta\theta$$
$$= \tfrac{1}{2}(OB - OA)(OB + OA)\delta\theta$$
$$= (AB)(OC)\,\delta\theta = AB \times CM$$
$$= \text{length } l \times \text{ elementary displacement of its mid-point at rt. } \angle\text{s to } l$$
$$= l\,d\sigma.$$

The total area $= \int l\,d\sigma$.

In the use of the actual instrument the generating line returns to its original position, so that $\int d\theta = 0$. The implication is that the integral motion at rt. $\angle$s to l may be measured at *any* point in l, say at D, for $\delta\sigma - \delta\sigma_1 = CD\,\delta\theta$, or $\delta\sigma = \delta\sigma_1 + CD\,\delta\theta$;

$$\therefore \int d\sigma = \int d\sigma_1 + CD \int d\theta = \int d\sigma_1.$$

The recording wheel is often placed on BA produced.

Base Circle.—Consider the arrangement in fig. 3: the pole P is at the centre of a circle, the plane of the wheel passes through P, and B is on the circumference of the circle. In these circumstances the recording wheel will not revolve; it will have a motion of translation only. The circle is called the Base Circle and its area is constant for any one type of planimeter.

One design of Polar Planimeter is shown in fig. 5. The arm AB is adjustable for length and can be used on either side of AP. Movement of the wheel is read on a vernier scale and complete turns are recorded by a dial geared to the spindle of the wheel.

108a₁. Planimeter—Linear.

These are dealt with on the next page because of their use as integrators.

108b. Prytz's Planimeter.†

This is a bent bar of constant length having one end pointed and, at the other, a curved knife edge. As the point B traces the curve, the edge A can move freely only along AB, which is always tangential to the curve traced by B. B is taken initially at a point near the centre of gravity of the area, and A is pressed to mark the paper. B is then moved along a radius, around the curve and back along the radius to the starting point. A is again pressed. The area is given approximately by the length l multiplied by the length of the arc AA.

EXERCISE

A simplified planimeter of the type indicated in fig. 1 can readily be constructed from odd material including watch parts. It will not give accurate results but it will demonstrate principles. The wheel can be calibrated by taking B around known areas.

---

* The names of Amsler and Coradi, of Switzerland, are inseparably connected with these instruments. Amsler's first planimeter appeared in 1854.
† Prytz of Copenhagen.

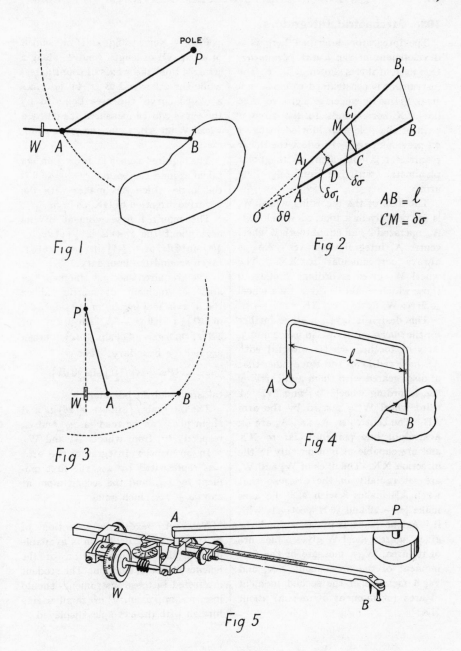

Fig 1

Fig 2

$AB = \ell$
$CM = \delta\sigma$

Fig 3

Fig 4

Fig 5

108c. Mechanical Integrators.

The integrator described here is a development of the *linear planimeter*. One form of this is shown in fig. 1. The instrument is constrained to move in a straight line by wheels in a groove. The line XX corresponds to the circular path of the pole arm (dotted in fig. 1 on previous page). In effect the linear planimeter is equivalent. to the polar planimeter with an infinitely long arm.

The axis of the recording wheel W_1 is set radially in a toothed wheel, axis A_2, operated by an equal toothed wheel centre A, integral with AB. AA_2 is always perpendicular to XX. The wheel W_1 gives recordings similar to those which would be given by a wheel such as W, dotted, on AB.

This design is taken a stage further for the *integrator*, shown in figs. 2 and 3. Arcs of toothed wheels integral with AB have radii two and three times that of discs geared with them and carrying the recording wheels W_1 and W_2. A third wheel W is carried by the arm AB. The centres A, A_1 and A_2 are on a straight line perpendicular to XX and are capable of motion only in the direction XX. The axes of W_1 and W_2 are set radially in the discs so that when AB makes θ with XX the axes make $\frac{1}{2}\pi - 2\theta$ and 3θ respectively with it. As the point B moves around the given figure, wheel W gives a measure of its area, W_1 a measure of the first moment of the area about XX, and W_2 a measure of the second moment of area (or moment of inertia) about XX.

Proof.—Refer to fig. 4. If one end A of a bar AB of length l moves along a straight line (the x axis of coordinates) while the other end B (x, y) describes a closed curve, the area bounded by the curve can be measured (as ω) by a wheel W on AB, as for the polar planimeter.

The required area is $\int y\,dx$ or $l\int \sin\theta\,dx$ taken around the boundary, where θ is the angle which AB makes with the negative direction of OX, as in fig. 4.

The required first moment of the area about the x axis is $\frac{1}{2}\int y^2\,dx$, or $\frac{1}{2}l^2\int\sin^2\theta\,dx$, or $-\frac{1}{4}l^2\int\sin\left(\frac{1}{2}\pi - 2\theta\right)dx$, taken around the boundary.

The required second moment of area, or moment of inertia, about the x axis is $\frac{1}{3}\int y^3\,dx$, or $\frac{1}{3}l^3\int\sin^3\theta\,dx$, or $l^3\left[\frac{1}{4}\int\sin\theta\,dx - \frac{1}{12}\int\sin 3\theta\,dx\right]$, or $\frac{1}{4}l^2\left[l\int\sin\theta\,dx - \frac{1}{3}l\int\sin 3\theta\,dx\right]$ taken around the boundary; that is

$$\frac{1}{4}l^2\omega - \frac{1}{12}l^2\left[l\int\sin 3\theta\,dx\right]$$

taken around the boundary.

The integrals $l\int\sin\left(\frac{1}{2}\pi - 2\theta\right)dx$ and $l\int\sin 3\theta\,dx$ can be read (as ω_1 and ω_2 respectively) from wheels W_1 and W_2.

In an Amsler Integrator, the area was represented by 2ω, the first moment by $4\omega_1$, and the second moment as $32\omega - 10\omega_2$ inch units.

Note.—An excellent collection of Instruments of Calculation is available in the Mathematics Section of the Science Museum, London. The student interested in these instruments should lose no opportunity of familiarizing himself with the examples displayed.

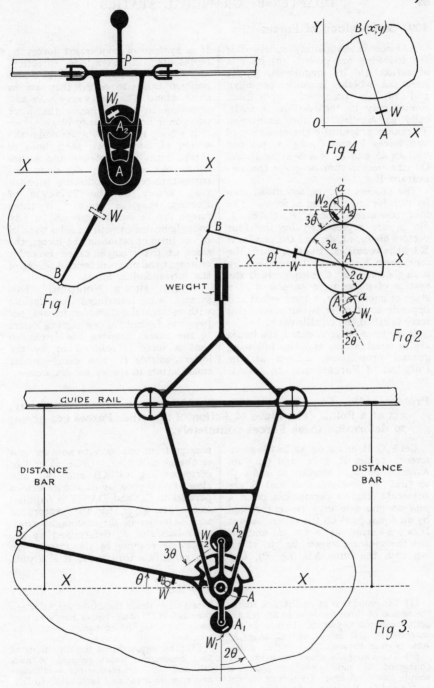

Fig 4

Fig 1

Fig 2

WEIGHT

GUIDE RAIL

DISTANCE BAR

DISTANCE BAR

Fig 3.

109. Composition of Forces.

A force is completely specified if the following are given: (a) its line of action, (b) its magnitude, (c) its sense (i.e. whether it pushes or pulls), and (d) its point of application. Forces may be regarded mathematically as *localized* vector quantities. To obtain graphically the resultant of two forces P and Q, fig. 1, set out vectors *ab* and *bc* representing P and Q; the vector sum *ac* gives the **resultant** R.

The process may be continued step by step for any number of forces. The four concurrent forces P, Q, S, T, fig. 2, are represented by the four vectors *ab*, *bc*, *cd*, *de*, and the resultant R by the vector *ae*. In other words the four given forces may be replaced by a single force R. Consequently if the system of forces is to be placed in a state of equilibrium, a force equal and opposite to R must be supplied: this force is called the **equilibrant**.

If the force polygon closes, the forces are in equilibrium, and the following general proposition, known as the **Polygon of Forces**, may be stated:

If a system of concurrent forces be represented in magnitude, direction, and sense, by the sides of a closed polygon taken in order, they are in equilibrium. The forces may have any direction in space, hence the force polygon is not necessarily plane.

If a body is in equilibrium under the action of *three* forces, their lines of action must be coplanar and either concurrent or parallel. For three concurrent forces the following proposition, known as the **Triangle of Forces**, may be stated: If three forces are in equilibrium and if any triangle be drawn with its sides parallel to the lines of action of the forces, the sides of the triangle taken in order represent the magnitudes and senses of the corresponding forces.

109(a). Bow's Notation.

This system was introduced for dealing with reciprocal figures for frames, see page 108 It consists in assigning letters to the spaces between the forces, so that a force is referred to by the letters astride it, and using similar small letters in the vector diagram.

Problem 110.—Four Forces, two of which are specified completely, act at a Point. The Lines of Action of the other Forces are given; to determine these Forces completely.

Let P, Q, R, and S, fig. 3a, be the given forces; values P, 8 lb., and Q, 5 lb. Arrange the force diagram as in fig. 3b so that the unknowns R and S are adjacent: this is permissible since a *pull* on one side may be represented by an equal *push* on the other. Adopt Bow's notation, as in fig. 3b, and set out the vector polygon, fig. 3c, starting with the force AB (i.e. P), and

passing from one force to another in a clockwise direction. Draw *ab* and *bc* corresponding to AB and BC, and close the polygon by *cd* and *da* drawn parallel to CD and DA. The required forces are given by the vectors *cd* and *da taken in the continuous directions indicated*, as determined by the clockwise reading of the forces. In fig. 3b, S is a thrust and R is a pull.

EXAMPLES

(1) The point O is in equilibrium under the action of three forces as in figure, AB acting vertically upwards. Determine the magnitudes and senses of BC and CA. Ans. BC 9·66 lb. push; CA 7·07 lb. pull.

(2) The two bars shown in figure are connected by pin joints and form a simple frame. A load of 3 tons is carried at the end O. Determine the forces in the

bars and indicate the directions in which they act at O. Ans. Upper bar, 4·91 tons away from O; lower bar, 5·98 tons towards O.

(3) The figure shows the directions of four concurrent forces referred to axes OX and OY. Determine the equilibrant and give its sense and inclination to OX. Ans. 8·75 at 343° 9′ to OX away from O.

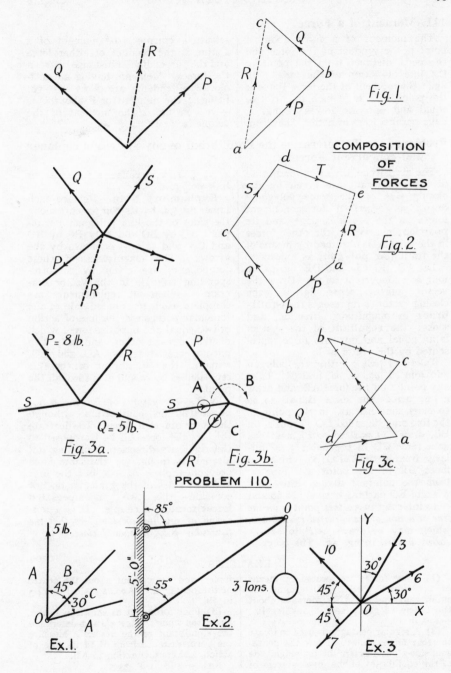

Fig. 1.

COMPOSITION
OF
FORCES

Fig. 2.

Fig. 3a.

Fig. 3b.

Fig 3c.

PROBLEM 110.

Ex.1.

Ex.2.

Ex.3.

111. Moment of a Force.

The moment of a force about a point is the product of the force and the perp. distance from the point to the line of action of the force. In fig. 1 the moment of the force F about the point P is F . *d* inch lb. If two equal and opposite forces F, fig. 2, have parallel lines of action, they con- stitute a **couple**: the moment of a couple is the product of either force and the perpendicular distance between the forces. Moments having an anti-clockwise tendency are taken as +ve: in fig. 1 the moment of F about P is —ve; in fig. 2 the moment of the couple is +ve.

Problem 112. To determine the Equilibrant of any System of coplanar non-concurrent Forces.

The magnitude, direction, and sense of the equilibrant are given by the closing side of the force polygon,* drawn as described for concurrent forces on the previous page, and its **position,** relative to the other forces in the system, is obtained by means of the **funicular polygon,** as follows.

Let AB, BC, CD, and DE, fig. 3*a*, be four non-concurrent forces. Draw the vector diagram *abcde*, fig. 3*b*; the closing vector *ea* represents the **equilibrant** in magnitude, direction, and sense: the **resultant** of the system is an equal and opposite force represented by the vector *ae*.

Take any pole *o* within the polygon and join *oa*, *ob*, *oc*, *od*, and *oe*. Take any point 1 on the force AB, and draw a line across the space B par¹ to *bo* to meet the force BC in the point 2; the line may be called BO. From 2, on BC, draw CO par¹ to *co* to meet the next force CD in the point 3. Similarly from 3 draw DO par¹ to *do* to meet DE in the point 4. Finally, from the points 1 and 4 draw lines AO and EO par¹ respectively to *ao* and *eo* to intersect in 5; *this point is on the line of action of the required equilibrant,* which may be drawn par¹ to *ea*—as shown dotted in fig. 3*a*. The polygon 1, 2, 3, 4, 5 is called a funicular or link polygon.

Explanatory Note. Replace each force, fig. 3*a*, by its components along the sides of the link polygon; i.e. replace AB by BO and AO; BC by BO and CO, and so on—as shown by the arrows. The magnitudes of these components are given by a corresponding triangle in the vector diagram. When all replacements are complete, each terminated side of the funicular polygon is the line of action of two equal and opposite forces which neutralize each other, and only two uncompensated forces, AO and EO, remain: the equilibrant of these is represented by *ea* and acts through the point 5.

N.B.—A system of forces is not necessarily in equilibrium although the force diagram closes. To illustrate this, let the force AE be transposed to the right, its direction remaining un altered, as in fig. 3*c*; the same force polygon applies, fig. 3*b*, but the first and last sides of the funicular do not coincide — the two uncompensated forces forming a couple. *It is a condition of equilibrium therefore that the funicular polygon should close.*

EXAMPLES

(1) Three forces are located by means of the square shown. Determine the equilibrant and state its moment about the centre of the square. Ans. 11 lb.; —10·5 in. lb.

(2) A regular pentagon is used to locate the four forces shown. Draw the pentagon, side 1″, and determine the magnitude of the equilibrant of the given system of forces, and its moment about the centre of the pentagon.† Ans. 2·45 lb.; +14·5 in lb.

(3) Four forces act as shown in figure. Draw the square, edge 3″, and determine the equilibrant of the system. Measure the perp. distance from B to its line of action, and its inclination to AB. Ans. 5·14 lb., 1·8″, 33·5°.

---

* This is readily seen if the forces are combined in pairs, the lines of action being produced as necessary to intersect. † Accurate working is essential in this question.

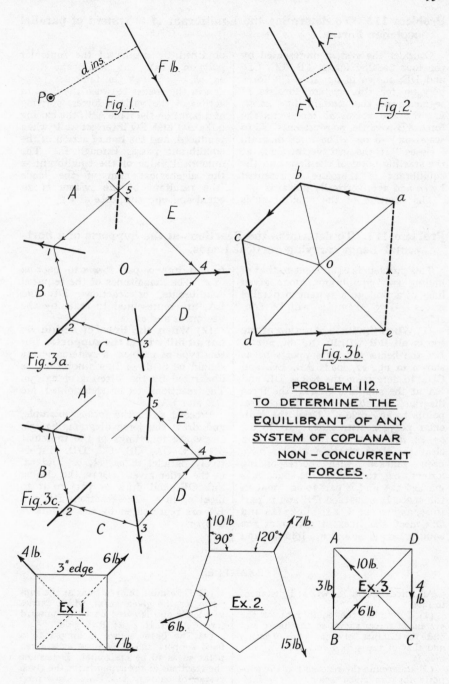

Fig. 1.

Fig. 2.

Fig. 3a.

Fig. 3b.

Fig. 3c.

PROBLEM 112.

TO DETERMINE THE
EQUILIBRANT OF ANY
SYSTEM OF COPLANAR
NON – CONCURRENT
FORCES.

Ex. 1.

Ex. 2.

Ex. 3.

Problem 113. To determine the Equilibrant of a System of parallel coplanar Forces.

Consider the system represented by the four parallel forces AB, BC, CD, and DE, shown in fig. 1a. The force polygon for this system consists of segments of the straight line *abcde*, fig. 1b; the vector *ab* represents the force AB to scale, *bc* represents BC to scale, and so on. The force diagram " closes " if the last vector ends at the starting point of the first, and the equilibrant is therefore represented by *ea* and acts vertically upwards.

The position of the equilibrant is obtained by means of the funicular polygon. Take any pole *o* and draw *oa, ob*, . . . Draw the funicular polygon in the spaces between the lines of action of the various forces, starting at a point on the force AB: the closing links AO and EO intersect at P when produced, and the line of action of the equilibrant passes through P. The numerical value of the equilibrant is the algebraical sum of the loads (the resultant of the system is an equal and opposite force at P).

Problem 114. To determine the Reactions at the Supports of a horizontal Beam carrying vertical Loads.

This problem is equivalent to that of finding two equilibrants along given lines of action, to a system of parallel forces. Two examples will be considered.

(1) When the lines of action of the loads all fall within the supports. Let the beam and its supports be as shown in fig. 2a, loads AB, BC, and CD: to determine the forces DE and EA at the supports. Draw the force diagram, fig. 2b, and join the several points to any pole *o*. Draw the funicular polygon with sides par¹ to *ao, bo, co, do*: to make the construction clear the sides of the funicular have been marked with corresponding letters, e.g. the link in the space A is marked OA and is par¹ to *oa*; that in the space B is marked OB and is par¹ to *ob*, and so on. Let the links OA and OD meet the lines of action of the equilibrants in *m* and *n*. Join *nm* and

from *o* draw *oe* par¹ to *nm* to meet *ad* in *e*. The magnitudes of the required equilibrants, or reactions, DE and EA, are represented to scale by the vectors *de* and *ea*.

(2) When the lines of action do not all fall within the supports. For this type of problem a systematic use should be made of Bow's notation, as illustrated by the lettering of fig. 3a. The reactions to be determined are DE and EA.

Proceed as in the former example, and draw the force diagram, fig. 3b. Draw the four links of the funicular polygon, OA, OB, OC, OD, respectively parallel to *oa, ob, oc,* and *od,* in the order given. Draw the closing link OE, and draw *oe* par¹ to it to meet *ad* in *e*. The reactions DE and EA are represented by the vectors *de* and *ea*.

EXAMPLES

All forces are in lb. wt. and distances in feet.

(1) Determine the equilibrant of the system of forces shown in the figure and state the distance between its line of action and that of the 5 lb. force. Ans. 23 lb.; 6·61 ft.

(2) Determine the reactions at the supports for the given beam. Ans. L.H. 692·6; R.H. 457·4.

(3) Determine the reactions at the supports for the given beam, which carries two loads and is acted on by an upward force. Ans. L.H. 4·32; R.H. 3·68.

(4) The beam shown is hinged to a fixed support and is propped at a given point so as to be horizontal. Determine the reactions at the supports for the given system of loading. Ans. Hinge 4·36; prop 8·64.

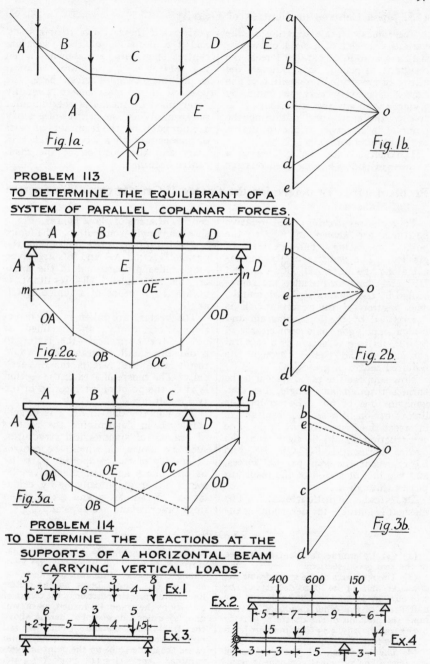

Fig. 1a.

Fig. 1b.

PROBLEM 113
TO DETERMINE THE EQUILIBRANT OF A
SYSTEM OF PARALLEL COPLANAR FORCES.

Fig. 2a.

Fig. 2b.

Fig. 3a.

Fig. 3b.

PROBLEM 114
TO DETERMINE THE REACTIONS AT THE
SUPPORTS OF A HORIZONTAL BEAM
CARRYING VERTICAL LOADS.

Ex. 1

Ex. 2

Ex. 3

Ex. 4

115. Mass Centres or Centres of Gravity.

The forces of gravity acting on the various particles of a rigid body constitute a system of parallel forces, the resultant of which acts through the centre of gravity. The position of the centre of gravity may be found by giving the body two positions, relative to the earth, and obtaining the point of intersection of the·resultants of the parallel forces.

The method may be applied to a number of bodies having fixed relative positions if the ratios of their masses, and the positions of their separate centres of gravity, are known. When the centres of gravity are coplanar, as in the examples given here, the position of the mass centre is readily determined by means of the funicular polygon. When the centres of gravity are not coplanar the resultants of pairs of masses may themselves be taken in pairs and the position of the mass centre obtained.

Problem 116. To determine the Centre of Gravity of a given System of Masses.

Four masses having fixed relative positions are shown in fig. 1, the points indicating their centres of gravity. The weights of the masses act along the lines AB, BC, CD, and DE, and their magnitudes are represented by the vectors *ab*, *bc*, *cd*, and *de*. The position of the resultant AE is determined by the force diagram and funicular polygon, as shown clearly in figure: its line of action is a vertical through *m* and passes through the required centre of gravity G.

Now suppose the mass system to be turned through any angle, for convenience one of 90°. Determine the line of action of a second resultant to intersect the first resultant in G. The construction should be clear from the figure: horizontals PQ, QR, RS, and ST represent the four parallel forces, and the line of action of the resultant passes through *n* .

Practical Applications. The method is suitable for determining the

centres of areas of plane figures. Fig. 2 shows an angle-iron section. The figure is divided into two rectangles, and the parallel vectors AB and BC are given magnitudes proportional to the areas of the rectangles: the resultant is obtained by means of a funicular triangle.

The vectors are taken in two directions, horizontally, and inclined at 60° to the horizontal—the latter to avoid the small diagram that would result were the vectors taken vertically. The centre of area of the section is at G, the point of intersection of the two resultants.

The construction in fig. 1 is largely employed in determining the centres of gravity of symmetrical structures, such as cranes, in which the centres of gravity of the component parts lie in one plane and can themselves be found either graphically or by calculation. Examples 3 and 4 are taken from practice and are typical.

EXAMPLES

(1) (2) Determine the centres of area of the two given sections.

(3) The members of the given frame are coplanar and of uniform density. The sectional areas of the diagonals are one half those of the other members. Determine the position of the centre of gravity of the frame. Ans. 3·73 feet from left-hand end, 1·32 feet from bottom.

(4) The positions of the centres of gravity of the principal component parts of a crane are shown in figure, their weights being: (*a*) counter-weight, &c.

8 tons; (*b*) machinery, 2 tons; (*c*) under carriage, 12 tons; (*d*) jib, 1 ton; (*e*) load, 5 tons. The gauge of the rails may be taken as 5' 0", the point *c* being on the centre line. Determine the position of the centre of gravity of the crane: (1) loaded, as shown, and (2) unloaded: give the height above the track, and distance from the rail remote from the load. What load would cause the crane to be on the point of overturning? Ans. (1) 5·44', 3·94'; (2) 4·46', 1·0'; 7·36 tons.

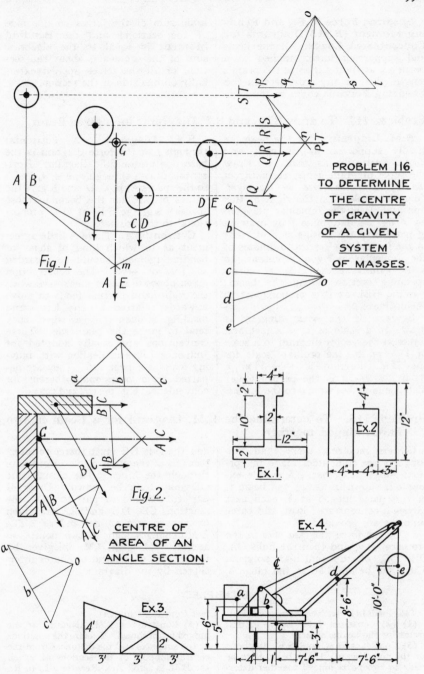

Fig. 1.

PROBLEM 116.
TO DETERMINE
THE CENTRE
OF GRAVITY
OF A GIVEN
SYSTEM
OF MASSES.

Fig. 2.

CENTRE OF
AREA OF AN
ANGLE SECTION.

Ex. 1.

Ex.2

Ex.3.

Ex. 4.

Shearing Force (S.F.) and Bending Moment (B.M.) Diagrams for Concentrated Loads. If the loads and supporting forces applied to a beam act at right angles to the beam's length, then at any cross section the Shearing Force is equal to the alge-

braic sum of the forces on one side of the section; and the Bending Moment is equal to the algebraic sum of the moments, about the section, of all the forces applied externally on one side of the section.

Problem 117. To draw B.M. and S.F. Diagrams for a given Beam.

B.M. Diagram. Let the beam be simply supported at its ends and loaded vertically, as in fig. 1. Draw the force diagram abcde, *beginning at the top with the force on the right.* Join to a pole o on the right of this line and draw the funicular polygon, as in fig. 2; determine f by drawing *of* parallel to the closing side (dotted) of the funicular. Vertical ordinates of the funicular polygon represent, to scale, values of the B.M. at corresponding sections along the beam; e.g. the B.M. at P is proportional to the ordinate p.

Scale. If the beam diagram is drawn to a scale of $1'' = h$ feet and forces in the vector diagram to a scale of $1'' = m$ lb., the ordinate scale for the B.M. diagram is $1'' = h \cdot m \cdot n$ ft. lb., where n is the perpendicular distance in inches from o to the line ae.

S.F. Diagram. A horizontal through f in the force diagram is the base line of the S.F. diagram. Horizontals drawn through a, b, c, d, and e to the spaces A, B, C, D, and E give the values of S.F. for the beam in those spaces; e.g. the S.F. at Q is proportional to the ordinate q.

Conventions. There is little agreement as to which kind of shear or bending moment should be regarded as +ve or −ve. The construction given opposite makes shear +ve when the right-hand portion tends to move upwards relative to the left; and bending moments +ve when they tend to make the beam sag. These conventions are usually adopted for Influence Lines in dealing with moving loads, see page 126 They are departed from occasionally herein for convenience, e.g. pp. 105 and 107

Problem 118. To determine the B.M. Diagram for a Beam in two Parts, hinged together.

Girders requiring more than two supports are sometimes made in portions hinged together. A simple example is shown in fig. 4; the beam is in two parts hinged at h, each part carries a concentrated load, and three supports are provided.

Draw the force diagram abco in the ordinary way, and then the links OA, OB, OC, of the funicular polygon. There can be no B.M. at the hinge h,

and the side OE must therefore intersect OB at the point m, on the ordinate through the hinge. The closing side OD may then be drawn. Draw *oe* par¹ to OE and *od* par¹ to OD: the reactions CD, DE, and EA are given by the vectors *cd*, *de*, and *ea*. The B.M. changes sign at two points, m and n. Values of B.M. are given by the lengths of vertical ordinates intercepted by the diagram.

EXAMPLES

Take loads in lb. wt. and distances in feet.
(1) (2) Construct B.M. and S.F. diagrams for the beams shown.
(3) Construct B.M. and S.F. diagrams for the given beam. If the beam is to be made in two parts hinged together determine the position of the hinge. Ans. 1·26

feet from R.H. support.
(4) Construct a B.M. diagram for the hinged beam shown. Measure the reactions at the supports and the distances from the central support of the sections at which the B.M. is zero. Ans. Reactions L. to R., 9·45 lb., 8·3 lb., 2·25 lb.; 1·65 feet, 7·8 feet.

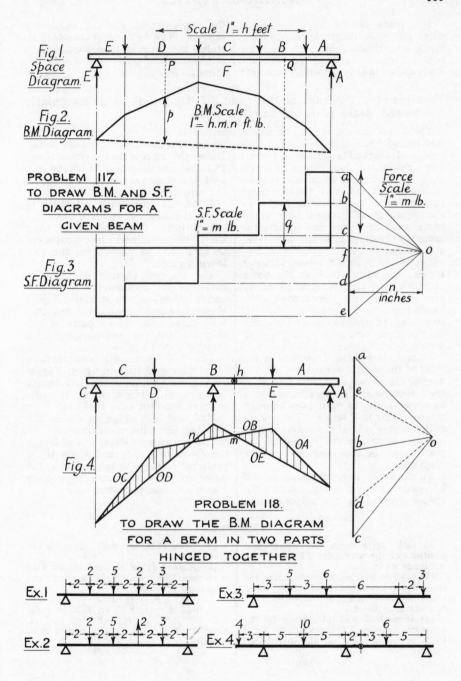

Fig.1. Space Diagram.

Scale 1" = h feet

Fig.2. B.M. Diagram.

B.M. Scale 1" = h.m.n ft. lb.

PROBLEM 117.
TO DRAW B.M. AND S.F.
DIAGRAMS FOR A
GIVEN BEAM

S.F. Scale 1" = m lb.

Fig.3 S.F. Diagram.

Force Scale 1" = m lb.

n inches

Fig.4.

OC   OD   OB   OA   OE

PROBLEM 118.
TO DRAW THE B.M. DIAGRAM
FOR A BEAM IN TWO PARTS
HINGED TOGETHER

Ex.1

Ex.2

Ex.3

Ex.4

By means of the force and funicular polygons, diagrams of Shearing Force and Bending Moment for horizontal beams carrying distributed loads are as readily drawn as those for the concentrated loads already considered. The following example illustrates the method of dealing with a combination of concentrated and distributed loads.

Problem 119. To draw the B.M. and S.F. Diagrams for an irregularly loaded Beam with overhanging Ends.

The beam shown, fig. 1, carries a distributed load of 7 tons, and concentrated loads of 4 tons and 3 tons. The distributed load may be dealt with by dividing it up into a number of small portions, and treating each portion as a concentrated load; in the figure it is divided into seven parts, each of 1 ton, but for greater accuracy the division should be carried still further.

Letter the spaces between adjacent forces; the spaces beneath the beam to which the letters A, M and L apply are indicated by the arrow lines. Draw the force diagram *ab* . . . *kl*, beginning at the top with the force on the right, and join each point to a pole *o*.

Starting at *p*, fig. 4, draw the various links of the B.M. diagram through the several ordinate spaces, finishing at *s*; then draw *pq* and *sr*, parallel to *lo* and *oa* respectively, to meet lines through the reactions in *q* and *r*. The line *rq* is the closing side of the diagram, and *om* drawn parallel to *rq* settles the position of the base line of the shear diagram.

To draw the shear diagram set off horizontals from *a*, *b*, *c* . . . across the spaces covered by the corresponding letters in fig. 1—the limits of these horizontals should be carefully noted. Draw the appropriate vertical lines and close the diagram. It will be seen that the shear values change sign at the supports.

Note.—It is usually possible to draw the S.F. curve for a beam by inspection. The B.M. curve may then be obtained by graphical integration, as shown on the following page.

The stepped portion of the S.F. curve is an approximation to the correct form—shown dotted. Fig. 3 shows, to a smaller scale, the corrected S.F. curve with those parts of the diagram for the overhanging ends brought to the base line.

The B.M. curve has been redrawn on a horizontal base in fig. 5, heights on the same ordinate for both figures being equal. The scale for the diagram is obtained as in Prob. 117.

B.M. and S.F. diagrams are usually required for the determination of maximum stress values, and although the forms of figs. 3 and 5 are to be preferred to those of figs. 2 and 4, the latter are sufficiently complete as they stand, for most purposes.

EXAMPLES

In each write down the values and positions of the maximum S.F. and B.M. All loads are in tons.

(1) (2) The two beams shown carry the same total loads but in different ways. Draw the B.M. and S.F. curves for both and compare them.

(3) Draw B.M. and S.F. curves for the cantilever shown. The 2-ton and 3-ton loads are superposed on the distributed loading.

(4) Draw B.M. and S.F. curves for the given beam.

(5) Answer (4) if the distributed load is increased to 3 tons/ft., other conditions remaining the same.

Ans. (1) S.F. 5·43, B.M. 6. (2) S.F. 4·07, B.M. 6. (3) S.F. 11, B.M. 36. (4) S.F. 3·5, B.M. 3. (5) S.F. 6·5, at L.H. support; B.M. 3·56, 1·75 ft. within R.H. support.

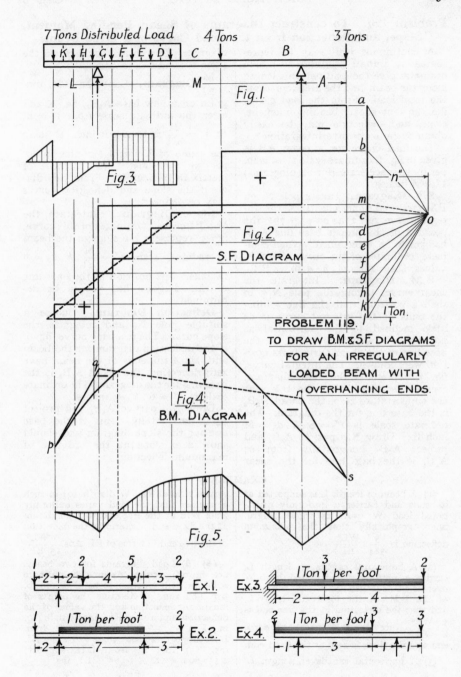

7 Tons Distributed Load 4 Tons 3 Tons

K H G F E D C B

L M A

Fig.1.

Fig.3.

Fig.2.

S.F. DIAGRAM

n''

o

a
b
m
c
d
e
f
g
h
k
l

1 Ton.

PROBLEM 119.
TO DRAW B.M. & S.F. DIAGRAMS
FOR AN IRREGULARLY
LOADED BEAM WITH
OVERHANGING ENDS.

Fig.4.
B.M. DIAGRAM

Fig.5.

Ex.1.
1 2 5 2
2 2 4 1 3

Ex.2.
1 1 Ton per foot 2
2 7 3

Ex.3.
3
1 Ton per foot
2 4

Ex.4.
2 1 Ton per foot 3 3
1 3 1 1

Problem 120. To construct Diagrams of Shear, Bending Moment, Slope, and Deflection from the Load Curve of a given Beam.

A continuous load may be represented by a **load diagram** in which ordinates give the load per unit length along the beam, and the area represents the total load. From the load curve, diagrams of shear, bending moment, slope, and deflection may be easily obtained by successive integration.*

The load curve for a beam AB is given in fig. 1, ordinate scale $1'' = w$ lb. per inch, space scale (for all diagrams) $1'' = h$ inches.

S.F. Diagram. Take a pole N, on BA produced, and construct the integral curve, fig. 2, as on page 76: the position of the proper base line A_1B_1 has yet to be determined. When complete, ordinates of the curve give S.F. values, scale $1'' = w . h . d$ lb. $= S$ lb.

B.M. Diagram. Integrate the shear curve, taking the pole N_1† in such a position that the line joining the ends of the integral curve is as little inclined to the horizontal as possible—N_1 would be taken *on* the base line of the shear curve if its position were known. The integral curve, fig. 3, represents the B.M. diagram. Since it is known that the B.M.s at the supports are zero, the chord A_2B_2 is the base line for the diagram. The ordinate scale is $1'' = S . h . d_1 = M$ inch lb. Draw N_1p parl to A_2B_2 and project A_1B_1 horizontally from p; A_1B_1 is the base line for the shear

diagram. Zero shear should lie on the same ordinate as maximum B.M.

$\dfrac{M}{EI}$ **Diagram.** Copy the B.M. diagram on a new base A_3B_3, fig. 4, and alter the ordinate scale from $1'' = M$ to $1'' = \dfrac{M}{EI}$ radians per inch, E being the value of Young's Modulus in lb. per sq. inch and I the moment of inertia in inch units. From this diagram the slope and deflection curves may be deduced.

Slope Diagram. Integrate the $M \div EI$ curve. Ordinates of this curve, fig. 5, represent the slope of the beam in radians, scale $1'' = \dfrac{M}{EI} . h . d_2 = \theta$ radians. The position of the base line of this diagram has yet to be determined.

Deflection Diagram. Choose a suitable pole N_3 and integrate the slope curve. The resulting curve, fig. 6, is the deflection diagram for the beam and, since the ends of the beam have zero deflection, the chord A_5B_5 is the base line for the diagram. The ordinate scale is $1'' = \theta . h . d_3$ inches.

Draw N_3q parl to A_5B_5 and project A_4B_4 horizontally: this is the base line for the slope diagram and should show zero slope at the ordinate of maximum deflection.

EXAMPLES

(1) A beam of length L is supported at its ends and carries a uniformly distributed load W over its whole length. Show graphically that the maximum deflection is $\dfrac{5}{384} . \dfrac{WL^3}{EI}$.

(2) A horizontal cantilever, length L, carries a uniformly distributed load W over its whole length. Show that the value of the B.M. at the fixed end is $\frac{1}{2}WL$, and that the deflection at the free end is $\dfrac{1}{8} . \dfrac{WL^3}{EI}$. Note: the B.M. is zero at the free end and the slope is zero at the fixed end.

(3) A horizontal cantilever, length L,

carries a total load W distributed in such a manner that the load varies uniformly from zero at the free end to a maximum at the fixed end. Determine the deflection of the free end in terms of EI. Ans. $\dfrac{1}{15} \dfrac{WL^3}{EI}$.

(4) (5) Load diagrams for two beams are shown opposite. Assume the ordinate scale for each diagram to be $1'' = 100$ lb. per inch run. Determine the points of maximum deflection and the values of the deflections in terms of EI. Ans. E. in lb./sq. in., I in (inch)⁴ units: (4) 3·04 ft. from L.H. support; 28,800 × 12³ × 1·23 ÷ EI, ins. (5) 10·3 ft. from L.H. support; 2,457,600 × 12³ × 1·385 ÷ EI, ins.

* The equations for these curves are: $S = \int w . dx$, $M = \int S . dx$, $\theta = \frac{1}{EI}\int M . dx$, $y = \int \theta . dx$.

† The conventional B.M. diagram (see p. 100) is given by taking the pole N_1 on the right, not the left.

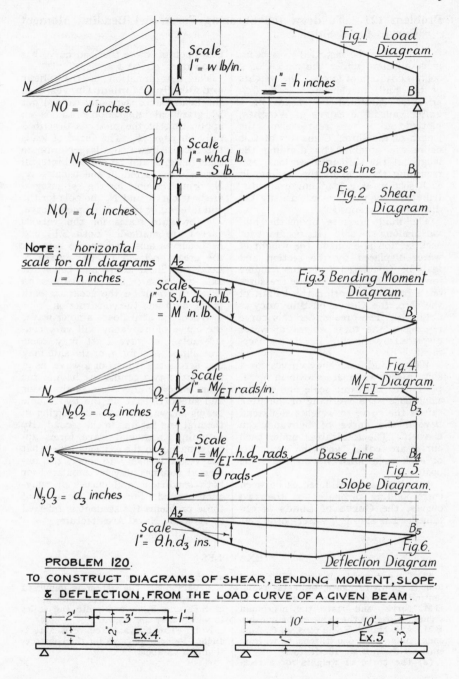

Fig.1 Load Diagram.

Scale
$1'' = w$ lb/in.

$1'' = h$ inches

N   O   A   B

NO = d inches.

Scale
$1'' = w.h.d$ lb.
$= S$ lb.

N_1 O_1 A_1 Base Line B_1
          P

Fig.2 Shear Diagram.

$N_1O_1 = d_1$ inches.

NOTE: horizontal scale for all diagrams $1 = h$ inches.

A_2

Scale
$1'' = S.h.d_1$ in.lb.
$= M$ in. lb.

Fig.3 Bending Moment Diagram.

B_2

Scale
$1'' = M/EI$ rads/in.

M/EI Diagram.

Fig.4

N_2 O_2 A_3 B_3

$N_2O_2 = d_2$ inches

Scale
$1'' = M/EI.h.d_2$ rads.
$= \theta$ rads.

N_3 O_3 A_4 Base Line B_4
          q

Fig.5. Slope Diagram.

$N_3O_3 = d_3$ inches

A_5

Scale
$1'' = \theta.h.d_3$ ins.

B_5

Fig.6. Deflection Diagram

PROBLEM 120.

<u>TO CONSTRUCT DIAGRAMS OF SHEAR, BENDING MOMENT, SLOPE,
& DEFLECTION, FROM THE LOAD CURVE OF A GIVEN BEAM.</u>

2'   3'   1'   Ex.4.   10'   10'   Ex.5.

Problem 121. To draw the Shearing Force and Bending Moment Curves for a Ship.

A ship may be regarded as a beam in equilibrium under the action of a complex system of forces. The weights of the hull, machinery, cargo, &c., are best exhibited by means of a graph, called the **Curve of Weights,** plotted on a base representing the length of the ship. Ordinates of this curve are obtained by dividing the length of the ship into sections, and regarding the sum of the weights in each section as a load uniformly distributed over the section: usually the diagram is " stepped "

The fluid pressure upon the hull varies along the ship. Its resultant for each section is equal to the weight of water displaced by the section and may be calculated from the volume immersed. By plotting these pressures on a base representing the length of the ship, the **Curve of Buoyancy** is obtained. The area under this curve represents the total weight of water displaced by the ship, i.e. its displacement.

Weight and Buoyancy curves for a ship in still water are shown in fig. 1, plotted to the same scale and on the same base; the overhang of the stern causes the curve of weights to extend beyond the curve of buoyancy on the left.. The total areas under each curve are equal, and the centroids of each area must lie on the same ordinate.

By plotting the differences between corresponding ordinates of the two curves, the **Curve of Loads** is obtained. The ship is " water-borne " at those points where the load curve has zero values, e.g. at p and q.

The determination of the **Shear and Bending Moment Curves** from the Load Curve is readily carried out by graphical integration,[*] and is an application of the methods described fully on page 76. The curve of loads has been redrawn to a larger ordinate scale in fig. 2, and 1st and 2nd integral curves constructed in the usual way, N_L being the pole for the 1st integral or shear curve, and N_S the pole for the 2nd integral or bending moment curve. The ordinate scales for the various curves are marked. Both S.F. and B.M. curves must end on the axis, for the areas enclosed by the curves of Load and Shear above and below the axis are equal: this constitutes an excellent check on the accuracy both of the data and the construction.

When the ship floats among waves the curve of buoyancy will vary considerably; a wave crest may come amidships, as in fig. 3, or the ship may rest across the trough of a wave, as in fig. 4, the two conditions having the same effect on the ship as the loads and supports have on the two simple beams shown, and causing hogging in the first and sagging in the second. By assuming trochoidal wave forms approximating to the extreme types that the vessel is likely to meet, the immersed volumes may be calculated for both conditions and curves of buoyancy plotted. For a full discussion of these problems the student is referred to books on Naval Architecture.

EXAMPLES

(1) The curve of loads for a ship 220' 0" long is shown in figure, ordinates representing tons per foot run. Draw S.F. and B.M. curves and state the maximum values. Also give the value of the midship B.M. Ans. Max. S.F., 20 tons, 40 ft. from ends; max. B.M., 630 ft. tons, 65 ft. from ends. Midship B.M., 282 ft. tons.

(2) The curve of weights for a rectangular crane lighter, 100 ft. × 40 ft. × 15 ft., is a parabola. The lighter is immersed to a depth of 10 ft. Draw the curves of S.F. and B.M., and state the scales to which they are drawn.

Note: The buoyancy curve is a rectangle. Refer to page 26 for the construction of a parabola.

---

[*] A mechanical integrator, page 90, is usually available in drawing offices.

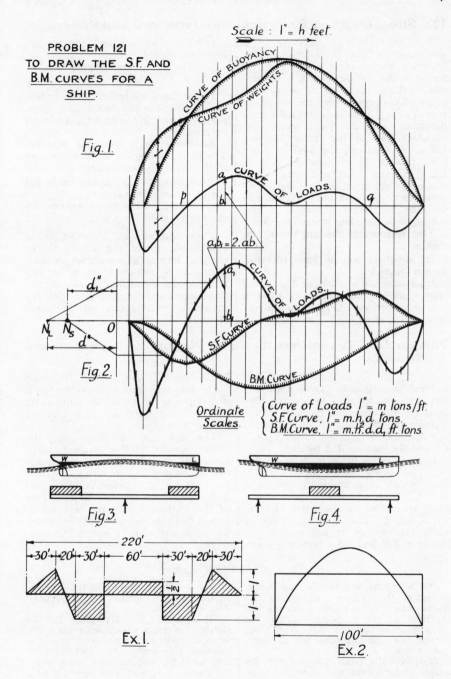

Scale : 1″ = h feet.

PROBLEM 121
TO DRAW THE S.F. AND
B.M. CURVES FOR A
SHIP.

Fig. 1.

CURVE OF BUOYANCY
CURVE OF WEIGHTS.

a CURVE OF LOADS.
b
p   q

$a_1 b_1 = 2 . ab$

a_1 CURVE OF LOADS.

d_1''

b_1

S.F CURVE

N_L N_S O

d″

Fig. 2.

B.M. CURVE.

Ordinate
Scales.
{ Curve of Loads 1″ = m tons/ft.
  S.F. Curve, 1″ = m.h.d. tons.
  B.M. Curve, 1″ = m.h².d.d₁ ft. tons.

W   L

Fig. 3. ↑

W   L

↑   Fig. 4.   ↑

220′
30′ 20′ 30′   60′   30′ 20′ 30′

Ex. 1.

100′

Ex. 2.

122. Stress Diagrams for Framed Structures in Equilibrium.

The forces in the members of a simple frame may be determined by applying the triangle or polygon of forces to the various joints in succession, taking them in a convenient order. This necessitates the construction of a number of disconnected vector diagrams, and the force vector for each member of the frame appears twice. In general, the various polygons can be fitted together into a single **Stress Diagram,** with considerable simplification. The construction of the separate polygons will first be considered, then their assembly in a complete stress diagram.

The following assumptions are made:

(1) Members are secured together by pin joints.*

(2) Loads are applied at the joints only. (Loads applied between the joints can be apportioned to give the equivalent joint loading.)

(3) The members and the applied loads are coplanar.

(4) The frames are " simply firm "— see fig. 1. Test: bars = (2 × joints) − 3.

Under these conditions the members are subjected to simple tensile or compressive forces in the directions of their lengths.

External and Internal Forces. The forces acting on a structure may be divided into two groups:

(1) External—the applied loads and the reactions at the supports.

(2) Internal—the forces exerted by the members on the joints.

The structure as a whole is in equilibrium under the action of the external forces; any member, or combination of members, is in equilibrium under the action of the external forces applied to it, together with the internal forces acting in the part selected.

Problem 123. To construct the Stress Diagram for a given Framed Structure.

Fig. 2 shows a loaded structure lettered according to Bow's notation. Draw the vector and funicular polygons for the external forces and determine the reactions at the supports; reaction CD is represented by vector *cd*, and the reaction DA by *da*.

Draw a force polygon (or triangle) for each joint of the frame, beginning at a joint which is determinate (joints I and III have each three unknowns); they are shown in the order II, III, IV, I, fig. 3. The △ of forces for II gives the vectors *be* and *ea* for the forces in BE and EA; knowing these

forces, joints I and III may be dealt with, for each member exerts equal and opposite forces on the joints at its ends. The polygon for joint I serves merely as a check.

The five force diagrams may be fitted together into a single figure— the stress diagram, fig. 4. Arrow heads are unsuitable in the stress diagram, for each vector represents two equal and opposite forces, and the directions of the various forces are best determined from the sequence of the lettering, as described fully on the following page.

EXAMPLES

(1) Determine graphically the forces acting in the various members of the frame shown without first obtaining the reactions at the supports. Draw the complete stress diagram. Ans. Top, 1·73; bottom, 2·6; sloping bars, left to right, 3·46, 1, 5·2.

(2) Draw the stress diagram for the simple girder shown and state the force

acting in each member. The triangles are similar and the common base is vertical. Ans. 500.

(3) Draw the stress diagram for the simple roof truss shown, and state the force acting in each member. Obtain the reactions graphically. Ans. Reading from left to right, rafters, 9·07, 7·07, 6·37; horizontals, 7·85, 3·2; inner bars, 3·5, 5·83.

* Rigid joints require another treatment: refer to that by Hardy Cross.

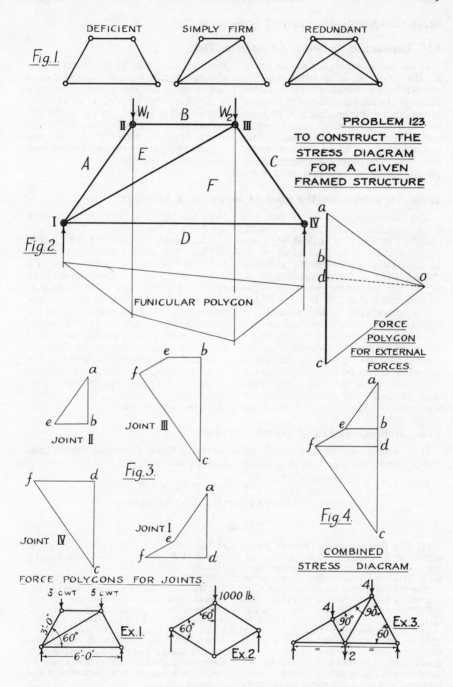

Fig. 1.

DEFICIENT SIMPLY FIRM REDUNDANT

PROBLEM 123.
TO CONSTRUCT THE
STRESS DIAGRAM
FOR A GIVEN
FRAMED STRUCTURE

Fig. 2.

FUNICULAR POLYGON

FORCE
POLYGON
FOR EXTERNAL
FORCES.

JOINT II

JOINT III

Fig. 3.

JOINT IV

JOINT I

Fig. 4.

COMBINED
STRESS DIAGRAM.

FORCE POLYGONS FOR JOINTS.

3 CWT 5 CWT
3'-0"
60°
6'-0"
Ex. 1.

1000 lb.
60°
60°
Ex. 2.

4
4
90°
90°
60°
2
Ex. 3.

Stress Diagrams for Framed Structures (*cont*.).

124. Distinction between Struts and Ties.

If the forces which act at the joints at the ends of a member tend to shorten it, the member is in compression and is called a **strut**: the internal forces oppose this tendency to shorten, and the member itself exerts equal and opposite forces at the joints. This is illustrated in fig. 1: the external forces at (*a*) produce the internal opposing forces at (*b*), and in a force diagram arrows as at (*c*) indicate that the member is a strut. Conversely if the forces at the joints tend to lengthen a member, the member is in tension, and is called a **tie**. In fig. 2 the external forces at (*a*) produce the internal opposing forces at (*b*), and the arrows at (*c*) indicate in a diagram that the member is a tie.

124a. To determine the kind of Stress in a Member.

The frame discussed on the previous page, together with its stress diagram, is shown in fig. 3. Consider the force diagram *abcd*—it is convenient to regard it as a narrow rectangle, as shown dotted. The **order of vectors** is *ab*, *bc*, *cd*, *da*, corresponding to an order of external forces AB, BC, CD, DA— i.e. a **clockwise order** around the frame. Take the letters around *all* the joints in the same order and insert arrow heads on each member in the direction or way given by the corresponding vector in the force diagram. The arrow gives the way of the force with which the bar in question acts on the joint. If the force is toward the joint the bar is a strut; if away from the joint, a tie.

As an example consider the joint I, the force polygon for which is thickened in. The characteristic order around I is clockwise; i.e. the members must be read as AE, EF, and FD (not as EA, FE, or DF); the vector *ae* (i.e. the direction *a* → *e*) gives the way of the arrow head on AE at I; the vector *ef* (i.e. from *e* → *f*) settles the way of the arrow on EF; and so for FD.

If the force diagram is drawn in the reverse order, i.e. taking the external forces in an anti-clockwise order, then the characteristic order for all joints must be anti-clockwise.

124b. Representation of Struts and Ties.

It is convenient to distinguish between the struts and ties in a frame by using thick, thin, and dotted lines as follows:

$\left\{ \begin{array}{l} \text{Struts—thick lines} \\ \text{Ties—thin lines} \end{array} \right\}$ As shown in fig. 3 and Ex. 1.

Members not stressed (see Ex. 4)—Dotted lines.

EXAMPLES

(1) The figure shows a Warren girder and its stress diagram for the given loading. Construct the stress diagram to a scale of 1″ = 1 ton, measure the forces in the various members, and confirm that the struts and ties are as marked. Ans. DE, DF, EF, 3·85; DH, DK, HK, 3·08; GF, GH, ·77; CE, 1·92; BG, 3·46; AK, 1·55.

(2) Draw the given roof truss, scale 1″ = 10′ 0″; the bars HK and MN bisect the main rafters at right ∠s. Draw the stress diagram, as shown, scale 1″ = 500 lb.; measure the forces acting in the various members and mark the members as struts or ties. Ans. BH, EN, 3850 s.; CK, DM, 3350 s.; HK, MN, 870 s.; KL, LM, 1470 t.; HG, NG, 3375 t.; LG, 2010 t.

(3) Draw the complete stress diagram for the given structure and specify the various stresses. Assume the crossed members to be pinned at P. Ans. AE, 3·07 s.; BF, 2·2 s.; CG, 3·95 s.; EF, DG, 1·55 t.; ED, FG, 1·2 t.

(4) Draw the complete stress diagram for the bridge girder shown and specify the various stresses. Show that when the loads are equal there is no stress in the diagonal member.

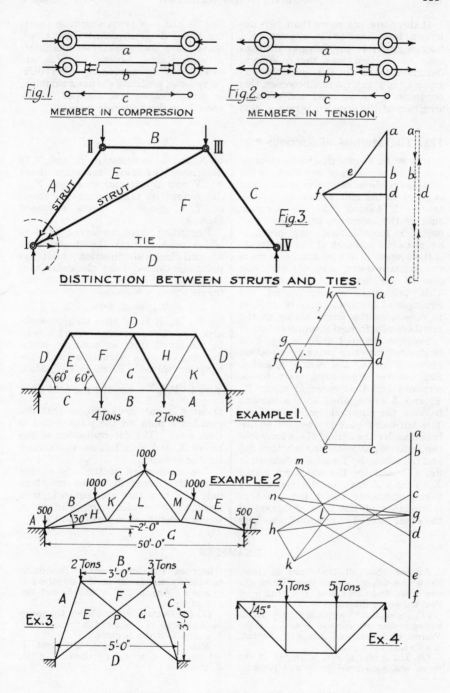

Fig. I.

MEMBER IN COMPRESSION.

Fig. 2.

MEMBER IN TENSION.

Fig. 3.

STRUT

STRUT

TIE

DISTINCTION BETWEEN STRUTS AND TIES.

60° 60°

4 TONS   2 TONS

EXAMPLE 1.

1000

1000   EXAMPLE 2

1000

500   30°   500

2'-0"

50'-0"

2 TONS   3 TONS
3'-0"

3'-0"

Ex. 3.

5'-0"

3 TONS   5 TONS

45°

Ex. 4.

If there are not more than two unknown forces acting at a joint in a framed structure the polygon of forces may be applied and the unknowns determined—as in the previous examples. In many frames, however, this condition is not satisfied and the construction of the stress diagram is held up: in the roof truss shown in fig. 1, the stress diagram cannot be started, there being more than two unknowns at each joint. Several methods are available for overcoming this difficulty; one, probably the most common, is given below, and another is discussed on the following page.

125. The Method of Sections.*

It may be shown that the algebraic sum of the moments of two forces in the same plane about any point in the plane is equal to the moment of their resultant.† It follows that the algebraic sum of the moments of a system of coplanar forces about any point is equal to the moment of the resultant; if the system is in equilibrium there is no resultant force, and the algebraic sum of the moments must be zero. This principle, usually called the Principle of Moments, may be applied to determine the forces acting in the members of a framed structure.

Suppose the roof truss in fig. 1 to be divided into two parts by a section plane SS cutting the bars a, b, and c. Suppose the right-hand part to be removed and that external forces X, Y, and Z are applied to the bars to balance the internal forces in them. The left-hand part is then in equilibrium under the action of six forces:— the loads W_1 and W_2, the reaction R_1, and the forces X, Y and Z, as shown in fig. 2. Each of the unknown forces X, Y and Z may be determined by taking moments about the point of intersection of the lines of action of the other two. For example, the lines of X and Z intersect in p, and Y is determined by taking moments about p; Y and Z intersect in q, and X is determined by taking moments about q; Z is given by taking moments about r.

Regarding those moments as +ve which tend to turn the structure in an anti-clockwise direction about the point considered, and clockwise moments as −ve, the equation of moments about p becomes:

$$-W_2 . m + Y . n = 0,$$

for W_1, R_1, X and Z have zero moments about p. The distances m and n may be measured from an accurate scale-drawing of the structure and the equation solved. From this equation

$$Y = (+)W_2 \frac{m}{n}: \text{ i.e. the moment of Y}$$

about p is anti-clockwise, so that the member b pulls on the joint d and is thus a tie. The determination of the forces X and Z is left as an exercise for the student.

The method of sections is of use when the forces in a few members only of a structure are required; it is specially convenient for girders of uniform depth, as in Exs. 1 and 4.

EXAMPLES

Assume that all loads are in tons. Draw the figures about three times the size given and check Exs. 1, 2, and 3 by drawing stress diagrams.

(1) Determine completely the forces acting in the members a, b, and c of the Warren girder shown. Ans. a, 9·24 strut; b, 2·89 tie; c, 7·79 tie.

(2) The girder shown is hinged at the upper end to a fixed point and rests on rollers at the lower end. Determine completely the forces in the members a, b, and c. Ans. a, 2·19 strut; b, ·006 tie; c, 2·49 tie.

(3) (4) Determine completely the forces in the members a, b, c of the truss, and a, b, c, and d of the N girder.

Ans. (3) a, 6·62 strut; b, ·053 strut; c, 6·1 tie. (4) a, 0; b, 14·5 strut; c, 4 tie; d, 11·66 tie.

---

* Ascribed to both Ritter and Rankine.   † Varignon's theorem.

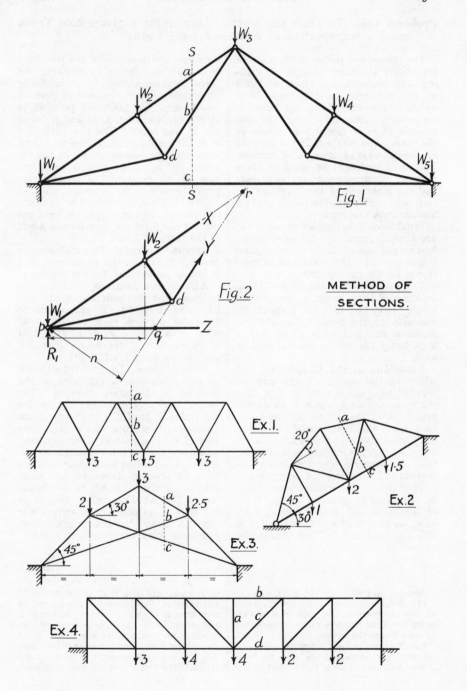

W_3

s
a

W_2

b

W_4

d

W_1

c
s

W_5

Fig. 1.

r

X

W_2

Y

Fig. 2.

d

W_1

p
R_1

m q Z

n

METHOD OF SECTIONS.

a

b

c

Ex.1.

↓3   ↓5   ↓3

a

20°

b

c ↓1·5

45°

↓2

30° ↓1

Ex.2

3

2   30° a 2·5

b

45°

c

Ex.3.

=   =   =   =

b

a c

d

Ex.4.

↓3   ↓4   ↓4   ↓2   ↓2

Problem 126. To draw the Stress Diagram for a given Roof Truss under a combination of Wind and Dead Loads.

The pressure of the wind upon a roof introduces a system of oblique loading, and one or both of the reactions at the supports must have a horizontal component. Large roof trusses are usually anchored at one end and allowed to move freely over rollers at the other; the reaction at the support for the rollers is assumed to be vertical. For small trusses it is usual to allow freedom of movement at both ends and the reactions at the supports are statically indeterminate unless certain assumptions are made.

Wind loads (which may be negative) are superimposed on the dead loads, first on one side of the roof and then on the other, and the members proportioned for the greater stress.

Two sets of forces are shown acting on the roof truss in fig. 1, the oblique forces being due to wind pressure on one side of the roof. The truss is anchored at F and is free to move over a perfectly smooth horizontal surface at A.

Reactions at the Supports. The whole of the applied loads may be reduced to a single resultant, so that the structure as a whole is in equilibrium under the action of this resultant load, the reaction at A—vertically upwards—and the reaction at F, unknown both in magnitude and direction. These three forces must be concurrent and are therefore determinate. To simplify the construction, begin by obtaining graphically the resultant of the wind and dead loads at each joint, as shown in fig. 2, lettered AB, BC, CD, and DE. Then obtain the

direction and position of the resultant of the applied forces by drawing the force polygon abcdef, fig. 3, and the funicular polygon: o is the pole in fig. 3, and the funicular polygon is clearly shown in fig. 2, pr and qr being the closing sides. The line of action sr of the resultant is parallel to af and intersects the line of action of the reaction AG in s: this settles the direction of the reaction GF which must pass through s. The magnitudes of FG and GA are determined by drawing the triangle of forces fga, i.e. by completing the force polygon for the whole of the external forces.

Stress Diagram. This may now be proceeded with in the usual way, and is left as an exercise for the student.

Alternative Method.—Having determined the resultant forces AB, BC, CD, and DE, fig. 2, the complete funicular polygon may be drawn at once, if a start is made at the point of intersection of the known load EF and the unknown reaction FG.

Special Case. The roof truss shown in fig. 4 presents a difficulty in the drawing of the stress diagram. The vector figures are readily drawn for the joints I, II, III, but there remain more than two unknowns at the joints IV and V. The method of sections may be employed, as on previous page. An alternative method is to substitute the form of bracing shown in fig. 5, to enable the force in AB to be determined, this force not being affected by the change. Once AB is known, the stress diagram for the original configuration may be proceeded with.

EXAMPLES

Assume that all loads are in lb. wt.

(1) Determine completely the forces in all the members of the roof truss shown in fig. 1. Take the directions of the wind forces perp. to the rafter.

(2) The truss in fig. 4 is symmetrical and the dead load system is the same for both sides. Assume the truss to be fixed at the

right-hand end and that the reaction at I is vertical. Apply wind loads normal to the rafter first on the left side, as in fig. 1, and secondly on the right side, of 700 lb. at each intermediate joint and 350 lb. at the extremities. Draw the complete stress diagram for both systems of loading and tabulate the two forces for each member.

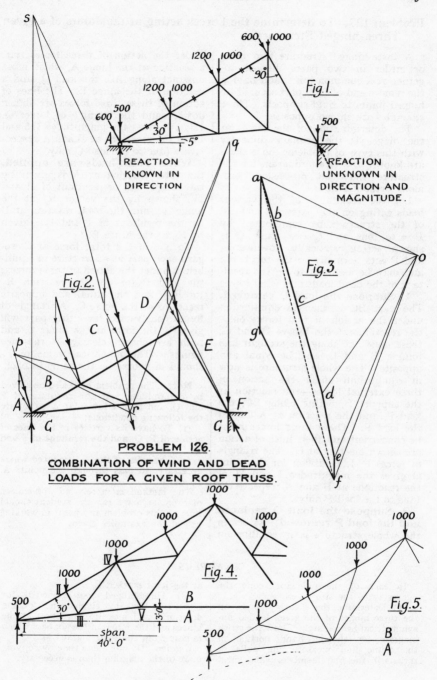

Fig. 1.

600 1000
1200 1000
1200 1000
600 1000

90°

30°

-5°

600 500

A

500

F

REACTION
KNOWN IN
DIRECTION

REACTION
UNKNOWN IN
DIRECTION AND
MAGNITUDE.

Fig. 2.

s

p

A
B
C
D
E
F
G
G

q

r

Fig. 3.

a
b
c
g
d
e
f

o

PROBLEM 126.
COMBINATION OF WIND AND DEAD
LOADS FOR A GIVEN ROOF TRUSS.

Fig. 4.

1000
1000
1000
1000
1000
500

II
IV
III
V
I

30°

3'0"

span
40'-0"

B
A

Fig. 5.

1000
1000
1000
1000
500

B
A

Problem 127. To determine the Forces acting at the Joints of a given Three-hinged Structure.

A three-hinged structure is a truss or girder in two parts, hinged together at a common joint, and having the remote end of each part secured by hinged joints to rigid supports. Three examples are shown opposite.

The determination of the forces at the hinges is the primary difficulty with this type of problem: once these are known, the stress diagram for the structure may be proceeded with along orthodox lines.

Let the resultants of the various loads acting on the parts AC and BC of the structure be represented by P = 10 tons and Q = 8 tons: Q is shown acting vertically downwards, and P acts normally to the top boom and may be assumed to be the resultant of the dead loads and wind forces

1. Suppose the load Q removed. The part BC is then in equilibrium under the action of two forces only, the reactions at the hinges B and C; these must act along the straight line joining B and C, and be equal and opposite. The whole structure is now in equilibrium under the action of three external forces—the reaction at the support B, acting along BC, the load P, and the reaction at A due to the load P. These three forces must be concurrent and their lines of action are shown meeting at O. The triangle of forces P, I, II drawn for the point O gives the magnitudes I and II of the reactions at B and A respectively (due to the load P only).

2. Suppose the load Q replaced and the load P removed. As before the whole structure is in equilibrium

under the action of three forces, viz. a reaction at the hinge A, which must now act along AC, the load Q, and a reaction at the hinge B. The lines of action of these three forces are shown dotted, and the triangle of forces Q, III, IV, gives the magnitudes III and IV of the reactions at B and A respectively (due to the load Q only).

When both loads are applied, the total reaction at A is given by the vector sum or resultant of II and IV, shown by the vector R_A at the hinge A; and the total reaction at B by the resultant of I and III, given by R_B at the hinge B.

To obtain the total force at C, regard each part as a structure in equilibrium under the action of three forces: the load P (or Q), the reaction R_A (or R_B), and the equal and opposite reactions at the hinge C. A triangle of forces constructed for either part will give F_C, the force at the hinge C, and both are shown clearly in the two figures. The numerical value of F_C should of course be the same in each.

Note. The problem may also be solved by determining first the resultant of P and Q, and then a funicular $\triangle$ to satisfy the following conditions:

(1) To have its vertices in the lines of action of P, Q, and the resultant of P and Q.

(2) The sides of the $\triangle$, produced where necessary, to pass through the points A, B, and C.

The method is merely an application of Prob. 11, page 12: the student should refer to this problem and use it in working one of the examples given.

EXAMPLES

In each of the following, copy the figures twice the size given.

(1) Determine the shearing forces on the three hinges of the given loaded unsymmetrical braced arch. Draw the stress diagram for the left-hand portion of the frame and measure the pulls or thrusts in the four members which meet

at the joint marked P.

(2) Three-hinged roof principal for a shed. For the given wind loading determine and measure the forces at the hinges and the forces in the six bars which radiate from the joint marked M.

Caution. Do not waste time by copying more of the diagram than is necessary.

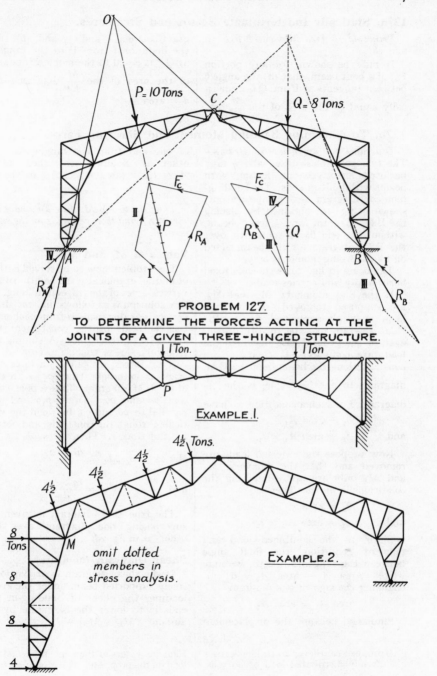

$P = 10\ Tons$

C

$Q = 8\ Tons.$

O

F_c

F_c

II

P

IV

R_A

R_B

Q

I

III

IV

A

B

I

R_A

II

III

R_B

PROBLEM 127.
TO DETERMINE THE FORCES ACTING AT THE
JOINTS OF A GIVEN THREE-HINGED STRUCTURE.

↓1 Ton.   ↓1 Ton.

P

EXAMPLE. I.

$4\tfrac{1}{2}$ Tons.

$4\tfrac{1}{2}$

$4\tfrac{1}{2}$

$4\tfrac{1}{2}$

$\dfrac{8}{Tons}$

M

omit dotted members in stress analysis.

EXAMPLE. 2.

8

8

4

127a. Statically Indeterminate Beams and Structures.

Principle of Area Moments: figs. 1a and 1b.

It may be shown, for any portion PQ of a bent beam, that (a) the angle θ between tangents at P and Q is numerically equal to the area of the $\frac{M}{EI}$ dia-gram between P and Q; and, (b), that the distance d from P to the tangent at Q * is equal to the moment about P of the area of the $\frac{M}{EI}$ diagram; i.e.

$$d = \text{area} \times l.$$

127b. To determine Bending Moment and Shearing Force.

Diagrams for an Encastré Beam.— The two principles stated above may be used effectively in dealing with beams with built-in ends. The bending-moment diagram is a composite one, shown in fig. 2d, obtained by placing the B.M. diagram for a " free " beam, without the end fixing moments, over the B.M. diagram for the beam acted on by the fixing moments only.

The beam in fig. 2 has its ends fixed horizontally and carries loads.

If the end moments, M_L and M_R, are supposed removed, so that the beam is simply supported, fig. 2a, the B.M. diagram may be readily drawn, even in the general case in which the loads are complex. If E and I are constant along the beam then the B.M. diagram, to another scale, is the $\frac{M}{EI}$ diagram. For such a condition we have

$$\theta_1 = \text{area } A_1$$

and $$d_1 = \text{area } A_1 \times l_1.$$

Now suppose the external loads are removed and that the moments M_L and M_R only are applied, giving the conditions in fig. 2b. Then

$$\theta_2 = \text{area } A_2$$

and $$d_2 = \text{area } A_2 \times l_2.$$

Since, in the combined load and moment condition, the final angle between the tangents is zero, we may write: Area A_1 + area A_2 = 0; or, changing the sign of one diagram,

$$\text{area } A_1 = \text{area } A_2.$$

Similarly, because the displacement of one end from the tangent at the other is finally zero, then (area $A_1 \times l_1$) = (area $A_2 \times l_2$); and hence

$$l_1 = l_2.$$

Thus the two B.M. diagrams have the same area and their centroids on the same vertical.

Values of M_L and M_R.

The problem now to be solved is that of finding graphically the lengths of the vertical sides of the trapezium in fig. 2b. One solution is as follows, fig. 2c: draw verticals at distances ⅓l from each end; the centroids of the constituent tri-angles lie on these lines. Assuming the triangular areas to be represented by vertical forces, we can use a force dia-gram, as on the left. Draw pq = area A_1 of the B.M. diagram. Take a pole o and join po and qo. Draw p_1o and q_1o, parallel to po and qo, between the verticals through the centroids, and obtain r_1o and thus, r. Then $pr = M_L \times l \div 2$.

Hence $$M_L = \frac{pr \times 2}{l}$$

and $$M_R = \frac{rq \times 2}{l}$$

The true B.M. diagram is given by superposing one diagram over the other, as in fig. 2d.

Shear Diagram (no figure).

If the shear diagram is drawn for a freely supported beam under the given loading, the effect of building-in the ends is to lower the base line by an amount $(M_L - M_R) \div l$.

EXAMPLES

A propped cantilever 10 ft. long carries a uniformly distributed load of 20 tons. Find the values of the max. B.M. and the load on the prop. Ans. 25 ft. tons; 7½ tons.

---

* In practice θ is small and the distance d may be measured as shown without appreciable error.

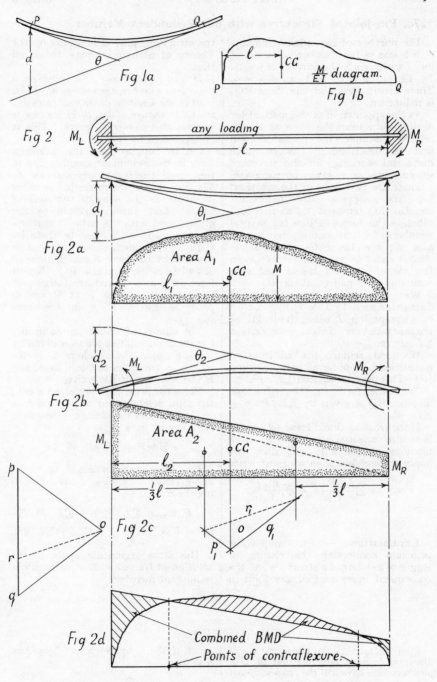

Fig 1a

$\frac{M}{EI}$ diagram.

Fig 1b

Fig 2   any loading

Fig 2a   Area A_1

Fig 2b

Fig 2c

Fig 2d   Combined BMD

Points of contraflexure.

127c. Pin-jointed Structures with one Redundant Member.

The number of redundant members in a frame is the number in excess of $2n - 3$ where n is the number of joints. In fig. 1, $n = 6$, and $2n - 3 = 9$. There are 10 bars and one (here RQ) is redundant.

As the application of the method for dealing with examples such as this is relatively simple, the procedure will first be described. The complete numerical working, for the structure shown opposite, is given on page 346.

There are two steps in the solution: (1) first suppose the redundant member RQ removed; then find and tabulate the forces (P) in the several members, for the loading given: (2) now suppose the external loads removed and ascertain the forces (p) in the several members due to *unit* force acting upon R and Q, along RQ.

We now require the addition of quantities such as $L \times P \times p$ for all bars except RQ, L being the length of the separate bars. This may be written $\Sigma L . P . p$.

We next require the addition of quantities $L \times p^2$ for all bars including RQ. This may be written $\Sigma L . p^2$.

Then the force in the redundant member RQ is given by $\Sigma L . P . p \div \Sigma L . p^2$.

If the cross-sectional areas (A) of the bars are unequal, and if the same modulus of elasticity (E) does not apply throughout, then we must write

$$\text{Force} = \Sigma \frac{L . P . p}{A . E} \div \Sigma \frac{L . p^2}{A . E}.$$

Explanation. — An important principle connecting the change of length of any bar in a structure and the consequent movement of any joint in the structure, proved in works on the Theory of Structures, may be stated thus:

If unit force at any joint produces a force p in any bar, the movement of that joint in the direction of the unit force due to ANY change of length of the bar is p times that change of length. This is illustrated in fig. 2.

Let us suppose that the unknown force in the redundant member due to the given loading is denoted by F. We may regard the forces in the other members as the sum of two sets of forces: first, those produced by the given loads, with the redundant member removed—let us denote these by P (fig. 3); second, those produced in the bars by the force F acting alone at R and Q in the direction RQ. If unit force at R and Q produces forces p in the bars, then force F at R and Q produces forces $F . p$ in the bars (fig. 4).

The change of length caused in any bar by the combined forces is given by $(P + F . p)L \div AE$, where L is the length, A the cross-sectional area, and E the Modulus of Elasticity.

From the principle enunciated above, the total displacement of R in relation to Q is p times the change of length in the various bars

$$= \Sigma p(P + F . p)L \div AE.$$

But the change of length of RQ is also given by $- F . L_{RQ} \div AE$. Equating these:

$$- F . L_{RQ} = \Sigma L . P . p + \Sigma L . p^2 . F,$$

i.e. $- F = \Sigma L . P . p \div (L_{RQ} + \Sigma L . p^2)$.

The same arguments apply to the solution of frames with more than one redundant member.

EXAMPLE

Find the force in the horizontal tie of the frame in fig. 5 assuming that the tension members have half the cross-sectional areas of the compression members. Ans. 5·56 tons.

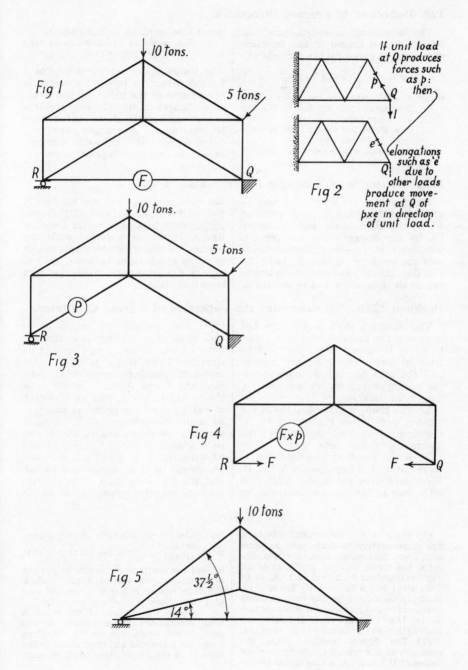

Fig 1

10 tons.

5 tons

R

Q

F

If unit load
at Q produces
forces such
as p:
then

p

Q

1

elongations
such as 'e'
due to
other loads
produce move-
ment at Q of
p×e in direction
of unit load.

e

Q

Fig 2

Fig 3

10 tons.

5 tons

R

Q

P

Fig 4

F×p

R → F

F ← Q

Fig 5

10 tons

$37\frac{1}{2}°$

14°

128. Deflection of Framed Structures.

The alteration in length produced when a bar is loaded in the direction of its axis is given by the formula

$$d = \frac{P}{A} \cdot \frac{L}{E} \quad \text{or} \quad d = f \cdot \frac{L}{E}$$

where d = alteration in length in inches,
„ P = total force applied to the bar in tons,
„ f = stress per square inch, in tons, produced by P,
„ A = total cross sectional area of bar, in square inches,

where L = length of bar in inches,
„ E = modulus of elasticity in tons per square inch.

The forces acting at the joints of a framed structure produce changes in the lengths of the bars, and the structure changes form. The displacement of one point relatively to another may be determined graphically when the deformation "d" of each member, caused by the load applied, is known.

128a. Deflection of a Simple Frame. Figs. 1 and 2.

Suppose that the effect of a load W applied at the joint C is to lengthen AC by the distance Ca, and to shorten BC by the distance Cb: the point C will drop to C_1—the intersection of arcs drawn from A and B, radii Aa and Bb. Actually the changes in length are small in relation to the lengths of

the bars, and the arcs are sufficiently flat to be regarded as straight lines perp. to the bars. The vector diagram, fig. 2, shows to a larger scale the method of obtaining the displacement of C: a comparison between the two figures CbC_1a should make the construction clear.

Problem 128b. To determine the Deflection of a given Cantilever.

The notation used in fig. 2 is not convenient in practice, and that shown in fig. 3 is largely used, and will be adopted here. The joints are lettered and the bars numbered; corresponding numbers and letters are used in the vector diagram.

Let the changes in the lengths of the various bars be as marked in fig. 3, negative quantities indicating the shortening of the struts.

Take the point ab as origin, fig. 4, representing the fixed points A and B. First deal with the frame ABC; set off $b1$ par^1 to bar 1 and representing to

scale the amount by which it is shortened, i.e. its *negative* stretch; draw $a2$ par^1 to bar 2 representing its stretch. Draw perps. $1c$ and $2c$: ac represents the displacement of C. Now take the frame ACE. Set off $c4$ = shortening in bar 4, and $a3$ = stretch in bar 3; draw the perps. $4e$ and $3e$—ae is the displacement of E. Finally set off $c5$ = shortening in bar 5, and $e6$ = stretch in bar 6; perps. $5d$ and $6d$ give the final point d. The displacement of D is represented by ad, and its horizontal and vertical components are represented by od and ao.

EXAMPLES

(1) Draw the displacement diagram, fig. 4, accurately to scale, and measure ao. Then determine, from a stress diagram, the forces, in tons, produced in the various members by a load of 1 ton at D. Take strut forces as −ve, tie forces +ve. Multiply each force by the corresponding change of length given, and add the six products. How does the result compare with the length of ao? Then refer to (b), page 124.

(2) The figures written along the members of the braced cantilever are the values of δl/P or l/EA, giving inches altera-

tion of length per ton force in bar. Determine $\frac{\delta x}{H}, \frac{\delta y}{H}, \frac{\delta x}{W}, \frac{\delta y}{W}$ for the point A. That is, find the component deflections per ton of horizontal and vertical load. Deduce the deflection at A due to a vertical load of 4 tons. Ans. $x = -\cdot032''$, $y = \cdot139''$.

Suppose the motion of the point A were constrained by a smooth vertical guide, what would now be the deflection under the 4-ton load and what would be the force H on the guide? Ans. $\cdot083''$, $7\cdot8$ tons.

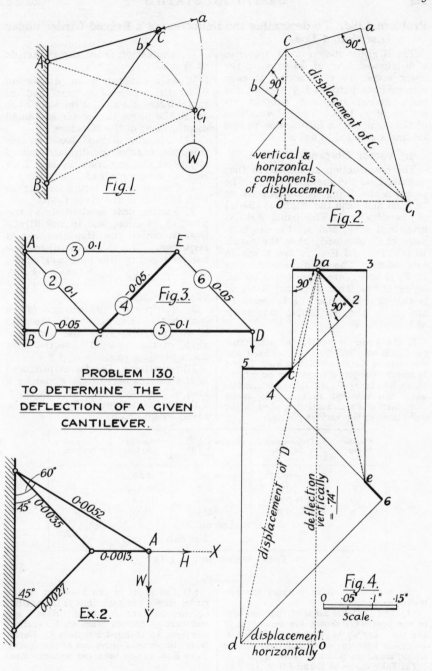

Fig.1.

Fig.2.

displacement of C

vertical & horizontal components of displacement.

Fig.3.

deflection vertically = .74"

displacement of D

displacement horizontally

Fig.4.

scale.

PROBLEM 130.
TO DETERMINE THE
DEFLECTION OF A GIVEN
CANTILEVER.

Ex.2.

Problem 128c. To determine the Deflection of a Braced Girder under a Distributed Load.

The N girder shown in fig. 1 carries a distributed load of 80 tons on the lower boom, 10 tons acting at each intermediate joint and 5 tons at the ends, giving resultant support reactions of 35 tons. The deformations of the members, in inches, due to this loading are stated in fig. 2.

(a) Vector Diagram Method.

The construction is similar to that described on the previous page. The points A and B (fig. 2) are assumed to be fixed and form the origin ab (fig. 4) of the diagram. The panel ABC is first dealt with, and ac, the displacement of C, obtained; then the panels BCD, CDE, DEF . . . are taken in succession. The displacements of D, F, H and L are represented by ad, af, ah, and al. Then ao (or bo) = 1·412″ is the total deflection of L above B, and ol is the horizontal movement of L relative to B.

If the girder is not loaded symmetrically, AB may still be taken as fixed and the relative displacements of both L and M found. The girder must then be rotated about B to bring M and L into a horizontal line: the rotation necessary is easily calculated and a correction applied to the deflections of the various points.

(b) Deflection from the Principle of Work.

When a structure is in equilibrium under the action of external and internal forces, the sum of the work done by all the forces is zero for any small displacement of the structure.

The following method, based on this principle, enables the displacement of any point in a structure, in any direction, to be determined.

1. Obtain the deformation (d) in each bar due to the given loading.

2. Assume **unit load** to act at the point in question, and **in the direction in which the displacement is required.** Find the force (F) in each bar produced by this unit load, acting alone. Take the ratio $\dfrac{\text{F tons}}{\text{1 ton}} \left(\text{or} \dfrac{\text{F lb.}}{\text{1 lb.}} \right)$ as a multiplier (r).

3. Multiply the deformation (d) in each bar by the ratio (r); the sum of the products gives the numerical value of the required deflection—see Ex. 1 previous page.

Illustration. The forces produced in the various bars of the girder by a load of 1 ton at B are shown in fig. 3. Tabulate as follows (labour may be saved by grouping):

Member No.	Deformation (d) Ins.	Ratio: $\dfrac{\text{F Tons}}{\text{1 Ton}}$ (r)	Product (dr) Ins.	Sum of Products
1	—0·05	— 2·0	0·10	
2	0·07	0·7	0·049	
3	—0·04	—0·5	0·02	
4	0·06	1·5	0·09	
		and so on.		
		For half structure		·706″

∴ **Total Deflection = 1·412″.**

EXAMPLES

Use both methods in working the following examples.

(1) The sectional areas of the members of the cantilever shown are such that f, the stress per sq. in., is 5 tons. Taking E = 13,000 tons per sq. in. determine the deflection of the point A. Ans. ·47″.

(2) Take values of E and f from (1) and determine the vertical deflection at the centre of the given girder.

(3) The areas of the members of the girder shown are such that f, in tons per sq. in., has the following values: top members 5; bottom members 6; vertical members 4; inclined members 5. Determine the vertical deflection at the centre. Take E = 12,000 tons per sq. in. Ans. 2·05″.

(4) Obtain the areas of members in (1), (2), and (3).

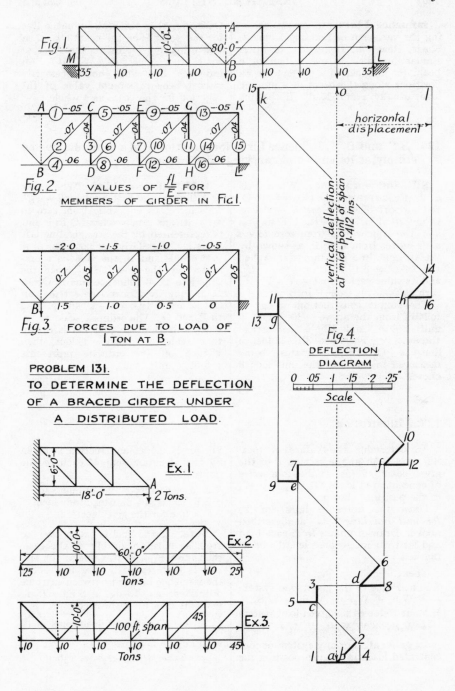

Fig.1.

M _35 ↓10 ↓10 ↓10 B↓10 ↓10 ↓10 ↓10 _35 L

10'-0" 80'-0" A B

Fig.2. VALUES OF $\frac{fL}{E}$ FOR MEMBERS OF GIRDER IN FIG.1.

A ①-.05 C ⑤-.05 E ⑨-.05 G ⑬-.05 K
.07 .04 .07 .04 .07 .04 .07 .04
② ③ ⑥ ⑦ ⑩ ⑪ ⑭ ⑮
B ④.06 D ⑧.06 F ⑫ H ⑯.06 L

Fig.3. FORCES DUE TO LOAD OF 1 TON AT B.

-2.0 -1.5 -1.0 -0.5
0.7 -0.5 0.7 -0.5 0.7 -0.5 0.7 -0.5
B 1.5 1.0 0.5 0

PROBLEM 131.
TO DETERMINE THE DEFLECTION
OF A BRACED GIRDER UNDER
A DISTRIBUTED LOAD.

Ex.1.
6'-0" 18'-0" A ↓2 Tons.

Ex.2.
10'-0" 60'-0"
↓25 ↓10 ↓10 ↓10 ↓10 ↓10 ↓25
Tons

Ex.3.
10'-0" 100 ft span 45
↓10 ↓10 ↓10 ↓10 ↓10 ↓10 45
Tons

Fig.4.
DEFLECTION DIAGRAM

horizontal displacement

vertical deflection at mid-point of span 1.412 ins.

0 .05 .1 .15 .2 .25"
Scale

Influence Lines are curves showing the variation of Bending Moment, Shear, Reaction, Deflection, or other similar function, *at one section* of a beam (or member of a girder) for all positions along the beam of a moving load, usually a unit load. If the line OCB (fig. 1) is the B.M. influence line for the point P on the beam OB, then the ordinate y represents the value of the B.M. *at P, when load* W *is at* A. An influence line does not necessarily indicate the maximum value of the function for the beam as a whole.

129. S.F. and B.M. Influence Lines for a section P of a beam supported simply at its ends and carrying unit load w.

S.F. Influence Line. When w is at A (fig. 2a) distant x ($< a$) from O, the support reactions are $(l - x)/l$ at O and x/l at B. Hence the S.F. at P is x/l: its value varies from zero to a/l as w moves from O to P, as shown by the straight line OE (fig. 2b). When x is $> a$ the S.F. at P is $-(1 - x/l)$; and its value varies from $-(1 - a/l)$ to zero as w moves from P to B.

The simple construction in fig. 2b follows from the above. BC = OD = unit length, and EF is a vertical through P. Then the S.F. influence line is OEFBO, ordinates being measured in terms of the unit length chosen.

B.M. Influence Line. When w is at A (fig. 2a) the B.M. at P is $x(l - a)/l$. This increases uniformly from zero to $a(l - a)/l$ as w moves from O to P and is represented by the straight line OK in fig. 2c. Similarly, for positions of w between P and B, the B.M. is represented by the line KB. Hence the complete B.M. influence line is OKB. It is quickly drawn by setting off OH = a and BG = $l - a$, and joining to B and O. The ordinate scale represents unit B.M.: e.g. if BG is drawn to a scale of 1 inch = 1 foot, then 1 inch on the ordinate represents 1 foot × unit load.

129a. Illustration.

To determine the values of total S.F. and B.M. at the section P of the given beam OB (fig. 3), due to a series of concentrated loads W_1, W_2, W_3, W_4 in the positions shown.

Draw the influence lines for P, *for unit travelling load*, as described above. Draw ordinates from each load and let their intercepted lengths be as marked.

Then:
S.F. at P due to the given load system
= $W_4S_4 + W_3S_3 + W_2S_2 - W_1S_1$.
B.M. at P due to the given load system
= $W_4M_4 + W_3M_3 + W_2M_2 + W_1M_1$.

In general. For any system of concentrated loads in a given position the SF. (or B.M.) at any section is given by the algebraic sum of each of the products:

load × corresponding ordinate to S.F. (or B.M.) unit-load influence line for the given section.

Note. If the load is taken over an area, as must inevitably be the case, then the S.F. or B.M. value at the section is represented by the *area* under the influence line and between extreme ordinates: e.g. in fig. 4, S.F. = (load per unit length) × (projected area) = $\dfrac{W}{l}$ × A. Distributed loads are dealt with on the next page.

Examples. See page 136.

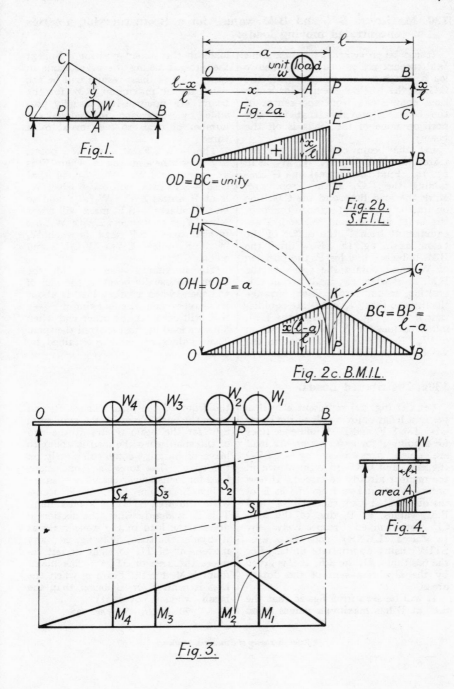

C

O   P   A   B

y

W

Fig. 1.

a

ℓ

O unit load B

w

$\dfrac{\ell-x}{\ell}$ x A P $\dfrac{x}{\ell}$

Fig. 2a.

$OD = BC = unity$

E

O $\dfrac{x}{\ell}$ +   P $\dfrac{x}{\ell}$ B

F

D

Fig. 2b.
S.F.I.L.

$OH = OP = a$

H

K

G

$BG = BP = \ell - a$

O $\dfrac{x(\ell-a)}{\ell}$ P   B

Fig. 2c. B.M.I.L.

W_4 W_3 W_2 W_1

O   P   B

S_4 S_3 S_2

S_1

W

ℓ

area A

Fig. 4.

M_4 M_3 M_2 M_1

Fig. 3.

130. Maximum S.F. and B.M. values for a Beam carrying a series of concentrated moving loads.

It may be proved that the maximum B.M. occurs under one of the heavy loads; further, that this load and the C.G. of the whole system of loads are then equidistant from the centre of the beam * (provided that for this position none of the loads is off the beam).

Let the beam carry five concentrated loads W_1, W_2, W_3, W_4, W_5, as in fig. 1a. Find the vertical line G containing the C.G. as in Prob. 116. Mark the load lines and the C.G. line on tracing paper and arrange this over the beam first so that W_1 and G are equidistant from C, the centre of the beam, as in fig. 1b. Now draw the B.M. influence line for P, the position of W_1, and obtain the value of the B.M. at P when the loads are in this position, as indicated. Next arrange the lines so that W_2 and G are equidistant from C (fig. 1c). Draw the B.M. influence line for Q, the new position of W_2. From this diagram obtain the value of the B.M. at Q for the five loads in the second position. Repeat the process taking the other loads in turn. One of these results will be the value of the maximum B.M. for the beam. In practice it is usually possible by inspection to reduce the number of trials to two, or at most three.

The +ve S.F. at any given point, say R, will be a maximum when W_1 is at R unless $\Sigma W > W_1 L/l_1$; in that case +S.F. max. will occur when W_2 is at R unless $\Sigma W > W_2 L/l_2$: and so on. Similarly, −S.F. max. will occur when W_5 is at R unless $\Sigma W > W_5 L/l_4$; in that case −S.F. max. is when W_4 is at R unless $\Sigma W > W_4 L/l_3$; and so on.

The maximum +ve S.F. for the beam will usually occur at the end of the span, when a heavy load is about to leave; and the maximum −ve S.F. will occur at the other end when a heavy load has just entered the span. Their values are readily obtained by trial.

130a. Distributed Loads.

Let CD (fig. 2a) represent a load w per unit length moving along the beam OB. The S.F. and B.M. influence lines for a point P for *unit concentrated* load are given respectively by OEFBO (fig. 2b), and OKB (fig. 2c) drawn in the manner already described. If now ordinates are drawn from the ends of the distributed load, then the S.F. and B.M. values, at P, due to the load CD are given respectively by ($w \times$ area LMNR) and ($w \times$ area STUV) using appropriate units. For the position $C_1 D_1$ the S.F. at P is given by the algebraic sum of the dotted areas.

It will be seen from fig. 2b that the S.F. at P has maximum values when the front of the load reaches P, and when the back leaves P; the maximum S.F. for the beam occurs at the end of the span when the load is about to leave or has just come completely on the span. The foregoing conclusions hold for loads longer than the span.

The maximum B.M. at P occurs when the area STUV is a maximum. This is evidently when the section P divides the load in the same ratio as it divides the span, as in fig. 2d; any movement of TU to right or left reduces the area of STUV. The maximum B.M. for the beam is when the load is central (or, if longer than the beam, when it covers it).

Examples. See page 136.

_____

* Refer to *Theory of Structures*, Morley.

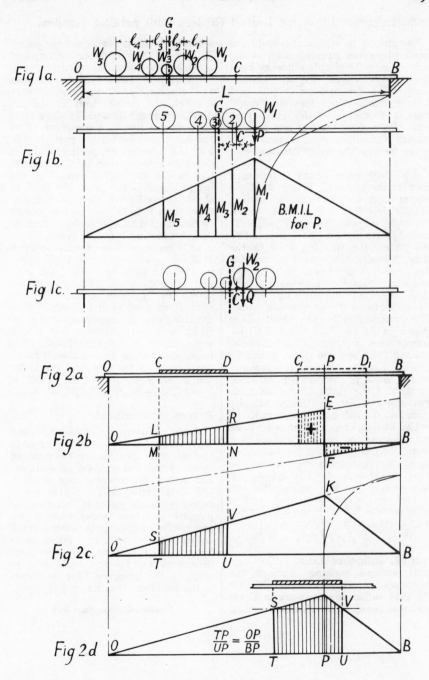

Fig 1a.

Fig 1b.

B.M.I.L.
for P.

Fig 1c.

Fig 2a

Fig 2b

Fig 2c.

Fig 2d

$$\frac{TP}{UP} = \frac{OP}{BP}$$

130b. Influence Lines for Braced Girders with parallel Flanges.

For girders in which the loads are applied by cross girders placed at the panel points the simple influence lines already considered require amendment. The construction of S.F. and B.M. influence lines for one typical truss, a Warren girder (fig. 1), is given here: for a treatment of non-parallel trusses and other special cases, see works on the Theory of Structures.

S.F. Influence Lines. Influence lines for the joints (C, E and G) of the loaded boom are drawn in the manner already described for a simple beam. For all sections between adjacent loaded joints (i.e. for all sections in any one panel) the shearing force is the same, for a given load position, for the load is transmitted only to the joints, not to any intermediate points. Hence the same influence line applies for all sections within the one panel. The construction for the panel EG is shown in fig. 2: the usual S.F. influence lines ak_1 and a_1k terminate at ordinates through the joints E and G, and the diagram (shaded) is completed by the line e_1g_2. Within a panel itself no part of the usual simple influence line applies; for the panel AC, for example, the influence line is ac_2k.

The whole of the S.F. influence lines for the truss can be incorporated in a single diagram without confusion, as in fig. 3. The influence line for the panel AC is ac_2k; for CE, ac_1e_2k; for EG, ae_1g_2k; for GK, ag_1k. For the section P, *or for any other section lying between E and G*, the S.F. for the four loads shown is given by

$$W_4S_4 + W_3S_3 + W_2S_2 - W_1S_1.$$

B.M. Influence Lines. For vertical sections passing through the joints of the loaded boom (C, E and G) the B.M. influence line is drawn as for a simple beam. For all other vertical sections (including those for the joints B, D, F, H) the influence line requires a slight modification. Consider e.g. the section P. Draw the influence lines as if P were a point on a simple beam (fig. 4), but break them at the ordinates through E and G; thus al is the influence line for the portion AE of the beam and km for the portion KG. For the intervening portion EG the influence line is given by the straight line lm, for the load is carried on the joints E and G and hence the B.M. at P varies uniformly with load movement from E to G (here lm happens to be horizontal). The complete influence line, for section P only, is $almk$.

It is convenient in practice to show B.M. influence lines for all panel joints on the one diagram. This has been done in fig. 5; if the construction shown in fig. 4 is applied by the student to each joint he should have no difficulty in interpreting fig. 5, which is merely an assembly of the seven separate influence lines. The influence line for the section through the joint F is the line kg_2e_1a, and for the given loading the total B.M. at F is $W_4M_4 + W_3M_3 + W_2M_2 + W_1M_1$.

Forces in Members.

From a knowledge of the S.F. and B.M. values for any panel, the forces in the members can readily be obtained by applying the method of sections; but with an appropriate scale, the S.F. and B.M. Influence Lines may themselves represent forces in inclined and parallel members respectively. An alternative method is to allow a small deformation to occur in a given member and to determine the consequent deformation of the structure: this, to scale, is the Influence Line for forces in the member. (See Ex. 7, p. 136.)

Examples. · See page 136.

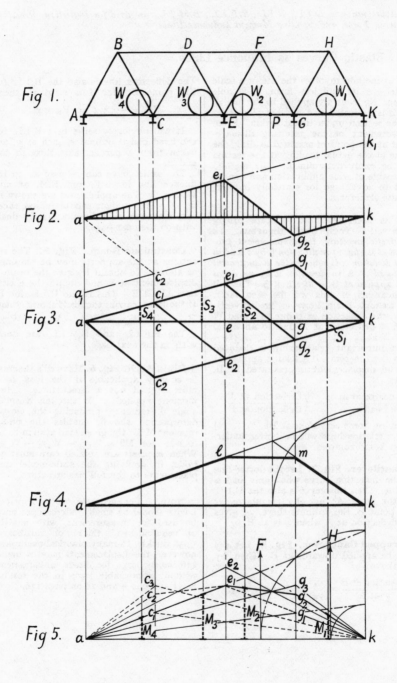

Fig 1.

Fig 2.

Fig 3.

Fig 4.

Fig 5.

Abbreviations: D.I.L., R.I.L., S.F.I.L., B.M.I.L., are used for Deflection, Reaction, Shearing Force, and Bending Moment Influence Lines.

131. Elastic Curves as Influence Lines.

This method involves the use of a scale model—in its simplest form, a flexible spline. No loads are applied to the model, but it is subjected to a small displacement at the section under investigation. From measurement of the resulting displacement at the point of assumed loading, the value of the actual loading at the section can be easily determined. The method is capable of wide application and can be used to advantage for statically indeterminate structures.

It is necessary first to state briefly **Maxwell's reciprocal theorem**. *In an elastic structure, the displacement produced at B (fig. 1) in direction β by a load at A in direction α is equal to the displacement produced at A in direction α by an equal load applied at B in direction β.* The full implications of this will be seen later. For a simple beam, however, loaded vertically, the deflection at B due to a load W at A = deflection at A due to an equal load W at B. . . . . . . (1)

An important relationship is deduced from the theorem: if a load is applied at A, and displacement is prevented at B, then:

$$\frac{\text{Constraint at B}}{\text{Load at A}} = \left(\frac{\text{Deflection at A}}{\text{Deflection at B}}\right)$$

when *any* force acts *alone* at B). . . (2) These relationships will be better understood from examples.

Cantilever. Fig. 2. Let the dotted line be the deflection curve when unit load *w* is at A. Then this curve is also the D.I.L. for the point A; for, by (1) above, at any point B, the ordinate there, y_B, gives the deflection at A when *w* is at B.

Propped Cantilever. Fig. 3. Let *any* load be applied upwards at B. Then, by (2) above:

$$\frac{\text{Reaction at B}}{w(\text{at A})} = \frac{y_A}{y_B}; \text{ or } R_B y_B = w y_A.$$

The deflection line is also the R.I.L. for the point B; for if *w* is at C, then:

$$R_B = \frac{w}{y_B} \cdot y_C; \text{ i.e. } R_B \propto y_C.$$

It will obviously serve as a R.I.L. for the fixed end if ordinates, such as *y*, are taken from a parallel axis through the end B.

The same curve can be used as an influence line for S.F. and B.M. at any section P. The applications are shown in fig. 4. The student should reason these out, although such lines are fully dealt with on the next page.

Continuous Beam. Fig. 5. The reactions at the supports may be obtained as above. To obtain R_B give the beam a displacement at B and obtain the elastic line AB_1CDE. This is the R.I.L. for B. If the beam carries the loads shown, then:

$$R_B \times y_B = W_1 y_1 - W_2 y_2 + W_3 y_3 + W_4 y_4.$$

The S.F.I.L. and B.M.I.L. are dealt with on the next page.

Rigid Arch. Fig. 6. Maxwell's theorem is equally applicable if the load is a moment and not a direct force. *Any moment applied at B through a small angle θ (radians), producing the elastic deformation shown, enables the fixing moment M for the given load system to be found;* for $M\theta = W_1 y_1 - W_2 y_2 - W_3 y_3.$ When *moments* are applied care must be taken in applying the scale-model displacements to the full size structure.

Note. For accurate work displacements should be small. Microscopes may be used for measurement, with models of celluloid (·1″ thick) or cardboard (·05″ thick). In many practical examples, however, the displacements may be made sufficiently large for direct measurement without appreciable error in the result. See Examples 9 and 10 on page 136.

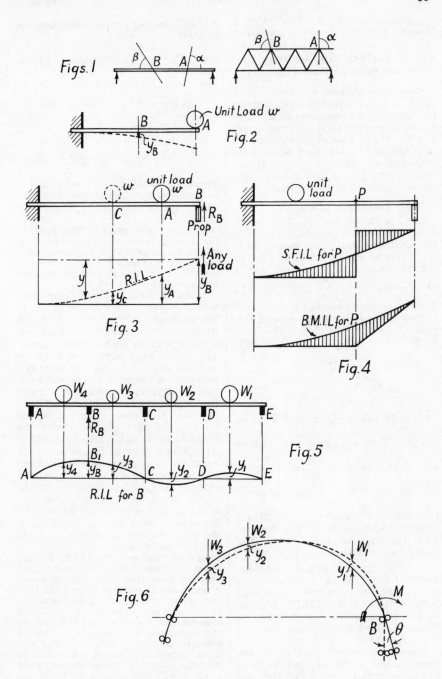

Figs. 1

Fig. 2

Unit Load w

Fig. 3

R_B
Prop
Any load
R.I.L

Fig. 4

unit load
P
S.F.I.L for P
B.M.I.L for P

Fig. 5

R.I.L for B

Fig. 6

Abbreviations: *D.I.L., R.I.L., S.F.I.L., B.M.I.L., are used for Deflection, Reaction, Shearing Force, and Bending Moment Influence Lines.*

131a. Further Applications of Elastic Curves as Influence Lines. Continuous Beam (Contd.)

Reaction Influence Lines. The R.I.L. for any support may be obtained by applying a displacement at that support, as shown on previous page. E.g. the R.I.L. at A is shown in fig. 2: it is the D.I.L. obtained when A is displaced y_A, while B, C and D are held. Then $R_A = (y_P/y_A) \times$ (unit load) for any position P of unit load.

B.M. Influence Lines.

(a) *For the support B.* The D.I.L. of fig. 2 is drawn again in fig. 3a. For all positions of unit load from B to D this curve is also the B.M.I.L. for B. E.g. the B.M. at B when unit load is at $Q = R_A \cdot AB = \dfrac{AB}{y_A} \cdot y_Q$; i.e. the B.M. $\propto$ ordinates such as y_Q.

Now consider the load to be at any point R *between* A and B. The B.M. at B is now $(1 \times RB) - (R_A \times AB)$

$$= RB - \frac{y_R}{y_A} \cdot AB$$

$$= \frac{AB}{y_A}\left(\frac{RB}{AB} \cdot y_A - y_R\right).$$

Join A_1B and mark S and T as in fig. 3b. Then $\dfrac{RB}{AB} \cdot y_A = RT$. Hence the B.M. at $B = \dfrac{AB}{y_A}(RT - y_R) = \dfrac{AB}{y_A} \cdot ST$; and B.M. values at B for load positions between A and B are given to scale by intercepts such as ST. The complete B.M.I.L. for B is shown by the shaded areas in fig. 3c.

(b) *For a point R between supports.* The D.I.L. of fig. 2 is repeated in fig. 4a. For *all* positions of unit load between R and D the B.M.I.L. is given by this curve, shown shaded in fig. 4a. When the load is at Q, the B.M. at $R = (y_Q/y_A)AR$.

Consider now unit load at V, between A and R (fig. 4b), (which is 4a to a larger scale). Join A_1R. Then the B.M. at $R = (R_A \times AR) - (1 \times VR)$

$$= \left(\frac{VX}{y_A} \cdot AR\right) - VR$$

$$= \frac{AR}{y_A}\left(VX - \frac{VR}{AR} \cdot y_A\right)$$

$$= \frac{AR}{y_A}(VX - VW) = \frac{AR}{y_A} \cdot WX.$$

Hence the B.M.I.L. between A and B is given by the area shaded (fig. 4b); the complete B.M.I.L. for R is shown in fig. 4c.

Alternative Method. The B.M.I.L. for B could equally well be obtained by applying a moment at B producing a displacement θ radians (fig. 5). Then if a load W acts at Q, and M is the moment produced at B, $M\theta = W \cdot y_Q$.

Built-in Beam. In applying Maxwell's theorem care must be taken to leave other constraints unchanged when one constraint is displaced. Consider e.g. the built-in beam (fig. 6). The R.I.L. for a support will not be as given in fig. 3, p. 133, for there is a restraint at the ends to prevent rotation as well as displacement: hence the displacement must be applied as in fig. 6a. This curve is of course the S.F.I.L. for B. Fig. 6b shows the B.M.I.L. for B and is self-explanatory. At a point C between the supports, the beam (or spline) may be imagined cut and the two free ends either displaced —to give the S.F.I.L. (fig. 6c)—or subjected to a moment—to give the B.M.I.L. (fig. 6d); in the former the beam ends must remain parallel; in the latter they should receive the same vertical displacement.

Braced Girder. The D.I.L. for the boom of a truss is easily obtained graphically by projecting from a displacement diagram of the type shown on page 125; such an influence line has all the implications of those discussed above. If, for example, unit load is applied at A (fig. 7), the displacement of each point can be obtained as on page 124; this displacement diagram projected on ordinates from the joints of the beam represents the D.I.L. as unit load moves along the boom.

Examples. See page 136.

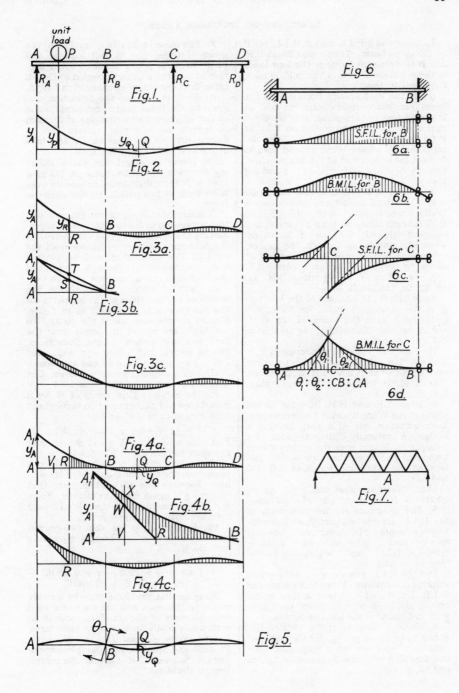

unit load

A P B C D
R_A R_B R_C R_D

Fig.1.

y_A y_P y_Q Q

Fig.2.

y_A y_R A R B C D

Fig.3a.

A_1 y_A T S B A R

Fig.3b.

Fig.3c.

A_1 y_A A V_1 R B Q C D y_Q

Fig.4a.

A_1 X W y_A A' V R B

Fig.4b.

R

Fig.4c.

A θ Q B y_Q

Fig.5.

Fig 6

A B

S.F.I.L. for B

6a.

B.M.I.L. for B

6b.

C S.F.I.L. for C

6c.

B.M.I.L for C

θ_1 θ_2

A C B

$\theta_1 : \theta_2 :: CB : CA$

6d.

A

Fig.7.

Examples on Influence Lines

1. Draw the S.F.I.L. and B.M.I.L. for P on the given beam. Then: (a) Obtain S.F. and B.M. values at P due to the four loads in the positions shown. (Ans. S.F. ·7 tons, B.M. 69·4 ft. tons.)

(b) If the load system travels over the beam from left to right obtain the B.M. at P as each load in turn passes over it; show that one of these is the maximum B.M. at P for any position of the group. Also determine S.F. values when an end load is 6″ from the nearer end of the beam. (Ans. 69·9, 73, 65·8 and 48 ft. tons; —·7 and 1·3 tons.)

2. A load of 8 tons is spread uniformly over a length of 10 feet and moves over a bridge AB 60 ft. long. Sketch the S.F. and B.M. influence lines for points D and C and compute mentally S.F. and B.M. values when the load is (a) just on the bridge, (b) central, and (c) when the front is just about to leave the bridge. (Ans. S.F. $+\frac{2}{3}$, -4, $-\frac{2}{3}$; $+\frac{2}{3}$, 0, $-\frac{2}{3}$ tons; B.M. 32, 48, 8; 20, 110, 20 ft. tons.)

3. (a) Find the position of the loads to give the maximum B.M. for the given girder, and give the value of the B.M. Determine also the value of the maximum B.M. at the centre.

(b) Find the value of maximum S.F. (Ans. Max. B.M. 942·2 ft. tons under centre load when it is 0·59 ft. from centre. Max. B.M. at centre 942 ft. tons. Max. S.F. 48·7 tons.)

4. Draw S.F. and B.M. lines for all the joints of the Warren and N Girders shown. Each girder is one of a pair forming a bridge. A uniformly distributed load of 2 tons per ft. (longer than the girder) moves over cross girders on the lower boom. Determine the maximum forces in AB, BC and AD.

(Ans. Warren: 36·4t and 2·3c; 55·4c; 46·2t tons; N: 25·4t and 2·8c; 48c; 32t tons.)

5. The equations for the two parts of the R.I.L. for the support A of the continuous beam ABC over two unequal spans are: $(a^2 + ab - bx - 1·5ax + ·5x^2/a) \div (aL)$; and $-(bx_1 - 1·5x_1^2 + ·5x_1^3/b) \div (aL)$.
Plot the R.I.L. when $a = ·5b$, taking any suitable dimensions. From this obtain the R.I.L. for B and C. Using a thin spline obtain displacement curves for each support and compare the pairs of curves.

6. Using a wooden spline obtain diagrams as in fig. 4, p. 133, for the propped cantilever. Re-plot on a horizontal base and compare with the diagrams in figs. 6c and 6d, p. 135.

7. The force on any bar, say Dd, of the given K truss, due to unit load at any point, say c, is proportional to the deflection at c due to unit compression of Dd, all other bars remaining unchanged in length.

(i) By first drawing the deflection diagram for the truss due to unit compression of Dd, all other bars remaining unchanged in length, or otherwise, obtain the influence line for the force on Dd as a load of 1 ton travels across the span.

(ii) Hence construct the curve showing the variation of the force on Dd as a load of W tons, distributed uniformly over the length of two panels, travels over the truss.

[Transfer the displacement diagram due to compression in Dd to a horizontal representing the span. Integrate this curve. If the difference between ordinates from the ends of the distributed load be plotted again on the ordinate through the front of the load, this curve gives the force in Dd.]

8. Using a wooden spline obtain the values of the reactions at the supports for the continuous beam loaded as in figure. Repeat for the case where the 20 and 10 ton loads are spread uniformly over the first and last panels. (Ans. Reactions, left to right: 6·78, 26·96, ·31, 6·90, and 4·06; 7·47, 25·3, 1·91, 5·87 and 4·44 tons.)

(Use thin xylonite for Q. 9 and 10. Cut with a fret-saw and finish with file.)

9. The value of H in terms of W for a portal frame of uniform section throughout with hinged ends is given by

$$H = (3 \text{ W.}a.b.) \div 2(2h^2 + 3hl).$$

Plot values of H as W (=1) moves over the span to give the R.I.L. for H.

Now take a model, place it over smooth paper, pin the two ends, and mark its outline. Remove one pin and displace that end, say $\frac{1}{4}$″. Again mark the outline. From the displacements plot the R.I.L. and compare with the former curve.

10. Repeat Q. 9 for the two-pinned circular arch rib. The corresponding formula is:

$$H = W[2 \cos\theta \cos\varphi + 2\varphi \cos\theta \sin\varphi + \tfrac{1}{2} \cos2\varphi - \theta \sin2\theta - 1 - \tfrac{3}{2} \cos2\theta] \div [4\theta - 3 \sin2\theta + 2\theta \cos2\theta].$$

Many similar problems may be quickly solved in this way, and the student should test out some of the standard cases dealt with analytically in works on structures. Refer also to *Journal Inst. C. E.*, 1936, Pippard and Sparkes, on " Simple Experimental Solutions of Certain Structural Design Problems ".

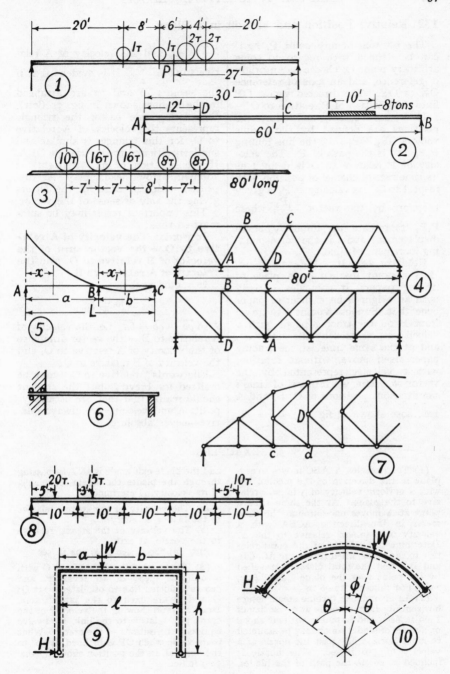

132. Relative Position and Motion in a Plane.

The position of any point P, fig.1, can be defined with regard to some arbitrary point O, chosen as an origin of reference, and an axis of reference OX or OY. The directed vector OP fixes the position of P relative to O.

If P has motion, and successive positions are defined by the radius vectors OP_1, OP_2 . . ., the line joining P_1P_2 . . . is the path of P. The velocity of P relative to O is defined as its time rate of change of position with respect to O; its velocity at P_1 is given therefore by the vector $\frac{P_1P_2}{t}$, where P_1P_2 represents an infinitely small displacement and t is the corresponding increment of time.

Relative Velocity. Refer to fig. 2a. B is a point marked on a piece of tracing paper; it coincides initially with an origin O on a sheet of paper beneath it. Suppose a point A to travel from B on the tracing paper with a velocity represented by the vector ba, and at the same time let the tracing paper itself move, without rotation, with a velocity represented by the vector ob. After an interval of time t simultaneous positions of O, B, and A are those shown in fig. 2b; $\frac{OB}{t} = ob$,

$\frac{BA}{t} = ba$, and the velocity of A relative to O $= \frac{OA}{t}$. It is evident that if the vectors ob and ba are arranged in the manner shown in fig. 2c (left), the closing side oa of the triangle represents the velocity of A relative to O, for this triangle is similar and similarly situated to the triangle OBA. If the notation of fig. 2c (right) is adopted arrow heads may be dispensed with, the order of the letters giving the way or sense of the vector.

This important result may be summarized thus:

In words: **The velocity of A relative to O = the vector sum of the velocity of B relative to O, and the velocity of A relative to B.**

In vector symbols:

$$\overline{oa} = \overline{ob} + \overline{ba}.$$

Also $\overline{oa} - \overline{ob} = \overline{ba}$; i.e. the velocity of A relative to B = the vector difference of the velocity of A relative to O, and the velocity of B relative to O.

The words " relative to O " may be omitted for brevity, but the student should realize that they are necessary; position and velocity are always relative—never absolute.

EXAMPLES

(1) The isosceles $\triangle$ ABC moves over a plane in the direction of the median DC with a uniform velocity of 3 in./sec. relative to the plane. At the same time a point at A moves once around the perimeter in the direction ACBA with a velocity of 2 in./sec. relative to the $\triangle$. Determine the velocity of the point relative to the plane, when traversing AC, CB, and BA, and the total distance travelled by the point over the plane. Ans. 4·92, 1·36, 3·61 in./sec.; 12·06 in.

(2) In an impulse turbine steam issues horizontally from a nozzle at a velocity of 1400 ft./sec., and impinges without shock on a series of blades which may be assumed to move horizontally past the nozzle at a velocity of 450 ft./sec. The nozzle is inclined at 20° to the path of the blades,

and the blade exit angle is 28°. In passing through the blades the steam loses 25% of its velocity. Determine:

(a) The velocity of the steam, relative to the blades, at entry.

(b) The velocity of the steam, relative to the nozzle, at exit.
Ans. (a) 989 ft./sec.; (b) 404 ft./sec.

(3) The crank OP rotates about O with a uniform speed of 10 revs./sec. and causes a slotted link to oscillate about Q; O and Q are fixed to the same frame. Draw a graph showing the velocity of the crank pin, relative to the link, for twelve equidistant positions of the crank. What is its value when OP is inclined at 30° to the vertical, in the position shown? Ans. 50·9 ft./sec.

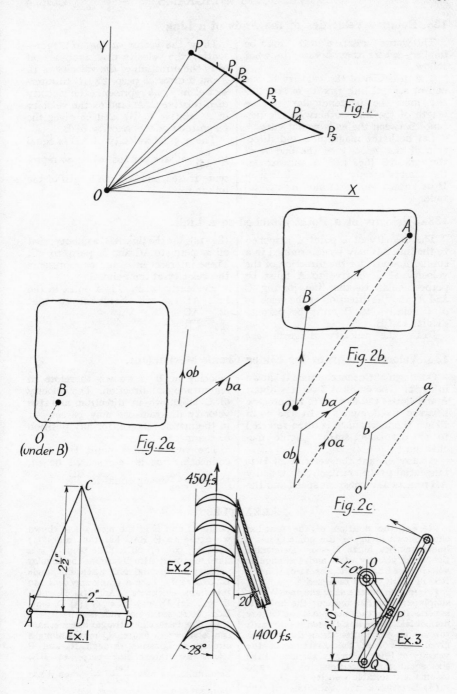

Fig. 1.

Fig. 2a.

O
(under B)

Fig. 2b.

Fig. 2c.

Ex. 1.

Ex. 2.

450 f.s.

1400 f.s.

28°

Ex. 3.

1'-0"

2'-0"

133. Relative Velocities of the Ends of a Link.

The words "relative to O" may be implied, where they obtain, in what follows.

The direction of the velocity of one end of a rigid link *relative to the other end* must be perpendicular to the length of the link—otherwise the distance between the ends would vary.

Let both the magnitude and direction of the velocity of the end A of the link AB (fig. 1a) be known; let the direction only of the velocity of B be known—to determine its magnitude.

Draw the vector *oa* (fig. 1b), representing the velocity of A, and set off *ob* in the direction of the velocity of B. From *a* draw *ab* perp. to AB to intersect *ob* in *b*; *ab* represents the velocity of B relative to A (and *ba* the velocity of A relative to B), and *ob* gives the magnitude of the velocity of B.

The angular velocity of AB is equal to $\dfrac{ab}{AB}$ radians per sec. when *ab* represents ft./sec. and AB the length of the link in feet.

133a. Velocity of a Point attached to a Link.

The velocity of a point C attached to the link AB may be determined in a similar manner, for the direction of the velocity of C relative to A must be perpendicular to the line joining C and A, and the direction of the velocity of C relative to B must be perpendicular to CB.

Draw the velocity diagram *oab*

(fig. 1c), for the link AB, as above. Set off *ac* perp. to AC and *bc* perp. to BC. These intersect in *c*, and *oc* represents the velocity of the point C.

Evidently, when C is a point in the link AB, *c* will divide *ab* so that $\dfrac{ac}{cb} = \dfrac{AC}{CB}$.

134. Velocity Diagram for Slider Crank Mechanism.

One form of this mechanism is shown in fig. 2. The crank AB rotates about A and causes the block C to reciprocate between fixed guides. In the form shown the mechanism is often referred to as the direct-acting engine mechanism.

At any instant the velocity of B is tangential to the crank circle. Usually AB rotates at a constant speed and the

velocity of B is known therefore in magnitude and direction. The velocity of C is known in direction and the velocity diagram *obc* may be set out in the manner shown, for any position of the mechanism.

The velocity of a point D on the connecting rod is represented by *od*, the ratio $\dfrac{bd}{dc}$ being equal to $\dfrac{BD}{DC}$.

EXAMPLES

(1) For the position of the four-bar chain shown in figure, the point Q has a linear velocity of 10 ft./sec. Determine the linear velocity of R and the angular velocities of RS and QR. Ans. 11·02 ft./sec.; RS, 44·1, QR, 18·72 rads./sec.

(2) In the mechanism shown (Watt's simple straight line motion) the link BD is vertical when the cranks AB and CD are horizontal. If the link DC swings upwards through an angle of 60° from the position given, show that the direction of the velocity of P is practically vertical. Take six equidistant positions of DC and assume any suitable velocity for D.

(3) Determine the velocities of the

points D and E in the mechanism shown when the angle BAC has the values (1) 120°, (2) 90°, (3) 60°. (The point F is a fixed fulcrum.) Also determine the angular velocity of the rod ED when the angle BAC is 120°. Take a crank speed of 120 revs./min. clockwise. Ans. (1) D, 7·63, E, ·94; (2) D, 9·43, **E**, 4·71; (3) D, 8·95, E, 13·66 ft./sec.; 9·92 rads./sec.

(4) The Peaucellier straight line mechanism is shown in figure. Q moves along a circular arc passing through R, and P traces a straight line perp. to RN. Determine the ratio $\dfrac{\text{vel. of P}}{\text{vel. of Q}}$ when θ has values 0 and 30°. Ans. 1·92, 2·03.

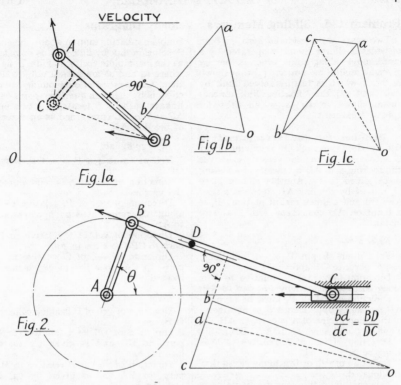

VELOCITY

Fig 1b

Fig 1c.

Fig. 1a

Fig. 2

$$\frac{bd}{dc} = \frac{BD}{DC}$$

VELOCITY DIAGRAM FOR SLIDER-
CRANK MECHANISM.

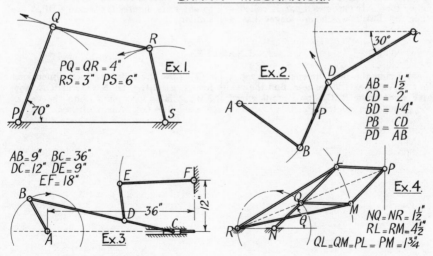

Ex.1.

$PQ = QR = 4''$
$RS = 3''$ $PS = 6''$

Ex.2.

$AB = 1\frac{1}{2}''$
$CD = 2''$
$BD = 1·4''$
$\frac{PB}{PD} = \frac{CD}{AB}$

$AB = 9''$ $BC = 36''$
$DC = 12''$ $DE = 9''$
$EF = 18''$

Ex.3.

Ex.4.

$NQ = NR = 1\frac{1}{2}''$
$RL = RM = 4\frac{1}{2}''$
$QL = QM = PL = PM = 1\frac{3}{4}''$

Problem 135. Sliding Members. Velocity Diagrams.

Two cases are considered here, in both of which a sliding member operated by a crank moves along a link which is thereby given an oscillating motion: (1) that shown in fig. 1 which is readily solved, link by link; and (2) the "floating link" mechanism shown in fig. 2 which calls for special treatment.

Solution for the arrangment in fig. 1.

Here the crank OP revolving about O with constant angular velocity ω, drives the rocking arm AB by means of a sliding piece joined to P. Another sliding piece D, moved by the link AD, receives a slow forward and a quick return motion: P_1 is a point on AB coincident with P at any instant.

Refer to fig. 1a.

Vel. of crank pin P (= vel. of slider) = op, perp. to OP and = OP.ω.

Direction of vel. of P_1 relative to P is along AB, represented by pp_1 par¹ to AB.

Direction of vel. of P_1 must be perp. to AB, i.e. along op_1. Hence p_1 is obtained.

Direction of vel. of A is perp, to AB, i.e. along op_1, and $oa = op_1$ (BA ÷ BP_1).

Direction of vel. of A relative to D is along ad perp. to AD.

Direction of vel. of D is along od, so that d is located.

This completes the velocity diagram.

Solution for "floating link". Fig. 2.

If a second crank is introduced at B, as in fig. 2, then AB becomes what is called a "floating link" which no longer has a simple oscillating motion about the end B. A solution link by link gets us only as far as the incomplete vector diagram in fig. 2a, where oa and ob are known only in direction. One method of overcoming the difficulty is by using an *auxiliary point* on the linkage such as C located by two perps. AC and PC; C is regarded as an extension of the link AB.

Refer to fig. 2b.

Vel. of crank pin P = ω.OP, perp. to OP.

Direction of vel. of A is represented by the vector oa.

Direction of vel. of P_1 relative to P is along AB_1 represented by pp_1 drawn par¹ to AB.

Direction of vel. of C relative to P_1 is perp to CP_1, so c lies in pp_1.

Direction of vel. of C relative to A is perp. to CA, so c lies in oa; hence c is located.

Refer to fig. 2c.

Direction of vel. of B is along ob perp. to O_1B.

Direction of vel. of C relative to B is perp. to CB and is given by cb, thus locating b.

Direction of vel. of B relative to BA is perp. to BA and is given by ba, thus locating a.

p_1 lies in ab and on pc produced. It will be found that p_1 divides ab in the ratio that P_1 divides AB and that cp_1 is perp. to ab; thus, figures cbp_1a and CBP_1A are similar.

EXAMPLES

(1) Using the inch dimensions in fig. 3 for the Walschaert valve gear, find the vel. of V when angle BAC is 60°; crank speed 200 r.p.m. Ans. 7·5 ft./sec.

(2) Take the following inch dimensions for fig. 2: O_1B 3, OP 6, O_1O 12, OA_1 10½, AB 24; ω = 200 r.p.m. Find the vel. of A when θ = 60°. Ans. 16 ft./sec.

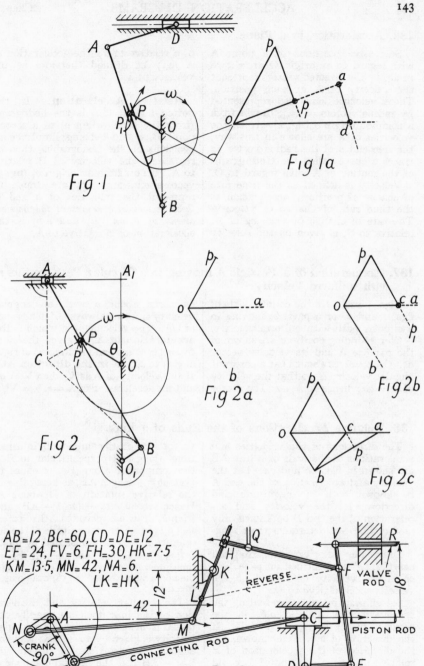

Fig.1

Fig 1a

Fig 2

Fig 2a

Fig 2b

Fig 2c

AB=12, BC=60, CD=DE=12
EF=24, FV=6, FH=30, HK=7·5
KM=13·5, MN=42, NA=6.
LK=HK

Fig 3

REVERSE

VALVE ROD

PISTON ROD

CONNECTING ROD

CRANK

90°

136. Acceleration in a Plane.

Successive positions of a point A with regard to an origin O are shown in fig. 1: the directed vectors represent the velocity of A at each position. These velocities are also represented by radius vectors oa . . . drawn from a common origin o (fig. 2). If all the velocities of A are shown in this way, the free end a of the radius vector oa traces a curve called the **Hodograph** of the motion of A with regard to O.

Velocity is defined as the time rate of change of position; acceleration as the time rate of change of velocity. The rate of change of velocity of A, relative to O, is given by the velocity of a relative to o; the acceleration of A may be defined therefore as the velocity of a.

Relative Acceleration. If the dotted line b......... is the hodograph (from the same origin o) of a second point B, a and b being simultaneous positions on the hodographs, then ab represents the velocity of B relative to A. If now from an origin o_1 (fig. 3) vectors o_1a_1 and o_1b_1 are drawn to represent the velocities of a and b, *these vectors also represent the accelerations of* A *and* B, and a_1b_1 is the acceleration of B relative to A.

137. Acceleration of a Particle A moving in a Circular Path, radius r, with uniform Velocity.

Figs. 4 and 5 are the counterparts of figs. 1 and 2 for a particle moving in a circular path with uniform velocity.

Corresponding positions are shown of the particle A and its velocity vector oa. OA revolves about O at a constant angular velocity ω, so that the velocity of A $= \omega r$: hence $oa = \omega r$. The vector oa turns about o with an angular velocity ω, being always 90° in advance of OA. The velocity of a, which is the **acceleration of A,** is given therefore by $\omega r \cdot \omega = \omega^2 r$, and is directed at right angles to oa, i.e. **in the direction AO.** If the velocity of A is V, then $V = \omega r$ and the acceleration of A $= \omega^2 r = V^2/r$.

138. Relative Accelerations of the Ends of a Link.

The directions of the velocities and accelerations of the ends of a link AB are shown in figs. 6a and 7a. Let the velocity and acceleration of the end A be known both in magnitude and direction: let the velocity and acceleration of the end B be known only in direction—to determine their magnitudes.

Draw the velocity ratio diagram oab (fig. 6b) as described on page 140: ob is the velocity of B and ab the velocity of B relative to A.

Set off o_1a_1 (fig. 7b) representing the acceleration of A, and o_1b_1 in the *direction* of the acceleration of B. If the direction of a_1b_1 were known, i.e. the direction of the acceleration of B relative to A, the position of b_1 could be determined. The vector representing the acceleration of B relative to A, however, may be represented as the vector sum of its components in directions along and perpendicular to BA: the component along BA is equal to (vel. $ab)^2 \div$ length AB, as being due to the relative rotation of B about A. Hence set off $a_1b_2 = (ab)^2 \div$ AB, and from b_2 draw b_2b_1 perp. to AB to intersect o_1b_1 in b_1; the magnitude of the acceleration of B is represented by o_1b_1.

If successive link positions at known time intervals are given (as in Ex. 1, p. 148) the hodograph method (§ 136) can be used effectively.

The angular velocity and acceleration of the link itself are readily obtained from figs. 6b and 7b. The angular velocity is given by $ab \div$ AB; and the angular acceleration is given by $b_1b_2 \div$ AB (i.e. the normal component of acceleration of one end relative to the other $\div$ length of link).

Examples. See page 146.

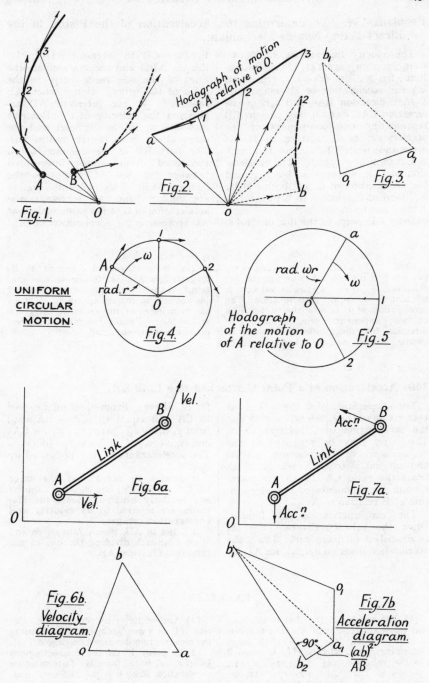

Fig. 1.

Fig. 2.

Hodograph of motion of A relative to O.

Fig. 3.

UNIFORM
CIRCULAR
MOTION.

Fig. 4.

Hodograph of the motion of A relative to O

Fig. 5.

Fig. 6a.

Vel.

Link

Vel.

Fig. 7a.

Acc.ⁿ

Link

Acc.ⁿ

Fig. 6b.
Velocity diagram.

Fig. 7b
Acceleration diagram.

$\dfrac{(ab)^2}{AB}$

90°

Problem 139.   To determine the Acceleration of the Piston in the direct-acting Engine Mechanism.

The velocity diagram, fig. 1b, follows from fig. 2 on page 141. For accelerations, fig. 1c, first draw o_1b_1 representing the acceleration of B—magnitude V_2/AB, direction BA—and o_1c_1 in the direction CA. Set off b_1b_2 parl to CB representing one component of the acceleration of C relative to B—its magnitude is $(bc)^2/BC$. Draw b_2c_1 perp. to CB to meet o_1c_1 in c_1. The acceleration of C is represented by o_1c_1, and the acceleration of C relative to B by b_1c_1 (and of B relative to C by c_1b_1). *Klein's construction.*

Draw AD perp. to the line of stroke.

Produce CB to intersect AD in D; the △s ABD and *obc* are similar, the sides of the one being perp. to the sides of the other. Hence when AB represents V, the intercept AD represents the velocity of C. Describe arcs one on BC as diameter and the other about B as centre, radius BD. The part EF of the common chord (produced if necessary) intercepted between BC and AC completes the figure ABEF. This figure and fig. 1c are similar: hence AF represents the acceleration on C to the same scale that AB represents the acceleration of B.

139a. If the velocity of the crank is *not uniform* but the crank has an angular acceleration α (when its angular velocity is ω), then fig. 1c requires modification. The acceleration of B is now the vector sum of two components: one, as before, in direction BA, magnitude $\omega^2.AB$ due to circular motion; and the other at rt. ∠s to it (i.e. in a tangential direction at B-magnitude $\alpha.AB$, due to angular acceleration.) If these two components are added (as vectors) to replace o_1b_1 in fig. 1c, then the remainder of the construction given still applies. Usually the tangential component is relatively small.

140. Acceleration of a Point C attached to a Link AB.

The acceleration of C (fig. 2a) is the vector sum of the acceleration of A and the acceleration of C relative to A. The latter may be regarded as the vector sum of components normal and tangential to CA, each of which bears the ratio CA/BA to the corresponding components of the acceleration of B relative to A.

The construction is as follows. Draw the velocity diagram (fig. 2b) as described on page 140. Draw the acceleration diagram $o_1a_1b_2b_1$ for AB (as for BC above). From a_1 set off a_1c_2 parl to CA and equal to $(CA \div BA)a_1b_2$; from c_2 draw c_2c_1 perp. to c_2a_1 (or CA) and equal to $(CA \div BA)b_2b_1$. Join o_1c_1. The acceleration of C is represented by o_1c_1.

It should be noted that the three triangles ABC, *abc*, and $a_1b_1c_1$ are similar and similarly situated. Hence the diagrams are referred to **as velocity and acceleration images.**

If C lies in AB, then *c* falls on *ab* and c_1 on a_1b_1, each dividing the line in the ratio that C divides AB.

EXAMPLES

(1) The crank AB (fig. 1a) is 12″ long and turns about A with a uniform speed of 120 revs./min. The rod BC is 48″ long. Determine the acceleration of C when θ has the values 0, 22½° 45°, 67½°, 112½°, 135°, 157½°, 180°. Ans. 197, 173, 111, 32, −88, −111, −118, −118 ft./sec.².

(2) Answer Ex. 1 by plotting the velocity of C on a displacement base and using the construction described on page 82, (1).

(3) A point D on the connecting-rod BC (fig. 1a) is 18″ from B. Determine its acceleration when $\theta = 30°$ and 60°. Ans. 153, 112 ft./sec.².

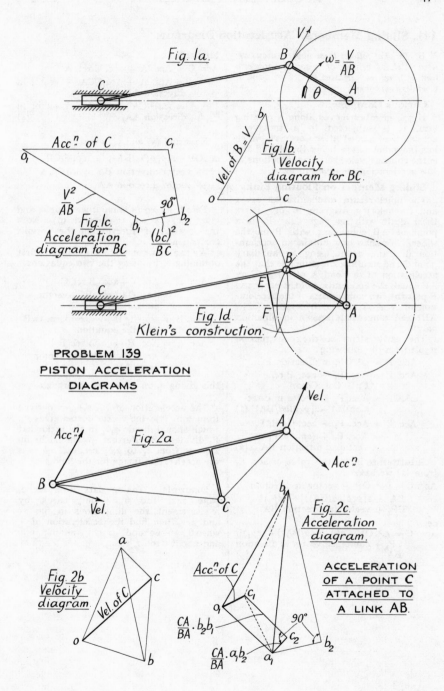

Fig. 1a.

$\omega = \dfrac{V}{AB}$

θ

Acc.ⁿ of C

Fig. 1b.
Velocity
diagram for BC.

Vel. of B = V

$\dfrac{V^2}{AB}$

Fig. 1c.
Acceleration
diagram for BC

90°

$\dfrac{(bc)^2}{BC}$

Fig. 1d.
Klein's construction.

PROBLEM 139
PISTON ACCELERATION
DIAGRAMS

Acc.ⁿ

Fig. 2a

Vel.

Acc.ⁿ

Vel.

Fig. 2c.
Acceleration
diagram.

Fig. 2b
Velocity
diagram.

Vel. of C

Acc.ⁿ of C

$\dfrac{CA}{BA} \cdot b_2 b_1$

$\dfrac{CA}{BA} \cdot a_1 b_2$

90°

ACCELERATION
OF A POINT C
ATTACHED TO
A LINK AB.

141. Sliding Members. Acceleration Diagrams.

N.B. *In what follows these abbreviations are used: Vel. is "velocity of"; acc. is "acceleration of"; rel. is "relative to"; Cor. Com. is Coriolis component.*

Coriolis Component.

If one member moves along a rotating arm, it is subjected to a transverse acceleration, named after Coriolis*, which has the value $2v\omega$, v being the sliding and ω the rotating velocity. The directions of this acceleration are given in fig. 3.

Sliding Member on Floating Link.

The quick-return mechanism in fig. 1 and its velocity diagram in fig. 2 are those dealt with on page 142: P_1 is a point on AB coincident with P on the slider. We shall now obtain an acceleration diagram, again using an auxiliary point C and recalling the principle that the acceleration of one end A of a rigid link AB equals the acceleration of the other end B plus the two components of the acceleration of A about B: centripetal $(=(ab)^2/AB)$ and tangential (known in direction only).

The early steps in the construction result from the following:

Acc. P = ω^2.OP, in direction PO.

Acc. P = Acc. P_1 + acc. P rel.
P_1 + Cor. Com.

∴ ω^2.OP = Acc. P_1 + vector in direction BA + $2[p_1p(ba/BA)]$ (1)

Acc. C = Acc. P_1 + acc. C rel. P_1
= Acc. P_1 + $(cp_1)^2/CP_1$ + vector in direction BA (2)

Substituting in (2) the value of acc. P_1 from (1) we have:

Acc. C = [ω^2.OP − vector in direction BA − $2\{p_1p(ba/BA)\}$] + $(cp_1)^2/$CP$_1$ + vector in direction BA.

i.e.

Acc. C = ω^2.OP − $2\{p_1p(ba/BA)\}$ + $(cp_1)^2/$CP$_1$ + combined vector in direction BA . . . . (3)

Similarly

Acc. C = Acc. A + acc. C rel. A
= Acc. A + $(ca)^2/$CA + vector in direction AA$_1$
= $(ca)^2/$CA + combined vector in direction AA$_1$. . . . (4)

Equating (3) and (4) gives:

ω^2.OP − $2\{p_1p(ba/BA)\}$ + $(cp_1)^2/$CP$_1$ + vector direction BA = $(ca)^2/$CA + vector direction AA$_1$

This is set out in heavy lines in fig. 4 and gives acc. C = o_1c_1 (not drawn in). Note the reversal of direction of the Coriolis component.

We can now complete the diagram, first obtaining b_1 by using the two equations:

Acc. B = Acc. C + acc. B rel. C
= Acc. C + $(bc)^2/$BC + vector perp. BC and

Acc. B = $(ob)^2/O_1$B + vector perp. O_1B
and finally using the equation

Acc. A = Acc. B + acc. A rel. B
= Acc. B + $(ab)^2/$AB + vector perp. AB

this giving a_1 on a horizontal through o_1.

The acceleration of P_1, o_1p_3, is derived from eqn. (1); the vector dotted gives p_3, which should divide a_1b_1 in the ratio that P_1 divides AB. Further, p_3 should lie on a perp. from c_1 to a_1b_1 because $a_1b_1c_1$ is the acceleration image for the linkage.

Example. Taking data from Ex. 2, page 142, and $\theta = -60°$, check by measurement the dimensions in figs. 1 and 2. Then find the acceleration of A when $\theta = -60°$ and $+120°$. Ans. 30 f./s./s. and about 1100 f./s./s.

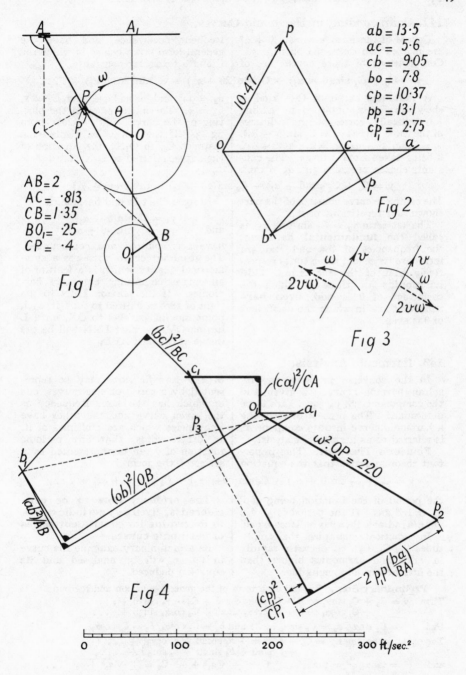

Fig 1

AB = 2
AC = ·8l3
CB = 1·35
BO₁ = ·25
CP = ·4

$AB = 2$
$AC = ·813$
$CB = 1·35$
$BO_1 = ·25$
$CP = ·4$

Fig 2

$ab = 13·5$
$ac = 5·6$
$cb = 9·05$
$bo = 7·8$
$cp = 10·37$
$pp_1 = 13·1$
$cp_1 = 2·75$

10·47

Fig 3

$2v\omega$

Fig 4

$(bc)^2/BC$

$(ca)^2/CA$

$(ob)^2/O_1B$

$(ab)^2/AB$

$(cp_1)^2/CP_1$

$\omega^2.OP = 220$

$2\,p_1p\left(\dfrac{ba}{BA}\right)$

0   100   200   300 ft./sec.²

142. Compounding of Harmonic Curves.

Curves of the form $y = c \sin(\theta + \alpha)$ are plotted and discussed on page 58. Combinations of these curves are of frequent occurrence, and have the general form given below, in which the C and α terms are constants.

$$y = C_0 + C_1 \sin(\theta + \alpha_1) + C_2 \sin(2\theta + \alpha_2) + C_3 \sin(3\theta + \alpha_3) + \ldots.$$

A harmonic curve of this kind is shown in fig. 1, obtained by adding (algebraically) corresponding ordinates of the three curves $y_1 = C_1 \sin(\theta + \alpha_1)$, $y_2 = C_2 \sin(2\theta + \alpha_2), y_3 = C_3 \sin(3\theta + \alpha_3)$, θ being taken from o to 2π. The constants chosen are $\alpha_1 = 30°$, $\alpha_2 = 120°$, $\alpha_3 = 90°$, and, to scale, $C_1 = 2, C_2 = 1$, $C_3 = \frac{2}{3}$. The initial steps in the plotting of the curves will be clear from fig. 3. If the x axis be displaced a distance C_0, to O_1X_1, the equation of the curve, referred to O_1X_1, is

$$y = C_0 + C_1 \sin(\theta + \alpha_1) + C_2 \sin(2\theta + \alpha_2) + C_3 \sin(3\theta + \alpha_3).$$

The complete curve consists of the part shown and repetitions of it.

The 1st term $y_1 = C_1 \sin(\theta + \alpha_1)$ is called the **fundamental** harmonic; the frequencies of the 2nd and 3rd terms are twice and three times respectively that of the 1st term. **Odd harmonics** are those in which the coefficient of θ is odd, **even harmonics** those in which the coefficient of θ is even.

In fig. 2 the two odd harmonics
$$y_1 = C_1 \sin(\theta + \alpha_1)$$
and
$$y_3 = C_3 \sin(3\theta + \alpha_3)$$
have been plotted and compounded. The resultant curve possesses a symmetrical quality which is a feature of all curves containing only odd harmonics. If the portion LMN to the right of 180° be shifted to the left, each point moving parallel to OX, until L lies on OY, the part LMN will be the image of the part ABL.

143. Harmonic Analysis.

In the converse problem, the resultant harmonic curve y is given, and the components $y_1, y_2, y_3 \ldots$ are to be determined. The process of resolving a harmonic curve into its components is referred to as harmonic analysis.

Fourier's Theorem. This important theorem states that the equation of *any periodic curve* may be represented by a series of sine curves, one of which has the same frequency as the given curve and the other have frequencies which are multiples of it. In other words, that any periodic function of θ can be represented by a series of the form

$$y = C_0 + C_1 \sin(\theta + \alpha_1) + C_2 \sin(2\theta + \alpha_2) + C_3 \sin(3\theta + \alpha_3) + \ldots,$$

the period of the function being 360° or 2π radians. If the period is $360°/n$ or $2\pi/n$ radians, then θ is replaced by $n\theta$.

In practical examples the amplitudes $C_1, C_2, C_3 \ldots$ decrease rapidly in value and harmonics higher than the fifth are usually negligible.

The problem now to be considered is, given *any* periodic curve, to determine its constituent series of harmonic curves.

As a preliminary example the curve in fig. 1 will be analysed and its equation deduced.

Preliminary Step. Expand each term of the general equation and regroup:

Then $y = C_0 + C_1 \sin\alpha_1 \cos\theta + C_2 \sin\alpha_2 \cos 2\theta + C_3 \sin\alpha_3 \cos 3\theta + \ldots$
$+ C_1 \cos\alpha_1 \sin\theta + C_2 \cos\alpha_2 \sin 2\theta + C_3 \cos\alpha_3 \sin 3\theta + \ldots.$

Put $a_1 = C_1 \sin\alpha_1, a_2 = C_2 \sin\alpha_2 \ldots$; and $b_1 = C_1 \cos\alpha_1, b_2 = C_2 \cos\alpha_2 \ldots.$

The equation becomes $y = C_0 + a_1 \cos\theta + a_2 \cos 2\theta + a_3 \cos 3\theta + \ldots$
$+ b_1 \sin\theta + b_2 \sin 2\theta + b_3 \sin 3\theta + \ldots,$

and $\dfrac{a_1}{b_1} = \tan\alpha_1, \dfrac{a_2}{b_2} = \tan\alpha_2 \ldots$; $C_1 = \pm \sqrt{a_1{}^2 + b_1{}^2}, C_2 = \pm \sqrt{a_2{}^2 + b_2{}^2} \ldots.$

For Examples, see page 156.

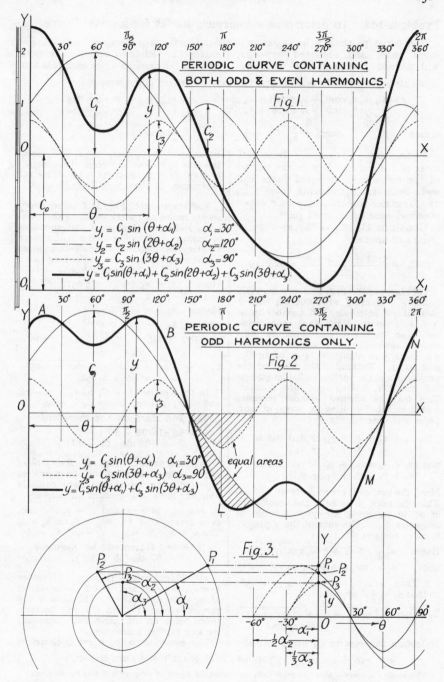

PERIODIC CURVE CONTAINING
BOTH ODD & EVEN HARMONICS.

Fig 1

$y_1 = C_1 \sin(\theta + \alpha_1)$ $\alpha_1 = 30°$
$y_2 = C_2 \sin(2\theta + \alpha_2)$ $\alpha_2 = 120°$
$y_3 = C_3 \sin(3\theta + \alpha_3)$ $\alpha_3 = 90°$
$y = C_1 \sin(\theta + \alpha_1) + C_2 \sin(2\theta + \alpha_2) + C_3 \sin(3\theta + \alpha_3)$

PERIODIC CURVE CONTAINING
ODD HARMONICS ONLY.

Fig 2

equal areas

$y_1 = C_1 \sin(\theta + \alpha_1)$ $\alpha_1 = 30°$
$y_3 = C_3 \sin(3\theta + \alpha_3)$ $\alpha_3 = 90°$
$y = C_1 \sin(\theta + \alpha_1) + C_3 \sin(3\theta + \alpha_3)$

Fig. 3

Problem 144. To determine a Fourier's Series for a given Curve.

The curve plotted in fig. 1 on previous page, and repeated opposite, will be analysed, and the constants C_1, C_2, C_3, α_1, α_2, α_3 obtained. *The student is advised actually to work through the example as he reads the text.*

As already shown, the general equation may be written in the form

$$y = C_0 + a_1\cos\theta + a_2\cos2\theta + a_3\cos3\theta + \dots \left.\right\} \text{ where } \begin{matrix} a_n = C_n\sin\alpha_n, \\ b_n = C_n\cos\alpha_n. \end{matrix}$$
$$+ b_1\sin\theta + b_2\sin2\theta + b_3\sin3\theta + \dots$$

Hence $\tan\alpha_1 = \dfrac{a_1}{b_1},\ C_1 = \pm\ \sqrt{a_1{}^2 + b_1{}^2}$, and so on.

Procedure. Take any base OX clear of the curve. Divide the period of the curve into 12 equal parts and erect ordinates, numbering them 0 to 11. Transfer the ordinates to a paper trammel—see fig. 1, next page.

Constant C_0. C_0 = the average of the 12 ordinates.

$$\therefore C_0 = \frac{5\cdot22 + \dots + 3\cdot55}{12} = 2\cdot69.$$

1st Harmonic. Fig. 2. Draw axes OX and OY. Starting with OX set out 12 radii at 30° intervals, and number them 0, 1, 2 ... to 11. Mark along consecutive radii the lengths of corresponding ordinates, transferring them from the trammel; e.g. mark 5·22 along O0, 4·42 along O1, and so on. Determine the sums of the projections of the various radial lengths on OX and OY, taking sign into account. This may be effected without separate measurement by using a strip of stiff paper and determining

$(x_0 + x_1 + x_2 + x_{11} + x_{10})$ minus $(x_4 + x_5 + x_6 + x_7 + x_8)$,

and $(y_1 + y_2 + y_3 + y_4 + y_5)$ minus $(y_7 + y_8 + y_9 + y_{10} + y_{11})$.

Here, the sum of the x projections = 6·02. This distance is = $6a_1$, the coefficient of a_n or b_n being always = number of ordinates ÷ 2. The sum of the y projections = 10·34 = $6b_1$.

Hence $a_1 = 6\cdot02 \div 6 = 1\cdot003$;

and $b_1 = 10\cdot34 \div 6 = 1\cdot72.$

$\therefore \tan\alpha_1 = a_1/b_1 = 1\cdot003 \div 1\cdot72 = \cdot58$; so that $\alpha_1 = 30°$ approx.

$C_1 = \pm\ \sqrt{a_1{}^2 + b_1{}^2} = \pm\ \sqrt{1\cdot003^2 + 1\cdot72^2}$
$= \pm\ 2$ approx.

To obtain the *sign* of C_1 use either the equation $b_1 = C_1\cos\alpha_1$ or $a_1 = C_1\sin\alpha_1$. Because b_1 is +ve and $\cos\alpha_1$ (i.e. cos 30°) is +ve, then C_1 must be +ve. **The 1st Harmonic is therefore $2\sin(\theta + 30°)$.**

2nd Harmonic. Fig. 3. Number every *second* radius and mark off along the *numbered* radii the lengths of corresponding ordinates. Each radius used will now carry two lengths. Determine the algebraic sums of the projections of the radial lengths on the axes OX and OY, as before.

Here $x_0 + x_1 + \dots = 5\cdot05$;
and $y_0 + y_1 + \dots = -3\cdot03.$
Hence $a_2 = 5\cdot05 \div 6 = \cdot84$;
and $b_2 = -3\cdot03 \div 6 = -\cdot505.$
$\therefore \tan\alpha_2 = a_2/b_2 = -\cdot84/\cdot505 = -1\cdot66$;
$\therefore \alpha_2 = -59°.$
$C_2 = \pm\ \sqrt{a_2{}^2 + b_2{}^2} = \pm\ \sqrt{\cdot84^2 + \cdot505^2}$
$= \pm\ \sqrt{\cdot96} = \pm\ \cdot98.$

In the equation $b_2 = C_2\cos\alpha_2$, b_2 is −ve, and $\cos\alpha_2$ is +ve; hence C_2 is −ve.

The 2nd Harmonic is therefore
$-\cdot98\sin(2\theta - 59°)$
$= \cdot98\sin(2\theta + 121°)$

3rd Harmonic. Fig. 4. Every *third* radius is now taken and the above process repeated. As the chosen radii coincide with the axes, a diagram is unnecessary.

$x_0 + x_4 \dots = 4\cdot02$; and $y_1 + y_5 \dots = 0$.
$a_3 = 4\cdot02 \div 6 = \cdot67$; $6b_3 = 0$, i.e. $b_3 = 0$.
$\operatorname{Tan}\alpha_3 = a_3/b_3 = \infty$; $\therefore \alpha_3 = 90°$. $C_3 = 67$.

The 3rd Harmonic is therefore
$\cdot67\sin(3\theta + 90°).$

4th Harmonic. (No fig.) Number every 4th radius and set off the ordinates along them in sequence. The sums of the projections on OX and OY will be zero, there being no 4th harmonic. Similarly for any further harmonics.

The equation to the curve is therefore

$$y = 2\cdot69 + 2\sin(\theta + 30°) + \cdot98\sin(2\theta + 121°) + \cdot67\sin(3\theta + 90°).$$

This result was actually obtained graphically and shows the degree of accuracy that may be expected.

For Examples, see page 156.

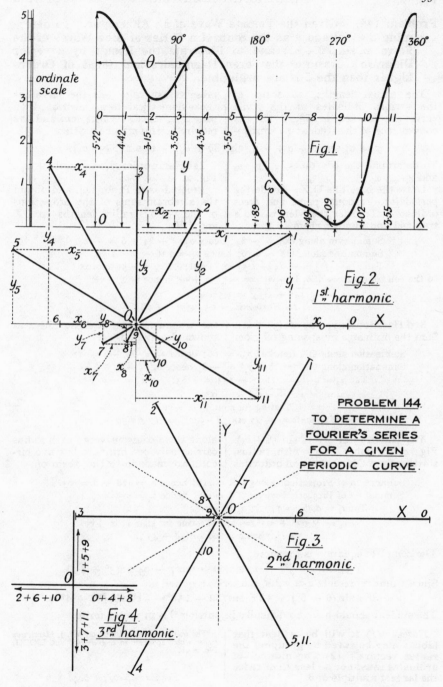

ordinate
scale

Fig.I.

Fig.2.
1st harmonic.

PROBLEM 144.
TO DETERMINE A
FOURIER'S SERIES
FOR A GIVEN
PERIODIC CURVE.

Fig.3.
2nd harmonic.

Fig 4.
3rd harmonic.

Problem 145.   Given the Voltage Wave of an Alternator. To determine the Voltage v as the Sum of a Series of Sine Waves of the Form $v_r \sin(r\theta + \alpha_r)$ and to illustrate the Result by a vector Diagram. Assume that even Harmonics and those of Orders higher than the 5th are negligible.

The actual lengths, to scale, of the various ordinates of the given curve are marked in fig. 1 for the convenience of the student: 1 unit represents 250 volts. As the curve has the symmetrical form mentioned on page 150, the C_0 term is zero and the required series may be written

$$y = v_1 \sin(\theta + \alpha_1) + v_3 \sin(3\theta + \alpha_3) + v_5 \sin(5\theta + \alpha_5).$$

To determine the constants, v_1, v_3, v_5 and α_1, α_3, α_5.

Draw the base line O_1X_1, divide the period into 12 equal parts and erect ordinates. Transfer their lengths to a trammel and number them.

1st Harmonic: $y_1 = v_1 \sin(\theta + \alpha_1)$. Fig. 2.

Proceed as in Prob. 144, and obtain the algebraic sums of the projections of the various radial lengths on OX and OY.

Summation along OX = −·84; hence $a_1 = -·84 \div 6 = -·14$.
Summation along OY = 9·16; hence $b_1 = 9·16 \div 6 = 1·53$.
$\text{Tan}\alpha_1 = a_1/b_1 = -·14 \div 1·53 = -·091$; $\therefore \alpha_1 = -5°$ approx.

In the eqn. $v_1 \cos\alpha_1 = b_1$, b_1 is +ve, cos − 5° is +ve; hence v_1 is +ve.

$$v_1 = \sqrt{·14^2 + 1·53^2} = 1·54.$$
$$\therefore 1st\ Harmonic = 1·54\ \sin(\theta - 5°).$$

3rd Harmonic: $y_3 = v_3 \sin(3\theta + \alpha_3)$. Sum the ordinates by spacing off from the trammel along straight lines, as shown clearly in fig. 3.

Summation along OX (marked x_3) = ·66; hence $a_3 = ·66 \div 6 = ·11$.
Summation along OY (marked y_3) = −·8; hence $b_3 = -·8 \div 6 = -·133$.
$\text{Tan } \alpha_3 = a_3/b_3 = ·11 \div -·133 = -·83$; $\therefore \alpha_3 = -40°$.
$$v_3 = \sqrt{·11^2 + ·133^2} = \pm ·17.$$
Testing for sign, v_3 is −ve.
$$\therefore 3rd\ Harmonic = -·17\ \sin(3\theta - 40°) = ·17\ \sin(3\theta + 140°).$$

5th Harmonic: $y = v_5 \sin(5\theta + \alpha_5)$. Fig. 4. Number every fifth radius, starting from OX, and set off ordinates along them consecutively; each radius carries only one ordinate but five circuits are made in setting them off.

Summation of Projections along OX = ·18; hence $a_5 = ·18 \div 6 = ·03$.
Summation of Projections along OY = ·18; hence $b_5 = ·03$.
$\text{Tan}\alpha_5 = a_5/b_5 = 1$; $\therefore \alpha_5 = 45°$.
$$v_5 = \sqrt{·03^2 + ·03^2} = ·04. \quad \text{Testing for sign, } v_5 \text{ is +ve.}$$
$$\therefore 5th\ Harmonic = ·04\ \sin(5\theta + 45°).$$

The complete equation is therefore
$$y = 1·54 \sin(\theta - 5°) + ·17 \sin(3\theta + 140°) + ·04 \sin(5\theta + 45°).$$
Since 1 unit represents 250 volts, the series required is
$$y = 385 \sin(\theta - 5°) + 42·5 \sin(3\theta + 140°) + 10 \sin(5\theta + 45°).$$
The student should have no difficulty in putting this in vector form.

Notes. (1) It will be evident that labour may be saved by grouping the radius vectors. **(2)** The number of ordinates must not be less than twice the largest multiple of θ.

The method given is based on Mr. J. Harrison's paper which appeared in *Engineering*, Vol. LXXXI, No. 2094.

For Examples, see page 156

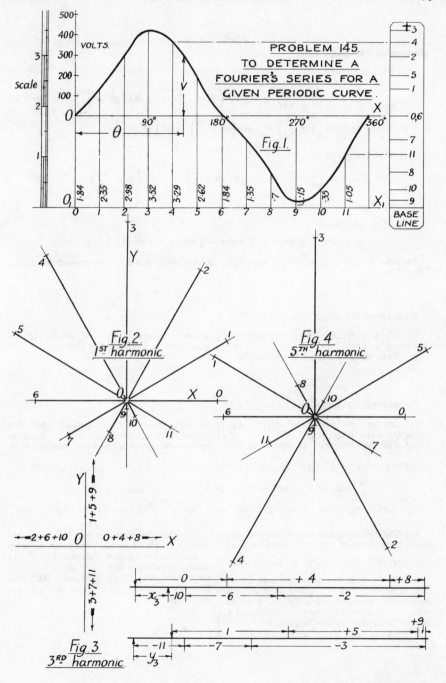

PROBLEM 145.
TO DETERMINE A
FOURIER'S SERIES FOR A
GIVEN PERIODIC CURVE.

Fig. I.

Fig. 2
1ST harmonic.

Fig 4
5TH harmonic.

Fig. 3
3RD harmonic.

EXAMPLES

Examples for page 150.

(1) Plot the graphs

(a) $\quad y = 3 \sin\left(\theta + \frac{\pi}{12}\right) + 1 \cdot 5 \sin\left(3\theta + \frac{\pi}{4}\right) + \cdot 75 \sin\left(5\theta + \frac{5\pi}{12}\right).$

(b) $\quad y = 3 \sin\left(\theta + \frac{\pi}{12}\right) + 2 \sin\left(2\theta + \frac{\pi}{6}\right) + \sin\left(4\theta + \frac{\pi}{3}\right).$

(2) Plot the graph

$$y = 3 \sin\left(\theta + \frac{\pi}{6}\right) - 1 \cdot 5 \sin\left(2\theta + \frac{\pi}{3}\right) - \cdot 5 \sin\left(3\theta + \frac{2\pi}{3}\right).$$

{NOTE: $-\sin\theta = \cos(\theta + 90°)$ or $\sin(\theta + 180°)$}.

(3) Plot the graph

$$y = 2\{\sin x - \tfrac{1}{2}\sin 2x + \tfrac{1}{3}\sin 3x - \tfrac{1}{4}\sin 4x + \tfrac{1}{5}\sin 5x\}.$$

Note that the greater the number of terms taken the more closely does the graph approach the straight line $y = x$.

Examples for page 152.

(1) Work out completely the example discussed in Prob. 144.

(2) Take the curves in Ex. 1b and Ex. 2 above. Assume the equations to be unknown and obtain the terms.

Examples for page 154.

(1) Obtain the series for Prob. 145 using the data in fig. 1, p. 155.

(2) The following are the values of two periodic functions, y, period 360°, for values of θ at intervals of 30°. Analyse each curve into its harmonic components.

(a) $\cdot 25 \quad 1\cdot 37 \quad 1\cdot 59 \quad 1\cdot 45 \quad 2\cdot 09 \quad 1\cdot 37 \quad -\cdot 25 \quad -1\cdot 37 \quad -1\cdot 59 \quad -1\cdot 45 \quad -2\cdot 09 \quad -1\cdot 37$

Ans. $y = 2 \sin\theta + \cdot 5 \sin(3\theta + \pi/6) + \cdot 12 \sin(5\theta + \pi)$.

(b) $\cdot 566 \quad 1\cdot 33 \quad 1\cdot 28 \quad 1\cdot 23 \quad \cdot 97 \quad \cdot 58 \quad 1\cdot 34 \quad -\cdot 43 \quad -\cdot 88 \quad -1\cdot 53 \quad -2\cdot 07 \quad -1\cdot 18$

(3) The following are values of a periodic function y (period 2π) for values of x at intervals of $\frac{\pi}{6}$ from o up to π; the value of y for $(2\pi - x)$ is the same as that for x, but negative. Plot the curve to scale and determine the terms of a Fourier's series which will give the curve. Harmonics higher than the 3rd are negligible.

o   63·3   65   33·3   21·6   20   o

(4) Plot the curve $y = 17 + 8x - 4x^2$ from $x = 0$ to $x = 3$. Determine a series of sine curves to give this portion of the curve, assuming that it is periodic, that the period $x = 0$ to 3 represents 360°, and that harmonics higher than the 5th are negligible.

PART II

SOLID OR DESCRIPTIVE GEOMETRY

The increasing use of 3rd Quadrant, or 3rd Angle, Projection for engineering drawings raises the question of the desirability of using this system in descriptive geometry. There is no difficulty in applying 3rd angle projection to solutions, but for the kind of pictorial view used herein, and strongly recommended as a preliminary to the solution of problems, 3rd angle drawings compare unfavourably with those of 1st angle. This will be clear from the examples on page 175, 177, 201, 205 and 215, particularly the last. As the primary aim of this book is the comprehension of solutions rather than their mechanical achievement, 1st angle constructions predominate. Nevertheless, the student is advised to acquire facility in the 3rd angle system by using it in the solution of problems.

146. Fundamental Principles of Projection.

Descriptive Geometry deals (a) with the representation on a plane surface of points, lines, and figures in space of three dimensions, in such a manner that their relative positions can be accurately determined; and (b) with the graphical solution of problems involving three dimensional space.

The axioms and theorems of Euclid XI are taken for granted in Descriptive Geometry, which is more concerned with graphical demonstration and use than with rigorous mathematical proof.

146a. Projection.

Any object may be regarded as an aggregation of points. If straight lines be taken from these points to meet a plane, the object is said to be **projected** on the plane, and the lines are called **projectors.** If the projectors converge to a point, **radial** or **perspective projection** is the result. If the projectors are parallel, and are normal to the plane, they give an **orthographic**, i.e. right line, **orthogonal** or rectangular projection; if they are parallel but inclined to the plane, they give an **oblique** projection.

146b. Planes of Reference.

Two principal planes are used in orthographic projection, one horizontal and the other vertical, intersecting and dividing space up into four **quadrants** or dihedral angles, numbered as in fig. 1. These planes are denoted here as the **H.P.** and **V.P.**; they are also called *x* and *y* planes and their lines of intersection the **xy** or **ground line.**

In descriptive geometry the line or object is supposed to be situated in one of the quadrants. It is represented by its orthographic projections on the H.P. and V.P., the views giving respectively its **plan and elevation.*** To show these views on a plane surface the H.P. and V.P. are opened out, or **rabatted,** about the *xy* line until they coincide; *the convention is that the 1st quadrant must always be opened out.*†

For projections in the 2nd, 3rd, and 4th quadrants the planes are assumed to be transparent, and projections are drawn looking on the rabatted 1st quadrant.

146c. Projections of a Point.

Refer to fig. 1, in which P is a point in the 1st quadrant. The projections *p* and *p*₁ are given by the projectors P*p* and P*p*₁, perp. to HP and VP respectively. When the 1st quadrant is opened out the projections *p* and *p*₁ lie in one plane and appear as in fig. 2.

Figs. 3, 4, and 5 illustrate the orthographic projections of points in the 2nd, 3rd, and 4th quadrants respectively. Only the 3rd is important.

Note. The following may be inferred from the figures given:

1. The line joining the plan and elevation of a point (*also called a projector*) is perp. to *xy*.

2. A point is above or below the H.P. according as its elevation is above or below *xy*.

3. A point is in front or behind the V.P. according as its plan is below or above *xy*.

4. The distance of P from the H.P. = the distance of its elevation from *xy*.

5. The distance of P from the V.P. = the distance of its plan from *xy*.

EXAMPLES

(1) Draw the projections of the following points: A, 2″ above H.P., 2½″ in front V.P.; B, 2″ above H.P., 2½″ behind V.P.; C, 2″ below H.P., 2½″ behind V.P.; D, 2″ below H.P., 2½″ in front V.P.; E, 2″ above H.P., in V.P.; F, 2″ below H.P., in V.P.; G, 2″ in front V.P., in H.P.; H, 2″ behind V.P., in H.P.

(2) The projections of a point coincide and both fall below *xy*. In which quadrant is the point situated?

* This system was invented by Gaspard Monge (1746–1818).

† It is convenient for blackboard demonstration to suppose the H.P. to swing about *xy* until it coincides with the V.P.; the student may, however, prefer to regard the H.P. as fixed and the V.P. as movable.

PROJECTIONS OF A POINT P
SITUATED IN THE 1ST QUADRANT.

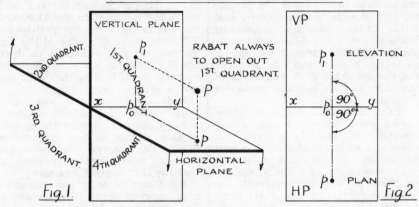

Fig. 1

Fig. 2

NOTE : { Pp & Pp_1 (DOTTED) ARE THE REAL PROJECTORS.
{ p_0p & p_0p_1 (BROKEN) ARE REFERRED TO AS PROJECTORS.

PROJECTIONS OF A POINT P IN THE

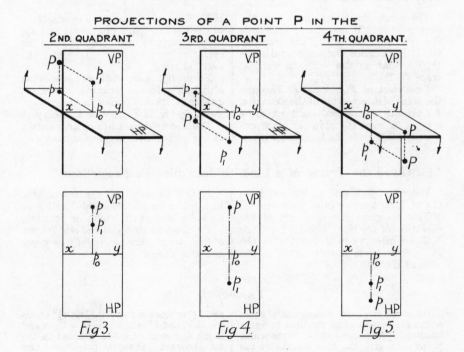

Fig 3.   Fig 4.   Fig. 5.

147. Projections of a Straight Line in the 1st Quadrant.

The projections of a straight line are straight lines joining the projections of its ends. Fig. 1 shows a line AB in the 1st quadrant, its projections being ab and a_1b_1. After rabatment, the orthographic projections appear as in fig. 2: a_0a_1 and b_0b_1 give the respective heights of A and B above the H.P., and a_0a, b_0b their distances from the V.P.

It should be noted that neither projection gives the true length of the line AB. This will only be given by a projection when the line itself is par^l to the plane on which it is projected.

The projections of a line in various positions in the 1st quadrant are shown in fig. 5. They are: (1) perp. to the H.P.; (2) perp. to the V.P.; (3) in the V.P. and par^l to xy; (4) in the H.P. and par^l to xy; (5) lying in xy; (6) par^l to both the H.P. and the V.P.; (7) par^l to the H.P. and inclined to the V.P.; (8) par^l to the V.P. and inclined to the H.P.; (9) inclined to, and with ends in, both the H.P. and the V.P.; (10) inclined to both the H.P. and the V.P., one end in xy; (11) perp. to xy.

Note. It is helpful to regard projected views in the following manner.

For Elevations: regard the xy line as an edge view of the H.P. from the front; i.e. suppose the planes of reference to be lifted to eye-level, so that the H.P. appears as a line.

For Plans: regard the xy line as an edge view of the V.P. from above. The student should test the method on fig. 4.

148. Traces of a Line.

These are the points in which the line, produced if necessary, intersects the H.P. and the V.P.; the trace on the H.P. is called the horizontal trace (H.T.), that on the V.P. the vertical trace (V.T.).

Construction. Figs. 2 and 4. Through the point of intersection of the elevation (or elevation produced) and xy, draw a perp. to meet the plan (or plan produced): this gives the horizontal trace, H.T. Through the point of intersection of the plan (or plan produced) and xy, draw a perp. to meet the elevation (or elevation produced): this gives the vertical trace, V.T.

From fig. 4 it will be seen that although a line is situated in the 1st quadrant its V.T. may be below xy; similarly the H.T. may be above xy.

The constructions are clearly shown in the pictorial views, figs. 1 and 3.

149. Given the Traces of a Line, to determine its Projections.

Refer to figs. 2 and 4. Project from the H.T. to meet xy in c; join c to the V.T.—this gives the direction of the elevation of the line. Project from the V.T. to meet xy in d; join d to the H.T.—this gives the direction of the plan of the line.

It will be seen that the traces of a line define its *direction* only, and give no indication of its length or position.

Problems on the traces and true lengths of lines are discussed more fully on pages 200 and 202 .

EXAMPLES

(1) Make freehand orthographic views of the projections of all the lines in fig. 5, inserting appropriate letters. State which projections give the true lengths of the lines.

(2) The projectors of two points A and B are $1\frac{1}{2}''$ apart. A is $\frac{3}{4}''$ above the H.P. and $2\frac{1}{4}''$ in front of the V.P.: B is $1\frac{1}{2}''$ above the H.P. and $1''$ in front of the V.P. A and B are the ends of a line situated in the 1st quadrant. Draw its projections and determine its traces. Measure the distance of each trace from xy. Ans. H.T. $3.52''$. V.T. $2.1''$.

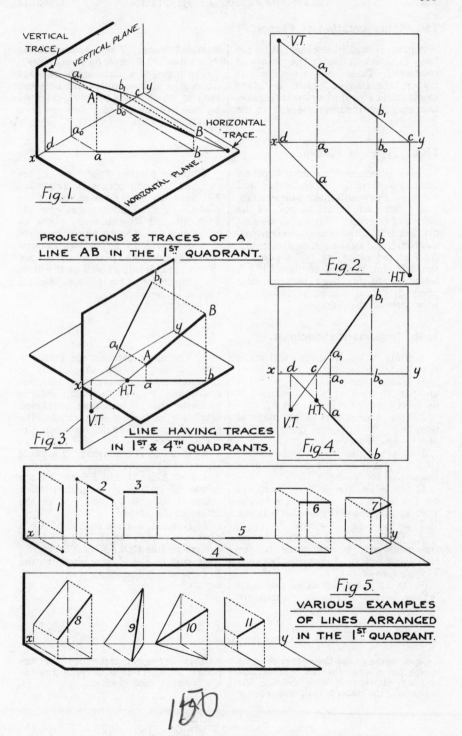

VERTICAL TRACE

VERTICAL PLANE.

HORIZONTAL TRACE.

HORIZONTAL PLANE.

Fig. 1.

PROJECTIONS & TRACES OF A
LINE AB IN THE 1ST QUADRANT.

V.T.

H.T.

Fig. 2.

Fig. 3.

LINE HAVING TRACES
IN 1ST & 4TH QUADRANTS.

V.T.   H.T.

Fig. 4.

Fig 5.

VARIOUS EXAMPLES
OF LINES ARRANGED
IN THE 1ST QUADRANT.

150. Representation of Planes.

A plane is usually represented by the lines in which it meets the planes of reference. These lines are called its traces; the line in which the plane meets the H.P. is the **horizontal trace**, that in which it meets the V.P. is the vertical trace. *The abbreviations H.T. and V.T. will be used here.*

The traces of a plane are not always accessible, and examples will be given later in which the plane is represented by the projections of points or lines contained by it.

150a. Types of Planes.

It is convenient to divide planes into two groups, perpendicular and oblique. **Perpendicular planes**, figs. 1 to 5, are perp. to one or both of the planes of reference. They may be subdivided into three types, *horizontal, vertical,* and *inclined,* as illustrated. An inclined plane, fig. 5, is perp. to the V.P. and inclined to the H.P. A feature common to all perpendicular planes is that one trace represents an edge view of the plane.

Oblique planes, figs. 6 to 11, are inclined to both planes of reference. Neither trace gives an edge view of the plane. In fig. 8 both traces coincide with *xy*, and to define the plane an additional trace on an auxiliary plane is necessary. In fig. 9 two oblique planes intersect the planes of reference in other quadrants as well as the first; the projections of their traces, fig. 10, should be noted.

150b. Important Principles.

A study of the diagrams will make clear the following important principles:

1. Unless the traces of a plane are par[l] to *xy* they will, produced if necessary, intersect in *xy*. See fig. 9.

2. If a plane is par[l] to one plane of reference it will have no trace on that plane—figs. 1 and 2.

3. A plane par[l] to *xy* but inclined to both planes of reference will have traces par[l] to *xy*—fig. 7.

4. (*a*) When the V.T. is perp. to *xy*, the inclination of the plane to the V.P. is given by the angle between H.T. and *xy*—fig. 4.

(*b*) When the H.T. is perp. to *xy*, the inclination of the plane to the H.P. is given by the angle between V.T. and *xy*—fig. 5.

(*c*) In all the other cases the angle between a trace and *xy* is *not* a measure of the inclination of the plane.

5. The angle between the traces in one quadrant is the supplement of the angle between them in an adjacent quadrant. Fig. 11 shows an oblique plane in the 1st quadrant continued into the remaining quadrants; the sum of adjacent angles between the traces = 180°.

6. If a line is contained by a plane its traces lie in the traces of the plane. The lines AB, CD, and EF, fig. 11, are contained by the given plane; their traces, indicated by dots, lie in the traces of the plane.

7. Horizontal lines contained by a plane are par[l] to the H.T. of the plane —e.g. the line CD, fig. 11. Similarly, lines par[l] to the V.P. are par[l] to the V.T. of the plane—e.g. the line EF, fig. 11.

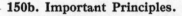

EXAMPLES

Copy, freehand, all the figures opposite except fig. 8, and make an orthographic drawing adjacent to each showing the positions of the traces of the plane. Choose suitable angles of inclination for the planes. In fig. 11 show the projections of the traces in each quadrant.

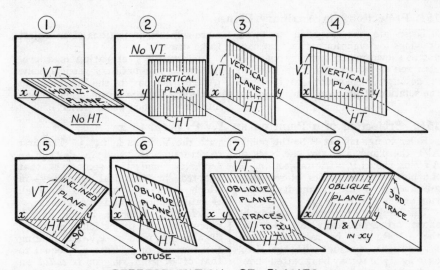

REPRESENTATION OF PLANES.
PERPENDICULAR PLANES, FIGS. 1-5 : OBLIQUE PLANES, FIGS. 6-11

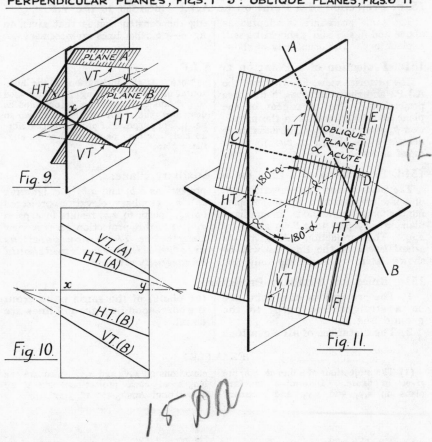

Fig. 9.

Fig. 10.

Fig. 11.

151. Projections on Auxiliary Planes.

These are necessary in technical drawings to give information that cannot be shown in the ordinary views of plan and elevation; they are required in descriptive geometry to facilitate the solution of problems and to enable more difficult projections of an object to be drawn.

An **auxiliary elevation** is a projection on any auxiliary vertical plane (A.V.P.) not par[l] to the V.P.; and an **auxiliary plan** is a projection on any auxiliary inclined plane (A.I.P.).

151a. Projection of a Point on an A.V.P. perp. to *xy*.

Refer to fig. 1. Let P be the point, AVP the plane, and x_1y_1 the trace of AVP on the H.P. The projections of the point on the planes of reference are given by p and p_1, and its projection on the A.V.P. by p_2.

This auxiliary projection p_2 may be shown orthographically in two ways: the A.V.P. may be rabatted about x_1y_1 until it coincides with the H.P., as in fig. 2; or it may be rabatted about its trace on the V.P. until it coincides with the V.P., as in fig. 3. The first construction is usual in descriptive geometry, and the second is that adopted in technical drawings—in which it is convenient to have all elevations upright and in alignment.

Referring to fig. 2, it will be apparent that the projection is given by drawing pp_2 perp. to x_1y_1 and making the distance of p_2 from x_1y_1 equal to that of p_1 from xy. p_2 *is called an auxiliary elevation.*

151b. Projection of a Point on an A.V.P. inclined to the V.P.

The same procedure is adopted as above and figs. 4 and 5 should be self-explanatory. The usual way of showing the construction is that given in fig. 6—no other lines are necessary.

151c. Projection of a Point on an A.I.P.

The pictorial view, fig. 7, shows the A.I.P., and the point P with its three projections. The rabatment of the plane about x_1y_1 results in the projection p_2 given in fig. 8. The actual lines necessary are shown in fig. 9. p_2 *is called an auxiliary plan.*

Note. If the inclination to the horizontal of the A.I.P. in fig. 7 be increased to 90° and the projection p_2 obtained as described, the result will be as shown in fig. 3. For purposes of projection therefore an end view may be regarded as an auxiliary plan.

151d. Projections of a Line on an Auxiliary Plane.

The plan ab and the elevation a_1b_1 of a line are shown in fig. 10. Applying the foregoing methods, an auxiliary plan on x_1y_1 par[l] to a_1b_1 is given by a_2b_2. This projection a_2b_2 gives *the actual length* of the line, as will be shown later. Considering only the projections a_1b_1 and a_2b_2 (i.e. ignoring ab), an auxiliary elevation projected on x_2y_2 perp. to a_2b_2 results in a *point* a_3b_3: i.e. this projection gives a view along the line. *These manipulations are of great importance and should be carefully studied.*

151e. Rules for Auxiliary Projections.

1. The projections of a point are on a straight line perp. to the ground line.

2. The distances of all elevations (or plans) of the same point from the corresponding ground lines are equal.

EXAMPLES

(1) The projections of a line ab, a_1b_1 are given in figure. Determine auxiliary plans on x_1y_1 and x_3y_3, and auxiliary elevations on x_2y_2 and x_4y_4. Measure the length of each projection: check by calculation. Ans. 3·47″, 1″, 3·47″, 1·8″.

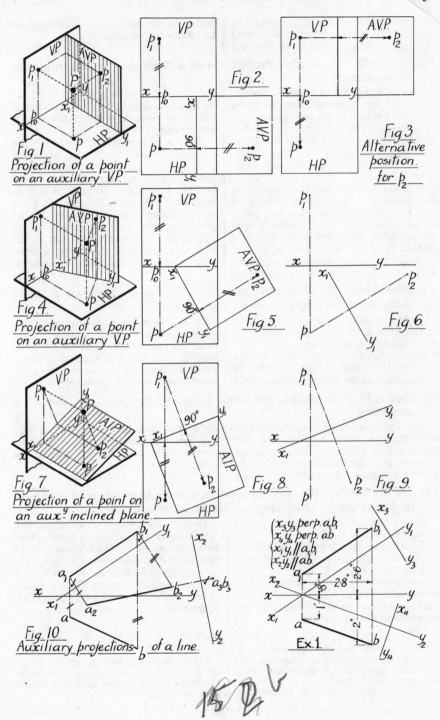

Fig 1
Projection of a point
on an auxiliary V.P.

Fig 2.

Fig 3
Alternative
position
for p_2

Fig 4
Projection of a point
on an auxiliary V.P.

Fig 5.

Fig 6.

Fig 7.
Projection of a point on
an aux.ᵞ inclined plane

Fig 8.

Fig 9.

Fig 10.
Auxiliary projections of a line

$\begin{cases} x_3y_3 \text{ perp. } a_1b_1 \\ x_4y_4 \text{ perp. } ab \\ x_1y_1 \parallel a_1b_1 \\ x_2y_2 \parallel ab \end{cases}$

Ex.1.

Projections of Plane Areas. To project any plane area, the principles of projection are applied to selected points on its outline, as illustrated in the following examples.

Problem 152. From the Projections of a Triangle to determine its True Shape.

The true shape of any plane area is given by its projection on a plane which is parallel to its own plane.

Refer to both pictorial and orthographic views. Let ABC be the given $\triangle$, and abc, $a_1b_1c_1$ its projections. Draw the projections of any horizontal line lying in the $\triangle$; a_1d_1 is par^1 to xy and is therefore horizontal and its plan is given by ad. Take an A.V.P., with its ground line x_1y_1 perp. to ad, and project an auxiliary elevation $a_2b_2c_2$ from the plan (i.e. project at rt. $\angle$s to x_1y_1 from a, b, and c, making the distance of a_2 from $x_1y_1 =$ the distance of a_1 from xy,

&c.). *The auxiliary projection of the $\triangle$ is now a line.* (If one projection of the $\triangle$ is given as a line, this step is unnecessary.)

Take a ground line x_2y_2 par^1 to the line $a_2b_2c_2$, and project an auxiliary plan $a_3b_3c_3$ (i.e. project from $a_2b_2c_2$ perp. to x_2y_2, making the distance of a_3 from $x_2y_2 =$ the distance of a from x_1y_1, &c.). The resulting $\triangle$ $a_3b_3c_3$ gives the required true shape: no other projection will give as large a $\triangle$ as $a_3b_3c_3$. The inclination of the $\triangle$ to the H.P. is given by θ, the angle between $a_2b_2c_2$ and x_1y_1—this is discussed later

Problem 153. To determine the Projection of a Circle on a Plane inclined at an angle φ to its own Plane.

Draw the simple plan and elevation of the circle shown in figure, and draw x_1y_1 inclined at φ to xy: x_1y_1 represents an A.V.P. inclined at φ to the plane of the circle. Take a number of points such as m, m_1 and obtain their

auxiliary elevations, as m_2. The fair curve joining points such as m_2 is the required projection. The result is an ellipse, and its construction is more easily carried out by locating its axes, as will be shown on page 178.

Problem 154. To obtain the True Shape of the plane Section of a Moulding.

Let the plan and elevation of a moulding be given by the line and figure shown opposite. Its true shape is given by a projection on x_1y_1 taken par^1 to the plan. Selected points such

as p, p_1 projected in the usual way give points such as p_2 on the required view. The construction should be clear from the figure.

155. NOTATION. To avoid confusion, a definite notation is essential in projection. That adopted herein is as follows:

Actual Lines (e.g. in pictorial views) —capitals, AB, CD, &c.

First Plans—small letters, ab, cd, &c.

First Elevations—small letters with suffix added, a_1b_1, c_1d_1, &c.

Auxiliary Projections—small letters with suffix added, a_2b_2, a_3b_3, &c., the number of the suffix indicating the sequence of the projection.

Ground Lines—xy, x_1y_1, x_2y_2, &c. If necessary the points of intersection between projectors and ground lines can be lettered a_0b_0 or hk, h_1k_1, h_2k_2, &c.

Angles—Inclinations to H.P., θ (thēta); inclinations to V.P., φ (phi); other angles, α (alpha), β (bēta), γ (gamma).

EXAMPLES

(1) Determine the true shape of the $\triangle$ given by the dimensioned projections in the figure opposite. Measure the sides. Ans. 3·71″, 3·37″, 2·96″.

(2) A 3″ diam. circle is inclined at 60° to the H.P. Determine its plan.

(3) A moulding section, at rt. $\angle$s to the length, is shown in figure. Determine the shape of the section as it will appear in a mitred joint, angle 45°.

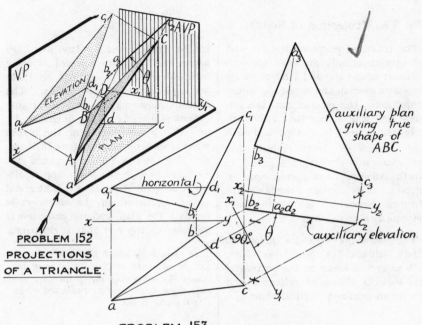

PROBLEM 152

PROJECTIONS

OF A TRIANGLE.

PROBLEM 153.

PROJECTIONS

OF A CIRCLE

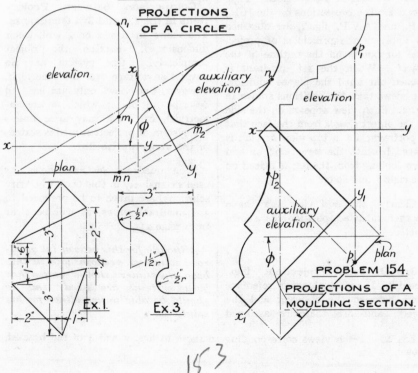

Ex.1.

Ex.3.

PROBLEM 154

PROJECTIONS OF A

MOULDING SECTION.

153

156. The Projection of Solids.

For practical purposes the 1st and 3rd quadrants only are used for projections; in the 2nd and 4th quadrants the views overlap and cause confusion. Projection in the 1st quadrant is traditional and was at one time universal. It is still current in Europe and to a large extent in Great Britain, but 3rd quadrant projection is used in North America, and increasingly in Britain. The British Standards Institution (Report No. 308) permits both systems as British standards.

The drawings opposite show an object situated in both quadrants, with projected views on the horizontal and vertical planes of reference and also on an auxiliary vertical plane.

1st Quadrant Projection. Figs. 1 and 2. The projections on the H.P., V.P., and A.V.P., fig. 1, are obtained by applying the principles of projection for points on the outline of the object. When the 1st quadrant is opened out until the planes coincide, the views take the positions shown in fig. 2. Each view appears on the side of the object *remote* from the face that it portrays; i.e. a top view or plan is placed beneath the elevation; an end view looking from the left is placed on the right; and so on.

Clearly, the end view may have another rabatted position, as shown dotted.

3rd Quadrant Projection. Figs. 3 and 4. Here the planes of reference come between the observer and the object, and are therefore assumed to be transparent. *Any auxiliary planes used are also regarded as being transparent,* and projections on them are viewed *through* the planes. The student may imagine that the transparent planes, fig. 3, are hinged at H, and that after drawing on the planes what is seen by looking through them, the H.P. and A.V.P. are opened out to coincide with the V.P. The resulting arrangement of views is that given in fig. 4, in which the elevation is beneath the plan, and the end view is *adjacent* to the end that it describes.

It should be noted that a third view can be obtained mechanically from two others, by applying the principles discussed on page 164, as indicated opposite for one point.

Comparison between Projections in the 1st and 3rd Quadrants. The views in figs. 2 or 4 will, when dimensioned, describe the object completely. One system has no natural advantage over the other. If, however, two views only are used to describe an object which is unsymmetrical, confusion may arise unless the system of projection used is stated, or the views are labelled.

First quadrant *pictorial* views, as used extensively in this book to clarify solutions, are much to be preferred to 3rd quadrant views, as will be clear from page 215.

Although for this reason 1st quadrant solutions predominate in this book, the student should realize that both systems are used, and he should be able to use either for his solutions.

Ex. 1. Sketch views corresponding to those in figs. 2 and 4 of the bracket shown.

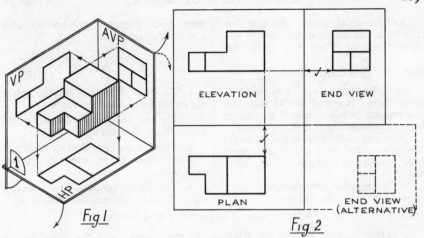

Fig 1

Fig 2

FIRST QUADRANT OR FIRST ANGLE PROJECTION

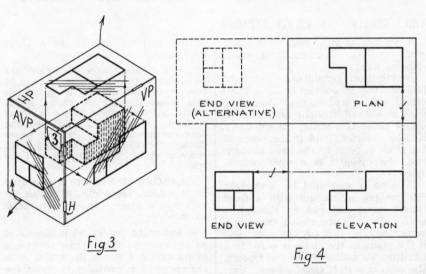

Fig 3

Fig 4

THIRD QUADRANT OR THIRD ANGLE PROJECTION

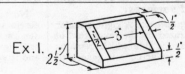

Ex.1.

The principal solids discussed in this book are illustrated opposite and defined below. They may be divided into two groups: polyhedra, and solids of revolution.

157. POLYHEDRA.

A **polyhedron** is a solid bounded by plane faces; it is **regular** if each face is a regular polygon, and if the angles formed between the faces are equal.

There are only **five regular polyhedra** *—refer to Euc. XIII—sometimes called the Platonic Solids. They are:

1. The regular **tetrahedron**: four equal faces, each an equilateral $\triangle$.

2. The **cube** (or regular hexahedron): six equal faces, each a square.

3. The regular **octahedron**: eight equal faces, each an equilateral $\triangle$.

4. The regular **dodecahedron**: twelve equal faces, each a regular pentagon.

5. The regular **icosahedron**: twenty equal faces, each an equilateral $\triangle$.

All the above may be inscribed in a sphere.

A **prism** has par[1] ends comprised of equal and similarly arranged polygons, joined by side faces which are par[ms]. A **parallelepiped** is a prism having par[ms] as ends.

A **pyramid** has a polygon as base; its side faces are triangles having a common vertex and the sides of the polygon for their bases.

Right regular prism and pyramid. These are regular if the end polygons are regular, and right if the axis is perp. to the ends or base. If the solid is not right it is **oblique**.

158. SOLIDS OF REVOLUTION.

A **sphere** is generated by a semi-circle revolving about its diam. which remains fixed.

A **cylinder** is generated by a straight line moving in contact with a fixed closed curve and always remaining par[1] to a fixed straight line. If the closed curve is a circle, the solid is a *circular cylinder*; and if the plane of the circle is perp. to the fixed straight line, the cylinder is a *right circular cylinder*.

A **cone** is generated by a straight line moving in contact with a fixed closed curve and passing through a fixed point. If the closed curve is a circle, the cone is a *circular cone*; and if the plane of the circle is perp. to a line from its centre to the fixed point, the cone is a *right circular cone*. Surfaces which are generated by straight lines are **ruled surfaces**.

Particular Definitions. A right circular cylinder is generated by the revolution of a rectangle about a side which remains fixed. A right circular cone is generated by the revolution of a rt. $\angle$d $\triangle$ about a side containing the rt. $\angle$.

Curves of contact between a Sphere and an enveloping Cylinder or Cone. If a right circular cylinder envelop a sphere, its axis passes through the centre of the sphere, and the curve of contact is a *great circle* of the sphere. If a right circular cone envelop a sphere, its axis passes through the centre of the sphere and the curve of contact is a *lesser circle* of the sphere.

An **anchor ring** or tore is generated by a sphere, the centre of which revolves in a plane circular path—refer to page 272.

A **spheroid** (no figure) is generated by the rotation of an ellipse about one of its axes: if about the major axis, the spheroid is **prolate;** if about the minor axis, the spheroid is **oblate.**

Note. When a solid is referred to herein simply as a prism, pyramid, cylinder, or cone, the words " right regular " or " right circular " are implied. Oblique and irregular solids will be so named.

* The Archimedian solids, 13 in number, have all the faces bounded by regular polygons which are not all congruent to each other; all the polyhedral angles are convex and congruent to each other. A large number of other interesting shapes occur in crystallography, and there are large numbers of semi-regular prisms and prismoids.

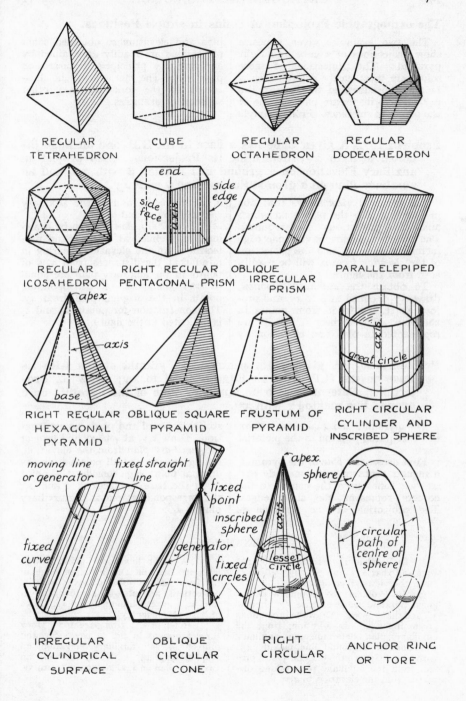

REGULAR TETRAHEDRON

CUBE

REGULAR OCTAHEDRON

REGULAR DODECAHEDRON

REGULAR ICOSAHEDRON

RIGHT REGULAR PENTAGONAL PRISM
end
side edge
side face
axis

OBLIQUE IRREGULAR PRISM

PARALLELEPIPED

RIGHT REGULAR HEXAGONAL PYRAMID
apex
axis
base

OBLIQUE SQUARE PYRAMID

FRUSTUM OF PYRAMID

RIGHT CIRCULAR CYLINDER AND INSCRIBED SPHERE
axis
great circle

IRREGULAR CYLINDRICAL SURFACE
moving line or generator
fixed straight line
fixed curve

OBLIQUE CIRCULAR CONE
fixed point
generator
fixed circles

RIGHT CIRCULAR CONE
apex
axis
inscribed sphere
lesser circle

ANCHOR RING OR TORE
sphere
circular path of centre of sphere

The orthographic Projection of Solids in simple Positions.

The two examples given opposite show projections of a cube and right pyramid. The projections of these solids are most easily drawn when a face of the cube and the base of the pyramid lie in, or are par¹ to, one of the planes of reference. From a simple plan and elevation so obtained other projections are readily determined by applying the principles discussed on page 164 The method will be illustrated in the following and other subsequent examples.

Problem 159. A given Cube has a Face in the H.P. and an Edge inclined at α to *xy*; determine its Projections. Also project an auxiliary Elevation on a ground line making β with *xy*, and an auxiliary Plan on a ground line making γ with *xy*.

The pictorial view shows the cube in position, also the planes on which auxiliary projections are required. The *plan* is a square having one edge inclined at α to *xy*, and the projection of the *elevation* from it will be readily seen from the figure.

To obtain the *auxiliary elevation*, draw x_1y_1 inclined at β to *xy*, and project at rt. ∠s to it from points in the plan. The corners A and B are represented in plan and elevation by a, a_1 and b, b_1, and in the auxiliary elevation by a_2 and b_2.

The *auxiliary plan* is given by drawing x_2y_2 inclined at γ to *xy* and projecting from the elevation, bearing in mind that the distances from x_2y_2 of points in the auxiliary plan are equal to the distances of corresponding points in the original plan from *xy*. The construction for points a_3 and b_3 is indicated in the figure.

Problem 160. A given right pentagonal Pyramid stands with its Base on the H.P. with an Edge parallel to *xy*; draw its Projections. Determine also an auxiliary Plan of the Pyramid on a ground line making γ with *xy*.

The pyramid and plane are shown in the required position in the pictorial view.

First drawn the *plan* of the pyramid, a regular pentagon with one side par¹ to *xy*. Lines from the centre to the corners represent the slant edges. The projection of the *elevation* is straightforward and needs no explanation. Draw x_1y_1 at γ to *xy* and project the *auxiliary plan* from the elevation, taking distances of all points from the original plan: e.g. consider the point A on the base, projections a and a_1; the corresponding point in the auxiliary plan is a_2.

EXAMPLES

(1) In Problem 159 assume a cube $1\frac{1}{2}''$ edge and that α = 30°, β = 45°, γ = 60°. Draw the views given opposite.

(2) In Problem 160 assume a pentagonal pyramid, edge of base $1\frac{1}{4}''$, height $3''$, and that γ = 60°. Draw the projections given opposite. In addition, from the auxiliary plan determine an auxiliary elevation on a ground line making 45° with x_1y_1. (Ignore the original plan; project from the auxiliary plan taking distances from the elevation to x_1y_1.)

(3) A right hexagonal prism stands with its base on the V.P., an edge of the base making 30° with *xy*. Draw its projections, and also an auxiliary elevation on a ground line at 45° to *xy*. Edge of base $1\frac{1}{4}''$, height $3\frac{1}{2}''$.

(4) Refer to Ex. 1 on page 168. Draw three views as in fig. 2, page 169, and project (a) an auxiliary elevation on x_1y_1 making 45° with *xy*, and (b) an auxiliary plan on x_2y_2 making 30° with *xy*.

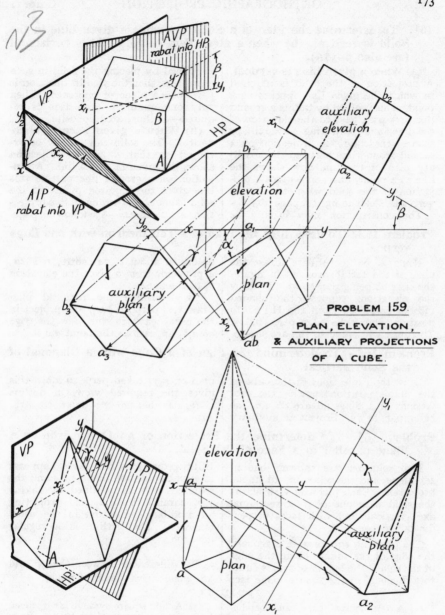

PROBLEM 159.

PLAN, ELEVATION,
& AUXILIARY PROJECTIONS
OF A CUBE.

PROBLEM 160.

PLAN, ELEVATION, & AUXILIARY PROJECTION
OF A RIGHT PENTAGONAL PYRAMID.

161.　To determine the Plan of a Solid (a) when a given Line in the Solid is vertical, (b) when a given Face of the Solid is horizontal (see also p. 176).

(a) When a given Line is vertical. A projection of the solid is required in which the given line appears as a point. This is given by taking a ground line first par[l], say, to the elevation of the line, and projecting an auxiliary plan of the solid. A projection on a second ground line perp. to the auxiliary plan of the line now shows the line as a point, and hence gives a projection of the solid when the line is vertical. Refer to fig. 10, page 165.

The construction may usually be simplified by placing the solid in such a position that the line in question is par[l] to one plane of reference. Then only one auxiliary projection is required—as illustrated opposite.

(b) When a given Face is horizontal. The solid can usually be arranged so that a simple projection shows the given face as a line. A projection on a ground line par[l] to this line gives the required plan. If the solid cannot be so arranged, two projections are necessary—see following page.

Problem 162.　To determine the Plan of a Tetrahedron with one Edge vertical.

Using a 60° set-square, draw the plan of the tetrahedron as in figure, the edge ab being par[l] to xy. To draw the elevation, suppose the shaded face to be raised from the H.P. into position, about the edge perp. to xy. The apex will traverse the arc shown,

described about c_1 as centre, rad. ca; a projector from b gives the elevation b_1 of the apex.

To determine the required plan draw x_1y_1 perp. to b_1a_1, and project in the usual way; in this view the edge ab a_1b_1 appears as the point a_2b_2.

Problem 163.　To determine the Plan of a Cube with a Diagonal of the Solid vertical.

Draw the projections of the cube in the simple position shown; the projections of a diagonal are ab a_1b_1, ab being par[l] to xy. Project an auxiliary plan on x_1y_1 taken perp. to a_1b_1: this gives the required view, in outline a regular hexagon. Refer to Art. 191.

Problem 164.　To determine the Projection of an Octahedron on a Plane parallel to a Face of the Solid.

Two solutions are shown: one in 1st angle, and the other in 3rd angle, projection. Each pictorial view shows the solid with one of its three equal axes vertical, and with faces perpendicular to the V.P.

The plan is a square with diagonals, the edge of the square being the edge of the solid; the length of a_1b_1 is equal to a diagonal of the square. One face

BCD, projections bcd, $b_1c_1d_1$, appears in elevation as the line $b_1c_1d_1$, and the required view is obtained by taking x_1y_1 parallel to $b_1c_1d_1$ and projecting an auxiliary projection in the usual manner. The outline is a regular hexagon.

The differences between the 1st and 3rd angle views should be noted.

EXAMPLES

(1) A tetrahedron, cube, and octahedron have edges 3″, 1¾″, and 2″ long respectively. Draw projections of each corresponding to those shown opposite.

(2) Draw the plan of a right hexagonal pyramid with a slant edge vertical. Edge of base 1¼″, height 3″. Use 3rd Angle Projection.

(3) A right square pyramid is suspended freely from a corner of its base.* Draw its plan. Edge of base 1¾″, height 3½″.

(4) Two views of a voussoir for an arch are given on p. 177. Draw them and project (a) an auxiliary elevation on x_1y_1, and (b) a plan when the line ab a_1b_1, is vertical.

* The C.G. of a pyramid is ¼ the distance along the axis from the base. A line from the corner of suspension through the C.G. will be vertical.

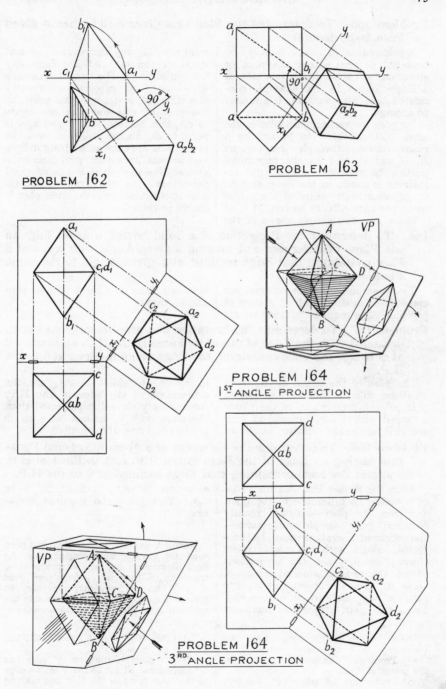

PROBLEM 162

PROBLEM 163

PROBLEM 164
1ST ANGLE PROJECTION

PROBLEM 164
3RD ANGLE PROJECTION

Problem 165. To determine the Plan of a given Solid when a given Face is horizontal.

A simple solid can usually be arranged so that either the plan or elevation will show the given face as a single line. In the example considered here, however, this is not easy to arrange.

Suppose the rectangular solid in the small pictorial view to be cut by a plane passing through the corners A, B and C, and let the pyramidal portion be removed. The solid remaining is shown in the lower sketch: to determine the plan of this solid when the face ABC is horizontal.

Draw any two simple views of the solid showing the projections abc and $a_1b_1c_1$ of the face ABC. Take any horizontal line in the face (if necessary draw one), and project an auxiliary elevation on a ground line perp. to the plan of this line; in the figure cb c_1b_1 is a horizontal line and x_1y_1 is perp. to cb. In the projected view the given face appears as a straight line, and an auxiliary plan projected on a ground line par^1 to it is the required projection. In the drawing x_2y_2 is par^1 to $a_2b_2c_2$ and the required plan is shown in thick lines.

166. To determine the Projection of a Solid having a given Edge in one Plane of Reference and making a given Angle with xy, and a Face containing that Edge inclined at a given Angle to the same Plane of Reference.

Two examples are given to illustrate methods of dealing with this type of problem. The required projections are obtained from the most simple plan and elevation.

Problem 167. To determine the Elevation of a given hexagonal Prism arranged with the Edge of one of its Faces in the H.P. and inclined at φ to xy, the Face containing the Edge being inclined at θ to the H.P.

The *plan* of the prism is unaltered whatever value is taken for φ. When φ is 90° the projections of the prism arranged with a face inclined at θ to the H.P. are readily drawn, as shown in figure—in which ab a_1b_1 are the projections of the edge in the H.P. An auxiliary elevation projected from the plan, on x_1y_1 inclined at φ to ab produced, gives the required view.

Problem 168. To determine the Elevation of a given hexagonal Pyramid having an Edge of the Base in the H.P. and inclined at φ to xy, and the Face containing that Edge inclined at θ to the H.P.

As in the former Problem the value of φ makes no difference to the plan, and from the plan—which is readily obtained from simple projections—the required elevation may be projected. Begin with the simple views shown; ab a_1b_1 are the projections of the edge in question and oab $o_1a_1b_1$ those of the face. Project an auxiliary plan on x_1y_1 inclined at θ to $o_1a_1b_1$. From this plan project an auxiliary elevation on x_2y_2 inclined at φ to a_2b_2; this gives the required elevation.

Note. A line mn, par^1 to x_1y_1, has been used for taking measurements for the final elevation; e.g. the distance of o_2 from x_2y_2 = that of o_1 from mn (not x_1y_1). This is frequently necessary to give a compact drawing and obviously has no bearing on the shape of the projection.

EXAMPLES

Using the following data, draw the views required in the above problems.

(1) Prob. 165: take dimensions from the pictorial view.

(2) Prob. 167: all edges $1\frac{1}{4}''$, θ = 35°, φ = 45°.

(3) Prob. 168: edge of base 1″, altitude 3″, θ = 50°, φ = 45°.

(4) A regular tetrahedron 3″ edge has a face inclined at 40° to the V.P., an edge of that face lying in the V.P. and making 35° with xy. Draw its plan.

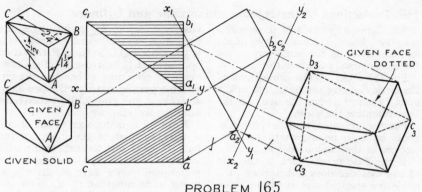

PROBLEM 165

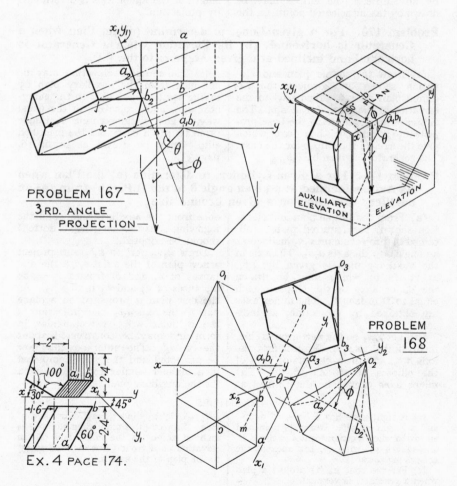

PROBLEM 167—
3RD. ANGLE
PROJECTION—

PROBLEM 168

Ex. 4 PAGE 174.

169. Projections of the right circular Cone and Cylinder.

Auxiliary projections of these solids may be obtained from simple plans and elevations in the same way as for the solids already dealt with.

For the cone, obtain projections of the circle forming the base, and of the apex; tangents from the apex to the base complete the view.

For the cylinder, outside par[l] tangents have to be drawn to the projections of the two circular ends.

If the base of the cone or the end of the cylinder does not appear as a circle or a straight line, it will be given by an ellipse. The ellipse may be drawn by taking selected points on the base, such as p, p_1, giving p_2 in the projection, as described in Prob. 153, page 166, and shown for the cone opposite; it is best constructed, however, by obtaining either the circumscribing parallelogram or the two axes, and using the standard methods for drawing the ellipse.

Note. The major axis of the ellipse is always equal to the diam. of the base (or end), and the axis of the solid bisects the major axis at rt. ∠s. The major and minor axes of the ellipse bisect each other at rt. ∠s, and the length of the minor axis is determined by projection.

Problem 170. For a given Cone, to determine (a) its Plan when a Generator is horizontal, (b) its Elevation when a Generator is horizontal and inclined at a given Angle φ to the V.P.

(*a*) Draw the simple plan and elevation shown and project the required plan on x_1y_1 taken par[l] to the elevation of the generator ab a_1b_1. The circumscribing square to the base projects into a rectangle, the sides of which give the axes of the ellipse for the base. The generator is given by a_2b_2.

(*b*) The required elevation may be obtained from the auxiliary plan by taking x_2y_2 inclined at φ to the generator a_2b_2, and projecting in the usual way. The circumscribing rectangle projects into a par[m] and the inscribed ellipse may be drawn as in fig. 6, page 33.

Problem 171. For a given Cylinder, to determine (a) the Plan when its Axis is inclined at a given angle θ to the H.P., (b) from (a) an auxiliary Elevation on a given ground line.

(a) From the simple plan and elevation shown, the required plan is obtained by projecting on x_1y_1 inclined at an angle θ to the axis a_1b_1. The axis in the auxiliary plan is given by a_2b_2, at rt ∠s to which may be drawn the major axes of the ellipses, each equal to D in length. The minor axes are obtained by projection, as indicated.

(b) Let x_2y_2 be the new ground line, inclined at β to x_1y_1. Project first the axis a_3b_3, and draw the major axes of the ellipses at rt. ∠s to it. The minor axes cannot be determined at once from the auxiliary plan, and the following construction (shown dotted) should be adopted.

Draw x_3y_3 par[l] to a_3b_3 and project a new plan of the axis, a_4b_4; the distances of a_4 and b_4 from x_3y_3 = the distances of a_2 and b_2 from x_2y_2. As the new plan is projected on a plane par[l] to the axis a_3b_3, the projection of the cylinder will be rectangular in form and may be constructed about the axis a_4b_4. The minor axes may now be projected and the view completed in a manner similar to that adopted for the auxiliary plan.

EXAMPLES

(1) A right cone has a base, 3″ diam., and a height of $3\frac{1}{2}$″. Draw its plan and elevation when a generator lies in the H.P. and makes 45° with xy, the apex of the cone being towards xy.

(2) For the cone in (1) project a plan when a generator is vertical.

(3) A right cylinder is $2\frac{1}{2}$″ diam. and 3″ long. Project (*a*) a plan when the axis is inclined at 30° to the H.P., and (*b*) an elevation on a ground line inclined at 45° to the plan of the axis in (*a*).

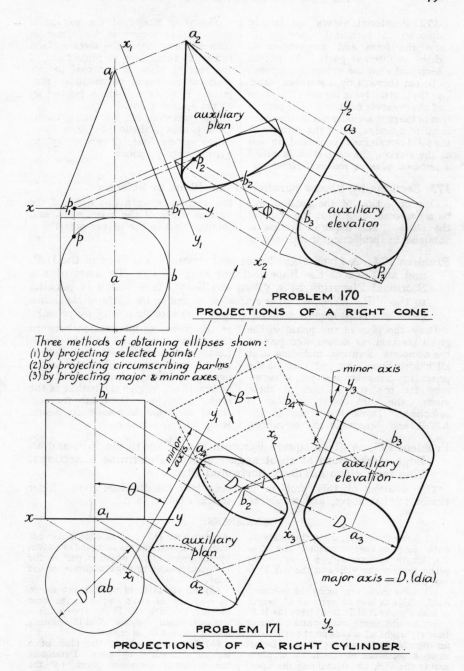

PROBLEM 170

PROJECTIONS OF A RIGHT CONE.

Three methods of obtaining ellipses shown:
(1) by projecting selected points!
(2) by projecting circumscribing parlms
(3) by projecting major & minor axes

PROBLEM 171

PROJECTIONS OF A RIGHT CYLINDER.

major axis = D.(dia)

172. Sectional views are largely employed in technical drawings to show the form and proportions of hidden or internal parts. To obtain a sectional view the object is supposed to be cut through by a suitable plane, and the portion lying between the plane and the observer removed. The projection of the part actually cut by the plane is called a *section*, but if the exterior of the solid remaining is projected, as well as the section, the projection is called a *sectional plan* or *sectional elevation*.

The **true shape** of the section is given in one projection, i.e. either in plan or elevation, if the section plane is par¹ to that plane of projection; if the section plane is not par¹ to one plane of reference, an auxiliary projection is necessary—on a plane par¹ to the section plane.

The material cut by the plane is usually shown shaded by thin lines, called *section lines*, drawn at 45° to *xy* and evenly spaced.

173. Sections by Planes parallel to one Plane of Reference.

One projection of the section will be a *line* coinciding with the trace of the plane. The required section is obtained by projecting the points of intersection between this line and the various edges of the given solid, and joining them in the proper order.

Problem 174. A given right hexagonal Prism has a Face in the H.P. and an Edge of the Base inclined at φ to *xy*. To determine a Sectional Elevation by a given auxiliary Plane which is parallel to the V.P. (*Note.—This is equivalent to finding the shape of the section given by a plane making an angle of* (90° − φ) *with the axis of the prism.*)

Draw the plan of the prism in the given position, as shown, and project the complete elevation, including in it all hidden lines. An end view of the prism, i.e. a regular hexagon, is necessary for the construction of these views: the part shown in plan is sufficient. Draw the trace of the A.V.P. and determine the elevations of the points of intersection between the edges of the solid and the plane. Letter the points in each view (letters have been omitted deliberately in the figures opposite); lines joining the points, in order, give the outline of the required section. Complete the sectional elevation as shown in figure. Refer to Ex. 1 below.

Problem 175. A given square Pyramid has a Face in the H.P. and an Edge of the Base at right angles to *xy*. Determine a Sectional Plan by a given auxiliary horizontal Plane.

The solution of this problem is similar to that above, and should be clear from the drawings given. Refer to Ex. 2 below.

EXAMPLES

(1) Solve Prob. 174, using the following data. Edge of base of prism 1″ (i.e. R = 1″), length of axis 2½″, φ = 60°. The prism touches the V.P. and the A.V.P. is 1·9″ from the V.P.

(2) Solve Prob. 175, using the following data. Edge of base of pyramid 2″, length of axis 3″; the A.H.P. is ·6″ from the H.P.

(3) A right hexagonal pyramid, edge of base 1″, height 3″, stands on the H.P. with an edge of the base at 45° to *xy*. Determine a sectional elevation by a plane par¹ to the V.P. and containing the apex.

(4) A cube, 2″ edge, is pierced centrally by 1″ square holes, as shown. The cube stands on the H.P. and an edge makes 60° with *xy*. Determine the sectional elevation by the given plane, which is par¹ to the V.P. and passes through a corner of the cube.

(5) An elevation of the bracket shown on page 169 is given opposite, the base making 30° with the H.P. Determine a sectional plan by an A.H.P. passing through a corner, as shown.

(6) The figure shows the plan of a regular octahedron 1¼″ edge. Determine the sectional elevation given by the A.V.P.

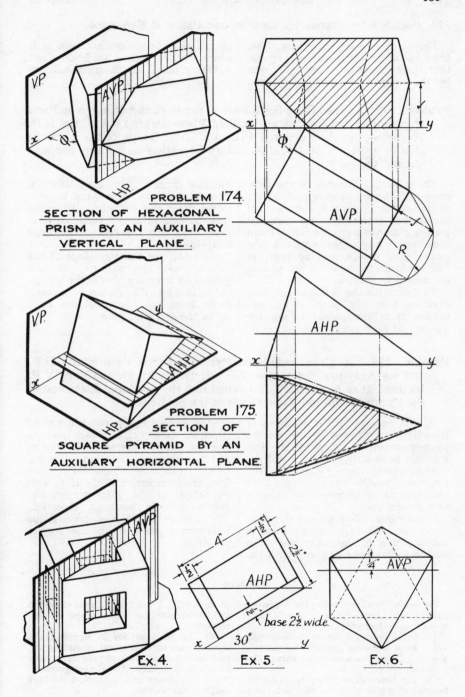

PROBLEM 174.
SECTION OF HEXAGONAL
PRISM BY AN AUXILIARY
VERTICAL PLANE.

PROBLEM 175.
SECTION OF
SQUARE PYRAMID BY AN
AUXILIARY HORIZONTAL PLANE.

Ex. 4.

Ex. 5.

Ex. 6.

176. Sections by Planes inclined to one Plane of Reference.

The sections considered here differ from those on the previous page in that their true shapes are not given by the ordinary sectional plans and elevations. To determine them it is necessary to obtain projections on auxiliary planes which are parallel to the plane of the section.

Problem 177. A given Cube has a Face in the H.P. and an Edge inclined at φ to *xy*. It is cut by an inclined Plane making θ with the H.P. To determine the Sectional Plan and True Shape of the Section; also to project a Sectional Auxiliary Elevation on a Plane perpendicular to *xy*.

The *plan and elevation* of the cube are readily drawn, as shown. The V.T. of the plane gives an elevation of the section, from which the sectional plan is obtained by projection. From these views any other sectional projection may be obtained by applying the fundamental rules of projection for auxiliary planes.

To determine the *sectional auxiliary elevation*, take x_1y_1 perp. to *xy*, and project from the plan, making the distance of each point from $x_1y_1 = $ its distance from *xy* in the elevation. (Note: the same view could be projected from the elevation; it would then be treated as a " sectional auxiliary plan ", but called a " sectional end view ".)

To determine the *true shape* of the section, take x_2y_2 par¹ to V.T. and project an auxiliary plan of the section only, making the distance of each point from $x_2y_2 = $ its distance from *xy* in the sectional plan.

Problem 178. A given hexagonal Pyramid has a Face in the H.P. and its Axis parallel to the V.P. It is cut by an auxiliary V.P. inclined at φ to the V.P. To determine the Sectional Elevation of the Pyramid and the True Shape of the Section.

In this example the projections of the solid in the required position (i.e. the position shown in the pictorial view) may not be readily obtained, and the student should begin by drawing the *simplest possible projections*, then obtaining those required by using a new ground line.

Draw the simple plan and elevation shown, and take x_1y_1 coincident with the elevation of the slant face. An auxiliary plan on x_1y_1 gives, with the original elevation, the two projections required. The original plan and *xy* line may now be disregarded entirely, and x_1y_1 taken as the normal ground line. Draw the plane VTH, and from the intersection of the H.T. with the edges of the plan project the points on the sectional elevation. The projection of the true shape follows the method of the former problem.

EXAMPLES

(1) Solve Prob. 177, using the following data. Edge of cube 2″, θ = φ = 30°; H.T. of plane is ·8″ from edge of cube.

(2) Solve Prob. 178, using the following data. Edge of base 1″, vertical height $3\frac{1}{2}$″, φ = 30°. The plane bisects the axis of the pyramid.

(3) The figure shows the plan of a pentagonal pyramid 3″ high, and the trace of an A.V.P. Project a sectional elevation and determine the true shape of the section.

(4) The solid shown in figure has a square base, an equilateral triangular top par¹ to the base, and sloping triangular faces. Project the sectional plan given by the A.I.P. shown and determine the true shape of the section.

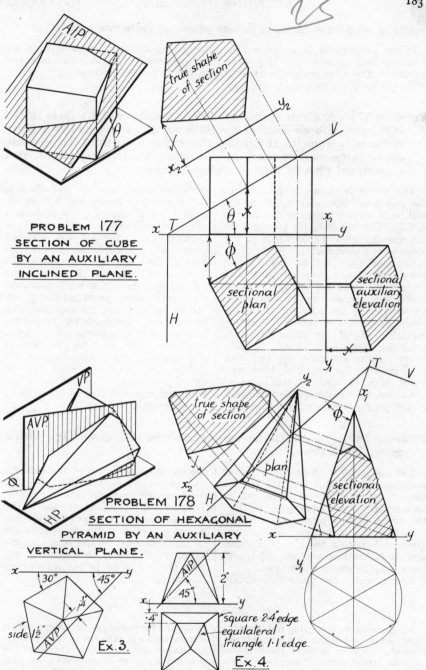

PROBLEM 177
SECTION OF CUBE
BY AN AUXILIARY
INCLINED PLANE.

true shape of section

sectional plan

sectional auxiliary elevation

true shape of section

plan

sectional elevation

PROBLEM 178
SECTION OF HEXAGONAL
PYRAMID BY AN AUXILIARY
VERTICAL PLANE.

30° 45°

side 1½"

Ex. 3.

45°

2"

·4"

square 2·4" edge
equilateral
triangle 1·1" edge.

Ex. 4.

Sections by planes inclined to one plane of reference (*cont.*).

If the projections of an edge of a solid are both perp. to *xy*, the plan of a point on that edge cannot be directly projected from its elevation, and vice versa: an auxiliary view is required to give the projection, as employed in the following problem.

Problem 179. A given hexagonal Prism stands with its Base in the H.P., and with an Edge of the Base parallel to *xy* and at a given distance D from it. It is tilted about that Edge until the opposite upper Edge touches the V.P. To determine its Projections, also a Sectional Plan by a given inclined Plane.

The prism is shown in position in the pictorial view, and the first step will be to obtain its projection on the A.V.P. The numbers (in brackets) in the drawing give the sequence of the steps in the construction.

Commence by drawing plan (1) of the prism standing on its base, and project elevation (2) on x_1y_1. These views are shown by broken lines. Draw *xy* perp. to x_1y_1 and at a distance D from the point a_2; rotate elevation (2) of the prism about a_2 until the edge a_2b_3 is at a_2b_2, b_2 lying in *xy*. From elevation (3) and plan (1), project the required plan (4). From plan (4) project the required elevation (5), making the distances of all points from *xy* in (5) = their distances from x_1y_1 in (3)—this is facilitated by projecting up to *xy* from (3) and swinging the projectors around O as shown.

Draw the trace of the inclined plane and proceed as follows to obtain the section. Consider the edges *cd* and c_1d_1; the plane intersects c_1d_1 in m_1. By projecting backwards, the corresponding point in (3) is given, viz. m_2 on c_2d_2, and the required plan *m* on *cd* is obtained by projecting from m_2. For this particular example, one other point must be projected in this manner; the other points on the section are obtained from (5) by direct projection. The outline of the section is shown dotted in (4) to leave the projection clear.

Problem 180. To determine the Intersection of a given Line with a given Polyhedron.

Let the given solid be the irregular pyramid OABC and let MN be the line. Suppose the solid to be cut by an auxiliary V.P. containing the line: the trace of this plane will coincide with the plan of the line. The section given will contain the points of intersection of the line and the solid, and these may be easily determined.

Construction. By projection obtain the sectional elevation of the solid by an auxiliary V.P., the H.T. of which coincides with *mn*, the plan of the line. This section is shown shaded. The elevations of the points of intersection, p_1 and q_1, are given by the intersection of the elevation of the line and the figure representing the section, and their plans, *p* and *q*, by projecting from p_1 and q_1.

EXAMPLES

(1) Solve Prob. 179, using the following data: edge of hexagon $1\frac{1}{4}''$, height $2\frac{1}{4}''$, D = 2''; trace of inclined plane makes 30° with *xy* and passes through the centre of the upper hexagonal end.

(2) An embankment, 30' high, is in the form of a triangular prism, the sides sloping at 30° and 60° to the ground. A pipe enters the embankment half-way up the short side and passes through to the long side. The plan of the pipe makes 30° with a long edge and the pipe rises at 5° to the ground. Find where it emerges and its height at this point. Scale: $1'' = 10'$. Ans. 19' 6".

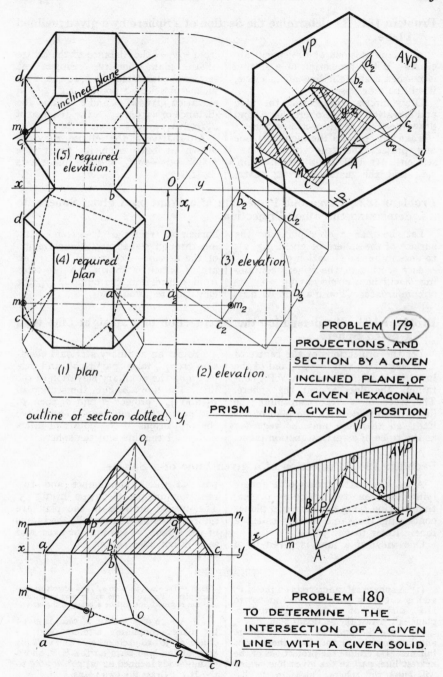

(5) required elevation.

inclined plane

(4) required plan

(1) plan

(2) elevation.

outline of section dotted

(3) elevation

PROBLEM 179

PROJECTIONS AND SECTION BY A GIVEN INCLINED PLANE, OF A GIVEN HEXAGONAL PRISM IN A GIVEN POSITION

PROBLEM 180

TO DETERMINE THE INTERSECTION OF A GIVEN LINE WITH A GIVEN SOLID.

Problem 181. To determine the Section of a Sphere by a given inclined Plane.

All plane sections of the sphere are circles; if the projection of a section is neither a straight line nor a circle, it will be an ellipse.

Draw circles representing the plan and elevation of the sphere, fig. 2, and draw the trace of the section plane.

Take x_1y_1 par[l] to the inclined plane and project the true shape of the section. It will be a circle, diam. c_1d_1, and the distance of its centre from x_1y_1 = the distance of the centre of the plan from xy. Project the sectional plan from the true shape by means of selected points; e.g. a_1b_1 in elevation gives a, a_2 and b, b_2, and the distance of a from xy = the distance of a_2 from x_1y_1; and so on.

Alternatively the ellipse may be drawn at once from its axes: the major axis = c_1d_1, and the minor axis = cd.

Problem 182. Given one Projection of a Point on a given Sphere, to determine the other Projection.

Let the plan p of a point on the surface of the sphere be given (fig. 2): to determine the elevation of p.

Any section of the sphere containing p will be a circle having p on its circumference. Choose a V.P. with its horizontal trace par[l] to xy and passing through p. Project the elevation of the section—a circle of rad. r, an arc of which is shown. A projector from p intersects the circle in p_1, the elevation required.

a needle

Problem 183. To determine the Intersection of a straight Line with a given Sphere.

Let the projections of the centre of the sphere be given by o, o_1 and of the line by ab, a_1b_1. Suppose a V.P., containing the line, to cut the sphere. The trace of a V.P. containing the line will coincide with the plan of the line; ab therefore may be regarded as the trace of a vertical section plane.

Project an auxiliary sectional elevation on x_1y_1 taken par[l] to ab, and project a_2b_2, the auxiliary elevation of the line. The line a_2b_2 cuts the section in p_2 and q_2; project p and q, then p_1 and q_1, in the usual way. These give the projections of the points of intersection of the line and the sphere.

Problem 184. Projection of a given Lune of a Sphere.

A zone is that portion of a sphere lying between two parallel planes; that portion lying between two planes containing the same axis is called a lune. Refer to fig. 3.

One view of a lune is given by a part of the circumference and two radii, as in the elevation of fig. 5. The curved outlines in the plan are semi-ellipses both having R, the radius of the sphere, as semi-major axes, and semi-minor axes oa and ob.

EXAMPLES

(1) A sphere $2\frac{1}{4}''$ diam. rests on the H.P. and is cut by a plane inclined at 60° to H.P., distance of centre of sphere from plane ·4″. Draw the sectional plan.

(2) Scale the projections of the sphere and line from fig. 4 and draw them twice full size. Determine the projections of the nearest line, par[l] to the given line, which will *touch* the sphere; measure its distance from the given line.

Hint.—Take x_1y_1 perp. to a_2b_2 and project an auxiliary plan. The given line will appear as a point, and the required line will be a point on the circle.

(3) A lune of a sphere, 2″ rad., is given by planes containing a diam. and inclined at 30° to each other. The straight edge of the lune rests on the H.P. and a plane face is inclined at (a) 20°, (b) 60° to the H.P. Draw the two plans.

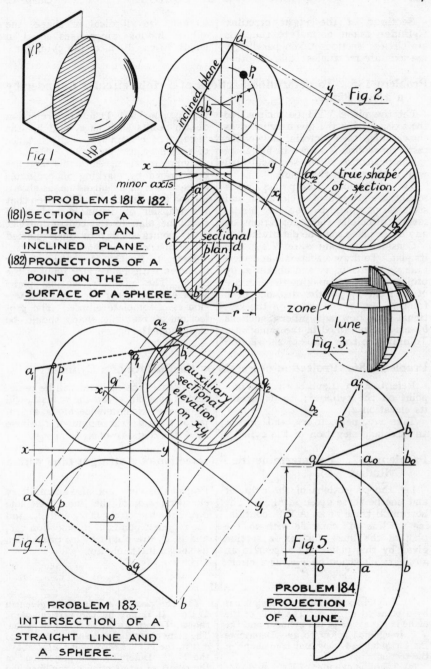

Fig 1

PROBLEMS 181 & 182.

(181) SECTION OF A
SPHERE BY AN
INCLINED PLANE.

(182) PROJECTIONS OF A
POINT ON THE
SURFACE OF A SPHERE.

Fig. 2.

true shape
of section.

Fig. 3.

zone

lune

Fig 4.

PROBLEM 183.
INTERSECTION OF A
STRAIGHT LINE AND
A SPHERE.

Fig 5.

PROBLEM 184.
PROJECTION
OF A LUNE.

Sections of the right circular Cylinder taken normal to the axis are circles; sections taken parallel to the axis are rectangles; and all other sections are elliptical in shape, the ellipses having minor axes equal in length to the diam. of the cylinder.

Problem 185. To determine Sections of a right circular Cylinder by a given Plane.

Let the plane VTH be inclined to the axis, cutting the given cylinder as shown in fig. 1. A pictorial view of the resulting section is shown in fig. 2.

The projection of the sectional plan will be clear from the figure.

Two projections of all points on the section are given by the elevation and sectional plan, and by applying fundamental principles their projections on any other plane may be determined.

Consider the point a_1 on V.T.; a is its plan. To draw a sectional end view take x_1y_1 perp. to xy and obtain a_2 by projection from a_1, making the distance of a_2 from x_1y_1 = the distance of a from xy. a_2 is one point on the sectional end view and others, such as b_2, may be obtained in the same way. To draw the true shape of the section take x_2y_2 par¹ to V.T., and by projection obtain points such as a_3 and b_3—the distance of a_3 from x_2y_2 = the distance of a from xy, and so on.

Symmetrical curves of this kind are best drawn by marking off ordinates on each side of a centre-line, as shown for the true shape. It will be seen that the transfer of distances such as d from the plan to the centre-line CL, gives points similar to those plotted by auxiliary projections.

Refer to fig. 3. Here the cylinder is shown with a generator lying in the H.P. The auxiliary vertical plane VTH cuts all generators, and the section is a complete ellipse. The projection of the true shape should be clear from the figure.

Problem 186. Projection of Points on the Surface of a Cylinder.

Refer to fig. 3. Let d be the plan of a point on the cylinder; to determine its elevation d_1.

Take x_1y_1 perp. to xy, and project an auxiliary elevation of the cylinder and point. The cylinder will be represented by a circle and the point d_2 will lie on its circumference along a projector from d: the height of d_1 above xy = the distance of d_2 from x_1y_1.

Problem 187. To determine the Intersection of a straight Line with a Cylinder.

Let the projections of the cylinder and line be those given in fig. 3. The horizontal trace of an A.V.P. containing the line will coincide with ab, the plan of the line; further, a section given by this plane will appear in an auxiliary elevation on x_1y_1 as a circle.

Project this view, and also an auxiliary elevation a_2b_2 of the line. The line a_2b_2 cuts the circle in c_2 and d_2, and the required points of intersection c, c_1 and d, d_1 are obtained by projection, as shown in the figure.

EXAMPLES

(1) A cylinder $2\frac{1}{4}''$ diam. $2\frac{3}{4}''$ long is cut by a plane as in fig. 1. The H.T. of the plane is $\frac{1}{4}''$ outside of the cylinder and the V.T. is inclined at 60° to xy. Determine the sectional end view and true shape of the section.

(2) Take the cylinder of Ex. 1 and project the true shape of the section given by a plane inclined at 45° to the axis.

(3) The plan of a cylinder is given in figure. It is cut by the vertical plane shown. Determine the sectional elevation. The plans of two lines ab and ac are shown: a is in the H.P., c is $3''$, and b is $2''$ above the H.P. Determine the projections of the points of intersection of the lines and the cylinder.

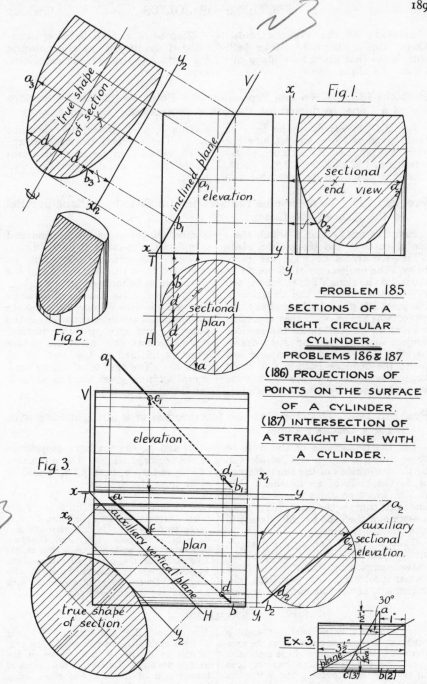

Fig.1.

true shape of section

sectional end view

inclined plane

elevation

sectional plan

Fig 2.

Fig 3.

V elevation

auxiliary vertical plane

plan

true shape of section.

auxiliary sectional elevation.

PROBLEM 185.
SECTIONS OF A
RIGHT CIRCULAR
CYLINDER.
PROBLEMS 186 & 187.
(186) PROJECTIONS OF
POINTS ON THE SURFACE
OF A CYLINDER.
(187) INTERSECTION OF
A STRAIGHT LINE WITH
A CYLINDER.

Ex. 3.

Sections of the right circular Cone. One section only will be dealt with here—that given by a plane par^l to the axis of the cone.*

The construction will be better understood if the projections of a point on the surface of the cone are considered first.

Problem 188. Given one Projection of a Point on the curved Surface of a Cone, to determine the other Projection. (*See also p. 250.*)

Let the projections of the cone be those given in fig. 2, and let b be the plan of a point on its surface: to determine b_1 the elevation of the point. Draw the plan oa of the generator containing b, and project its elevation o_1a_1. The required elevation of the point must lie on o_1a_1 and on a projector from b, and is therefore given by b_1.

Problem 189. To determine the Section of a Cone by a Plane parallel to the Axis of the Cone.

Refer to figs. 1 and 2 which show the given cone and the section plane VTH—the traces of VTH are perp. to xy. The projections of all points on the section lie along VT in elevation, and HT in plan; further, the elevation of the point b is given by b_1, and as these projections are on VTH they represent a point on the outline of the section. Project an end view on x_1y_1, drawn par^l to VT, regarding it as an auxiliary plan; the distance from x_1y_1 of b_2 on the curve = the distance of b from xy, and so on. Determine a sufficient

number of points on the section and complete the drawing as in the figure.

Note: The sectional view could be projected from the plan instead of the elevation, as in problem 190.

Objections to the method. The angle between o_1a_1 and VT may be too acute for the accurate determination of the point b_1, and this generator method should be used only when the section plane is inclined to the axis, as on page 263 The method of taking horizontal sections given there is suitable for the above example.

Problem 190. To determine the Intersection of a straight Line with a right Circular Cone.

Let the projections of the cone be those given in fig. 3, and let ab, a_1b_1 be the projections of the line. Regard ab as the H.T. of an auxiliary V.P. A V.P. which contains the plan of the line must contain the line itself; hence a section of the cone by this plane will also contain the line.

Take a ground line x_1y_1 par^l to ab, and project an auxiliary sectional elevation, employing the method of the former problem. An auxiliary elevation a_2b_2 of the line cuts the section

in p_2 and q_2, the auxiliary projections of the required points. The projection of p, q and p_1, q_1 from p_2, q_2 should be clear from the drawing.

Alternative Construction — *dotted lines.* Project the cone and line on x_2y_2 taken perp. to a_2b_2. In this view the line appears as a point a_3b_3, and the two generators o_3c_3 and o_3d_3 on which it lies must contain the points of intersection. Project these generators back to the original views: their intersections with ab, a_1b_1 give the required points.

EXAMPLES

(1) A cone 3″ diam. base, 3$\frac{1}{2}$″ high, is cut by a plane par^l to and $\frac{1}{4}$″ from the axis. Determine the true shape of the section.

(2) The cone in Ex. 1 has its base on the H.P. and touching the V.P. The projectors of a line AB are 3$\frac{1}{2}$″ apart;

B is 2″ above H.P. and 3″ from V.P.; A is $\frac{1}{4}$″ from each plane of reference. The projectors of A are 1$\frac{1}{2}$″ from those of the axis of the cone. Determine the points of intersection of the line and cone, using both the methods shown in fig. 3.

* The section given is a hyperbola. This section and others are fully discussed on pp. 262 and 266.

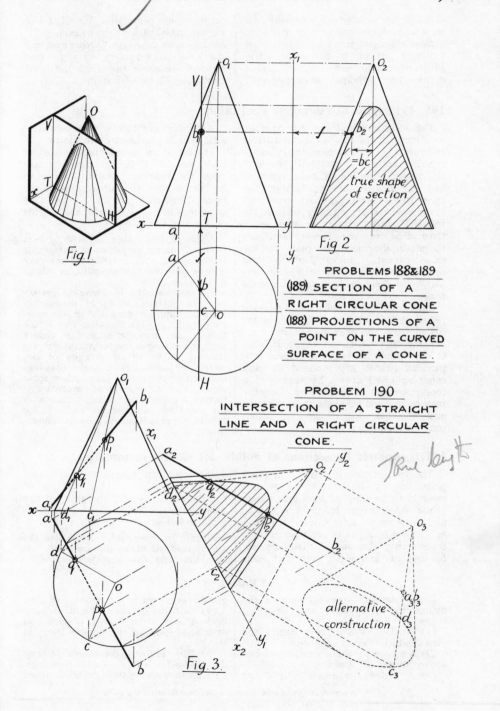

Fig 1

Fig 2

V

T

=bc

*true shape
of section*

PROBLEMS 188&189

(189) SECTION OF A
RIGHT CIRCULAR CONE

(188) PROJECTIONS OF A
POINT ON THE CURVED
SURFACE OF A CONE.

PROBLEM 190

INTERSECTION OF A STRAIGHT
LINE AND A RIGHT CIRCULAR
CONE.

True length

*alternative
construction*

Fig 3.

An object is more readily visualized from a pictorial view than from its orthographic projections. A system of pictorial projection, called Isometric Projection, will now be considered: it has the principal advantages of orthographic projection, viz. that the views can be quickly drawn and may be used as scale drawings. Other systems are dealt with on pages 298–305. The term *Axonometric Projection* has been applied to all these systems.

191. Principles of Isometric Projection.

Fig. 1 shows the plan of a cube when a diagonal of the solid is vertical (refer to Prob. 163). All faces of the cube are equally inclined to the H.P., each of the upper three faces being par[l] to one of the lower three; further, the three edges meeting at the upper and lower corners are equally inclined to the H.P. (although at another angle than that of the faces), and are therefore shortened in plan to the same extent. *Lines par[l] to these equally inclined edges will have the lengths of their projections to the same scale.*

Any rectangular solid may be projected in this way, and its dimensions scaled from the drawing. Three concurrent edges such as OA, OB, and OC are taken as **axes,** and their projections (which are inclined to each other at 120°) drawn by means of a tee-square and 30° set-square. Lengths may be marked off along them directly and the other edges of the solid drawn par[l] to these lines.

Any plane par[l] to two of the isometric axes is called an **isometric plane.**

It should be noted that *only lines which are par[l] to the axes can be scaled from the drawing*: the difference in length between the projections of the equal diagonals OD and BC, fig. 1, is considerable; compare also the lengths of AB and CD in fig. 2.

Although the system is used to best advantage in the projection of rectangular. solids, solids with curved surfaces may also be projected isometrically, as will be shown later.

The **advantages** of isometric projection are illustrated in fig. 2: the column base is completely described in one view. Refer to Ex. 5—the isometric view of the timber joint shows its form more clearly than the corresponding orthographic projections. The **disadvantages** of isometric projection are that the drawings take longer to prepare than orthographic views; that an element of apparent distortion is introduced,* which often makes the projection unattractive; and that the drawings are not always easily dimensioned.

192. Isometric Projections of Solids not wholly rectangular.

The projection of the hexagonal prism, fig. 3, illustrates the general method to be adopted in simple cases. Draw the hexagon *bchgfe*, produce *bc* and draw perps. to it from *e, f, g,* and *h.* Draw $a_1b_1c_1d_1$ at 30° to a horizontal, transferring the actual distances *ab*, *bc*, and *cd*, and erect perps. at each point equal in length to the corresponding heights of the corners of the hexagon. Set off the sides of the prism, also inclined at 30°, and complete the view as in the figure.

Note. In this problem actual lengths are marked off along the axes and this should be done in working the examples.

EXAMPLES

Draw isometric views of each of the figures shown opposite and described below.

(1) Frustum of a square pyramid, the plan of which is given. Full size.

(2) A square pyramid, standing on two square plinths. Full size.

(3) A rectangular block with a vee groove, as shown in elevation. Full size.

(4) A flight of steps, shown in plan. Scale ½″ = 1 foot. A is 1′ 0″ high and each step rises by the same amount to E.

(5) The timber joint shown by two dimensioned views. Scale, half size. Arrange the views as shown in the solution.

* An isometric view of a St. Andrew's cross will illustrate this point.

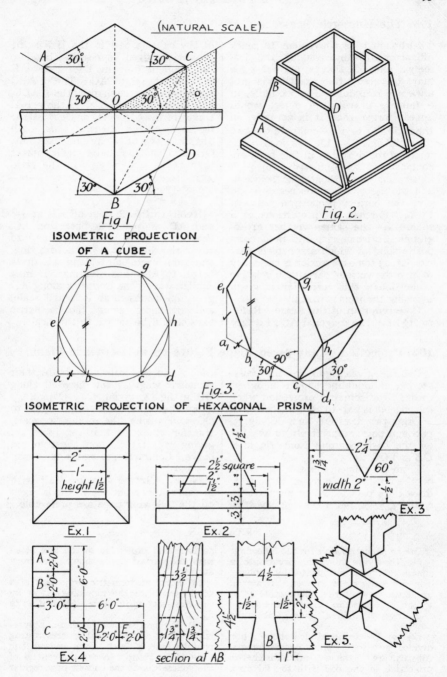

(NATURAL SCALE)

Fig 1

Fig. 2.

ISOMETRIC PROJECTION
OF A CUBE.

Fig. 3.

ISOMETRIC PROJECTION OF HEXAGONAL PRISM

Ex.1

2"
1"
height 1½"

Ex.2

2½" square
2"
1½"
1½"
3·3

Ex.3

2¼"
3¼"
60°
width 2"
½"

Ex.4

A 2'·0"
B 2'·0"
3'·0"
6'·0"
6'·0"
C 2'·0"
D 2'·0" E 2'·0"

section at AB.

3½"
4½"
1¼" 1¼"

A
4½"
1½" 1½" 2"
B
1"

Ex.5.

193. The Isometric Scale.

Although it is convenient to mark off actual distances along the iso-metric axes so that a natural scale may be used, the practice is not always permissible. For example, in a full-size drawing an object would appear larger than it is actually, in the ratio $\sqrt{3}$ to $\sqrt{2}$—compare the two views in fig. 3 on the previous page; many of the drawings in this book are comparative, an isometric projection illustrating an orthographic solution, and the use of a scale is necessary to give the correct proportion between them. Further, all projections of a sphere are the same, whether ortho-graphic or isometric, and in a com-posite solid of which a sphere is part (see Ex. 1) some correction is necessary to give the various parts their relative proportions: this correction is given by using the isometric scale.

Construction of the Scale. Refer to fig. 1. The diagonal AC of a face of the cube is par[l] to the H.P.; the projection given represents therefore the true length of this line. Draw AE and CE at 45° to AC. The △AEC represents the true shape of the △ADC, and either AE or EC gives the actual length of the edges AD or DC. *This is the basis of the isometric scale.* All distances set off along the axes OA, OB, and OC should be made shorter than the actual lengths in the ratio

$$\frac{AD}{AE} = \frac{\sqrt{2}}{\sqrt{3}}.$$

Refer to fig. 2. Set off AE at 45° and AD at 30° to a base line AC. Graduate AE in inches (or cms.) to cover the dimensions of the drawing, and from each point along AE draw perps. to AC, thus dividing AD in a similar way. The divisions along AD give dimensions to an isometric scale, and distances along the isometric axes should be set off to this scale.

194. Projection of any given Plane Figure on an Isometric Plane.

Let the given figure be the cross-section of moulding shown in fig. 3: to draw its isometric projection when the edge hk is par[l] to an isometric axis.

Draw two axes OB and OC, fig. 3, par[l] and perp. respectively to hk. Set off the corresponding isometric axes OB and OC, fig. 4.

From a sufficient number of points on the outline, such as m, n, o, p, k, draw lines par[l] to OC to meet OB in c, d, e, f, g. Transfer the distances Oa, Ob ... to the scale and obtain their isometric lengths; set these off along OB in fig. 4, giving Oa_1, Ob_1 .... From these points draw lines par[l] to OC and using the isometric scale, transfer the lengths cm, dn ... giving m_1, n_1 ... on the required curve. Complete the figure as shown.

Note. Circles may be dealt with in this manner but the methods given on the following page are preferable.

EXAMPLES

Use an isometric scale for the following:
(1) The figure shows the elevation of a square gate pillar surmounted by a sphere. The dimensions are in feet. Draw an isometric view of the pillar, scale 1″ = 1 foot. (Locate the centre of the sphere and draw a circle 2″ *diam.* about the point.)

(2) (a) The sectional view shows a box divided into four similar compartments. All parts are ·2″ thick. Draw an isometric projection of the box with the lid open at 90°.

(b) As a more difficult exercise draw an isometric projection of the box fitted with the curved cover shown, and open at 120°.

(3) Draw an isometric projection of a short length of the moulding given by the cross-section in figure. Show it as in fig. 4.

(4) The figure shows the plan of a pyra-midal stack of four equal spheres resting on a flat base, ¼″ thick. Draw an isometric view of the group. (*Note*: the centres of the spheres lie at the corners of a regular tetrahedron, 1¼″ side.)

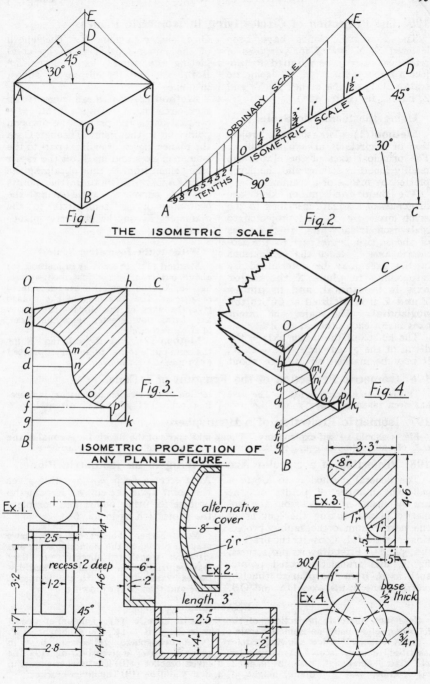

THE ISOMETRIC SCALE

Fig. 1   *Fig. 2*

Fig. 3.   *Fig. 4.*

ISOMETRIC PROJECTION OF
ANY PLANE FIGURE.

Ex. 1.

recess ·2 deep

Ex. 2.
alternative cover

Ex. 3.

Ex. 4.
base ½" thick

195. The Projection of Circles lying in Isometric Planes.

The isometric planes have been lettered X, Y, and Z for purposes of reference. Let it be required to project a circle, actual rad. R, isometric scale rad. r, on the planes X, Y, and Z, having its centre at O.

Using the Isometric Scale.

Method (1). The isometric projection of a circle is always an ellipse. The principal axes of the ellipse are readily found as follows and the curve plotted by means of a trammel.

The major and minor axes are diagonals of a square inscribed to the given circle, fig. 1a, and projected to scale isometrically, fig. 1—the sides of the square being par[l] to the isometric axes. Hence the inclinations of the axes may be determined by inspection: **in plane X the major axis is horizontal, and in planes Y and Z it is inclined at 60° to the horizontal.** The major and minor axes bisect each other at rt. ∠s.

The length of the major axis = the diam. of the given circle, i.e. = 2 . R ; it may be marked off at once about

O, as shown in plane Y. The length of the minor axis' is given by completing the square, as in plane Z.* Both axes and the ellipse are shown in plane X.

Method (2). With rad. r describe a semi-circle, fig. 2a. Divide it into, say, 6 equal parts and project the division points on to the diam. Through O on the planes, fig. 2, draw lines par[l] to the isometric axes and mark off the rad. r along them from O, thus obtaining the points a, b, c, d. Obtain further points on the curve by transferring the ordinates of the semi-circle to the diameters bd on the isometric planes —as shown.

Without the Isometric Scale.

Method (1). Proceed as in method (1) above but transfer the inscribed square, fig. 1a, to the isometric planes *without reduction*. The diagonals of the square give the axes of the ellipse—the major axis being now greater than the diam. of the given circle.

Method (2). Use the full rad. R for the semi-circle in fig. 2a and then proceed as in method (2) above.

196. Isometric Projection of the Frustum of a Cone.

This is clearly shown in fig. 3. The base has been taken in plane Y and the projection of the axis of the frustum is therefore inclined at 30° to the horizontal.

197. Isometric Projection of a Hemisphere.

Fig. 4 should be self-explanatory. The rad. of the outline is equal to the semi-major axis of the ellipse, i.e. is equal to the full radius of the given hemisphere.

198. Projection of a circular Arc not lying in an Isometric Plane.

The general method is to locate a sufficient number of points on the arc by means of perps. from the isometric planes, taking the lengths of the perps. from orthographic projections of the solid. Consider the bracket, figs. 5 and 6. First draw its projections, fig. 5, and transfer selected points, such as a, from the plan to the isometric plane, using axes OA and OB.

At a erect a perp. $= a_0a_1$; this gives the point a_1 on the curve. Repeat the process for other points and complete the projection as in fig. 6.

Note. Either the orthographic views may be drawn to the isometric scale and distances transferred directly; or they may be drawn to a natural scale and distances transferred first to the isometric scale and then to the drawing.

EXAMPLES

Construct isometric projections of the following solids, using an isometric scale.

(1) A cube $2\frac{1}{2}''$ edge showing circles inscribed in each face. (2) A cylinder, $2\frac{1}{4}''$ diam., height 3″, base in plane Z. (3) A cone, base $2\frac{1}{2}''$ diam., height 3″,

base in plane X. (4) A frustum of a cone, as in fig. 3. R = $1\frac{1}{2}''$, $r = \frac{3}{4}''$, height = 2″. (5) A hemisphere, $2\frac{1}{2}''$ diam., base in plane Y. (6) The given bearing. (7) The given bracket. (8) The bracket shown in figs. 5 and 6. (9) The given bracket.

* Length of minor axis = ·58 (diameter of given circle).

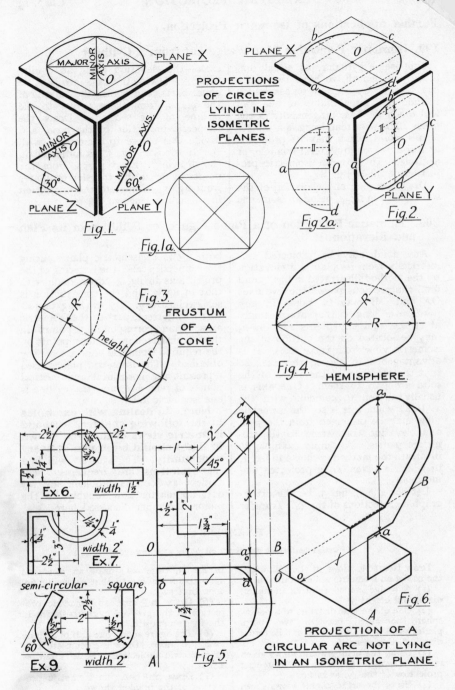

PLANE X

MAJOR AXIS
MINOR AXIS
O

MINOR AXIS O
30°

MAJOR AXIS O
60°

PLANE Z

PLANE Y

PROJECTIONS
OF CIRCLES
LYING IN
ISOMETRIC
PLANES.

Fig. 1.

Fig. 1a.

PLANE X

b   O   c

a   d

b
I-I
II-II
O

a   O

d

Fig 2a.

b
I
II
O

a

c

d

PLANE Y

Fig. 2.

Fig. 3.

R

height

r

FRUSTUM
OF A
CONE.

R

R

Fig. 4.

HEMISPHERE.

2½"   2½"
¾"
¼"
⅛"
½"   ¼"

Ex. 6.   width 1½"

1"
¼"
¼"
3"
2½"   width 2"   Ex. 7.

semi-circular   square
2½"
2"   ½r
¼"
½"
60°
Ex. 9.   width 2"

a₁
2"
45°
1"
½"   2"
1¾"
0   B
a₀

o   a
1¾"

A   Fig 5.

a₁
√
B
a
1
o   o
A

Fig. 6.

PROJECTION OF A
CIRCULAR ARC NOT LYING
IN AN ISOMETRIC PLANE.

Further applications of Isometric Projection.

199. Isometric Projection of a Wedge cut from a Cylinder.

Suppose a cylinder, of equal diameter and length, to have two parts cut from it by section planes which include the same diam. at one end and touch the base tangentially at the other. The part remaining is a wedge-shaped solid which may be projected as a square, an equilateral triangle, or a circle. To draw an isometric projection of this solid.

Draw the plan and elevation of the solid shown in fig. I, using an isometric

scale—the elevation being an equilateral △. Divide the base into, say, 12 equal parts, and obtain the lengths of the perps. from these points to the outline of the section. Draw the isometric projection of the base, and divide this ellipse to correspond with the divisions in fig. I, as indicated in the drawing. Erect perps. from the points in fig. 2 and transfer ordinates from fig. I. Complete the projection in the manner shown.

200. Isometric Projection of a Plane Figure, or Solid, from its Plan and Elevation.

Any solid may be projected isometrically from its plan and elevation by the method illustrated in figs. 5 and 6 on the previous page: the two axes OA and OB may be taken in any position as long as they are mutually at rt. ∠s, and any points such as a_1 may be located on the outline of the isometric view from these axes. It is advantageous to choose the axes so that the majority of the edges of the solid are par[l] to them. One axis is usually chosen to coincide with the xy line; when this is so, the isometric view may be projected from the plan and elevation transferred to the isometric planes. A simple example to illustrate the method is shown in fig. 4, in which a given △ is projected isometrically.

Let abc, $a_1b_1c_1$, fig. 3, be the orthographic projections of the △. Transfer

both views to isometric planes, using the isometric scale: the lettering of the projections in fig. 4 corresponds with that in fig. 3, and the construction is self-explanatory. By drawing projectors to intersect, par[l] to the isometric axes, from corresponding points in these transferred projections, points on the required isometric view are readily obtained. The isometric planes used represent the horizontal and vertical planes of reference, and the xy line is one isometric axis.

Note. In dealing with examples in the following chapter, freehand isometric views of the kind shown in fig. 4 should be drawn at every opportunity. The student will find that solutions are frequently self-evident as soon as problems are set out, no matter how roughly, in the pictorial manner discussed here.

EXAMPLES

(Further examples are given on pages 274 and 276.)

Draw isometric views of the following: the use of an isometric scale is optional.

(1) A wedge-shaped solid as in fig. 2; diam. and height 3″.

(2) One of the four quarters of a hollow sphere contained between two perp. planes passing through a diam. External diam. 4″, thickness ¾″.

(3) Using the orthographic views given in Ex. I, page 167, draw an isometric projection of the △, as in fig. 4.

(4) Using the orthographic views given

in Ex. 3, page 241, draw an isometric view corresponding to that shown in fig. 2 on that page.

(5) Draw, full size, an isometric projection of the grooved cylinder shown in the figure opposite.

(6) The figure shows one half of a stone arch 3′ 0″ deep. Draw the complete arch in isometric projection. Scale I cm. = I foot.

(7) Draw, full size, an isometric projection of the bracket shown.

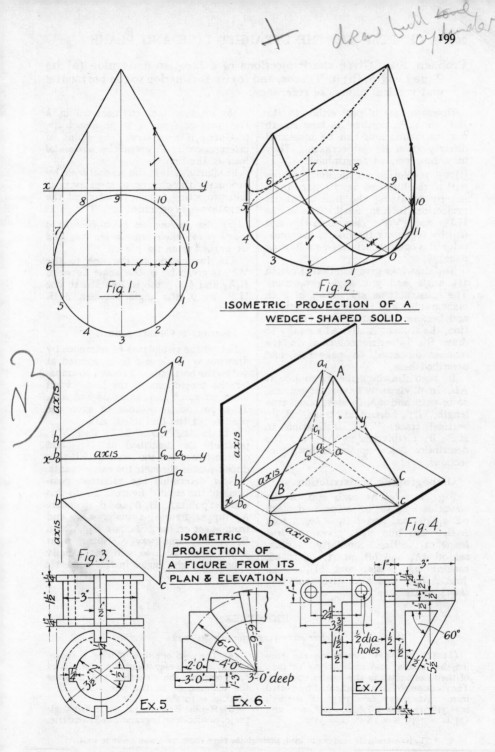

Fig. 1.

Fig. 2.

ISOMETRIC PROJECTION OF A
WEDGE-SHAPED SOLID.

Fig 3.

Fig. 4.

ISOMETRIC
PROJECTION OF
A FIGURE FROM ITS
PLAN & ELEVATION

Ex. 5.

Ex. 6.

Ex. 7.

Problem 201.   Given the Projections of a Line, to determine (a) its True Length, (b) its Traces, and (c) its Inclination to the horizontal and vertical planes of reference. *(found by GUX Projections)*

Unless a line is par[l] either to the H.P. or V.P., neither its true length nor its inclinations can be measured directly from its projections. They may, however, be determined:—

(i) by projecting on auxiliary planes par[l] to the line (see fig. 10, page 165),
(ii) by rabatting the line about its projections in turn, until it lies in the H.P. and V.P. (see below), and
(iii) by regarding the line as the generator of a conical surface (see following page).

The drawings give solutions in both 1st angle and 3rd angle projection. The constructions are shown in both orthographic and isometric projection, and because the latter aids visualization, the student is advised actually to draw the isometric solutions in the manner discussed on page 198 and described here.

In both drawings the given line is AB. In all views we have: projections, *ab* in plan and a_1b_1 in elevation; true length, L; horizontal trace, H.T.; vertical trace, V.T.; inclination to H.P., θ; inclination to V.P., φ. The description below applies to both projections.

Orthographic Construction.

Fig. 2.   (a) At each end of the elevation a_1b_1, set off perps. and mark off $a_1A_2 = aa_0$ and $b_1B_2 = bb_0$. Join A_2B_2; this line A_2B_2 gives the *true length* of the line. Similarly, in plan, set off aA_1 and bB_1 at rt. ∠s to *ab*, making $aA_1 = a_1a_0$ and $bB_1 = b_1b_0$. Join A_1B_1; this also gives the *true length* of the line.

(b) Produce .a_1b_1 to meet *xy* in *h* and project from *h* to intersect *ab*, produced if necessary; the point of intersection H.T. gives the *horizontal trace* of the line.

Similarly, obtain the *vertical trace* by producing the plan *ba* to meet *xy* in *v*, and projecting from *v* to intersect the elevation b_1a_1 produced in V.T.

(c) *The inclination of a line to a plane is the angle between the line and its projection on the plane.*

The inclination of the line to the V.P. is given by φ, the angle between B_2A_2 and b_1a_1; the inclination to the H.P. by θ, the angle between A_1B_1 and *ab*.

Isometric Construction.

Set off the two planes of reference by drawing verticals and lines inclined at 30° to the horizontal. Proceed to transfer the projections of the line. First mark off a_0b_0,[*] and obtain *ab* and a_1b_1. Then set off projectors to give the position of the line itself, AB.

Obtain the traces H.T. and V.T. precisely as described at (b), and demonstrate that the real line AB produced passes through the same points.

Now determine the rabatted positions of the shaded figures. To do this transfer points A_1, B_1, A_2 and B_2: to locate A_1 (upper figures) transfer a_0m, and from *m* set off mA_1. Locate B_1, A_2 and B_2 in the same way. (Note: right angles such as A_1ab will not usually appear as right angles in the isometric view.)

(Margin, right side: T R A C E S)

EXAMPLES

(Before answering refer to page 204)

(1) (2) and (3) Determine the true lengths, traces, and inclinations of each of the lines shown in the figures opposite. The solution to Ex. 3 is indicated by dotted lines. Ans. (1) $L = 3.13''$, $\theta = 18.5°$, $\varphi = 31°$. (2) $L = 2.87''$, $\theta = 20°$, $\varphi = 20°$. (3) $L = 2.5''$, $\theta = 18.5°$, $\varphi = 31°$.

(4) The plan of a line, 3″ long, is shown in Ex. 4. The elevation of one end is at b_1; complete the elevation and measure the inclinations of the line. Ans. $\theta = 41.5°$, $\varphi = 19.5°$.

(5) Repeat Exs. (1), (2), (4) in 3rd angle projection, both orthographic and isometric.

_____

* The isometric scale need not be used, although the views shown have been drawn to scale.

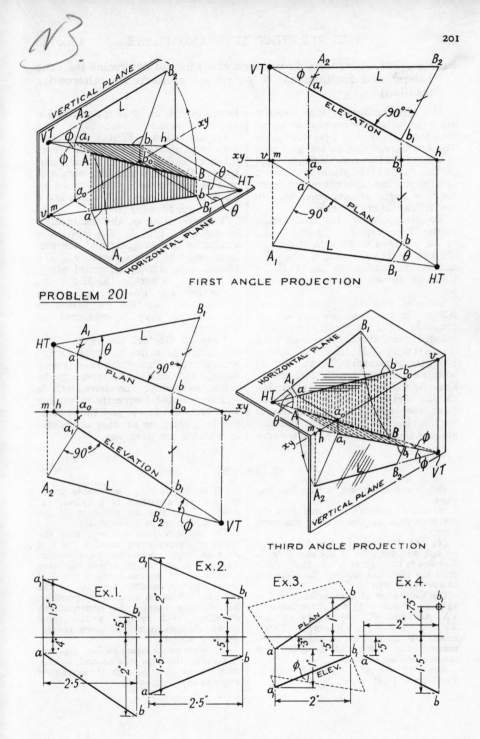

FIRST ANGLE PROJECTION

PROBLEM 201

THIRD ANGLE PROJECTION

Ex.1.

Ex.2.

Ex.3.

Ex.4.

Problem 202. Given the Projections of a Line, to determine its True Length and Inclinations to the planes of reference (alternative solution).

The constructions given opposite are sometimes more convenient to use than those given on the previous page. The line is regarded as the generator of a right circular cone having its base parallel to one of the planes of reference and the problem is solved by obtaining the true length of this generator and the base angle of the cone.

To determine θ, the inclination to the H.P. Fig. 1. If a line is par[l] to the V.P., the true length of the line is given by its projection on the V.P., and its inclination to the H.P. by the angle between the projection and xy.

Consider first the pictorial view; AB is the line, and ab a_1b_1 are its projections. If the plane figure ABba (shaded) turn about Bb as axis until it is par[l] to the V.P., i.e. in the position BCcb, the projection of BC on the V.P. will give the true length, b_1c_1, of the line. As the figure revolves about Bb, the line AB traces out a conical surface; the length of the generator of the cone is given by b_1c_1, and the base angle of the cone by θ; b_1c_1 is the true

length of AB, and θ is the inclination of AB to the H.P.

Construction. With b as centre and ba as radius, describe an arc to intersect bc drawn through b and par[l] to xy. Project from c and obtain c_1 on a horizontal through a_1. Join b_1c_1; b_1c_1 is the true length of the line and θ is the required inclination.

To determine φ, the inclination to the V.P. Fig. 2. The solution is similar to the foregoing and the pictorial view is self-explanatory. The shaded area ABb_1a_1 is turned about Bb_1 until it is par[l] to the H.P.; the projection of BC gives the true length of the line, and the inclination to the V.P., φ, may be measured, as indicated.

Uses of Conical Surfaces. The conception of a line in space as the generator of an imaginary cone is very useful, and enables solutions to be visualized readily. The above methods will be adopted frequently in dealing with problems herein and the student should apply them as often as possible in working out examples.

EXAMPLES

Before answering refer to the following page.

(1) Solve examples 1, 2, and 3 on the previous page by the method discussed above.

(2) The elevation of a line AB is given by a_1b_1 and the plan of the end B by b. The line is inclined at 30° to H.P. Complete the plan and measure the distances of d from xy. Ans. $-1''$, $6''$.

(3) Each edge of a pyramid VABC is $3\frac{1}{2}''$ long. Points P, Q, and R are taken on VA, AB, and VC respectively so that VP $= 1''$, BQ $= \frac{3}{4}''$, VR $= 2\frac{1}{4}''$. Determine the true shape of the $\triangle$ PQR, measure the lengths of its sides, and measure the angle PQR.

Ans. PR $= 1 \cdot 95''$, RQ $= 2 \cdot 8''$, QP $= 2 \cdot 62''$, angle PQR $= 43°$.

(4) With radius $1\frac{3}{4}''$, a circle is drawn on the curved surface of a cylinder 4'' diam. Determine an elevation of the curve struck out by the compasses.

The problem resolves itself into the following: given the plan of a line and its true length, determine its elevation. The plan is ab, and its length will vary from o to $1\frac{3}{4}''$, always being a chord of the circle representing the plan of the cylinder. Select a number of positions for ab, and obtain the corresponding elevations, such as a_1b_1; a fair curve drawn through points b_1 gives the required elevation.

(5) Refer to the figure for Ex. 4, p. 182. Determine the true lengths and inclinations to the H.P. of each of the sloping edges of the given solid.

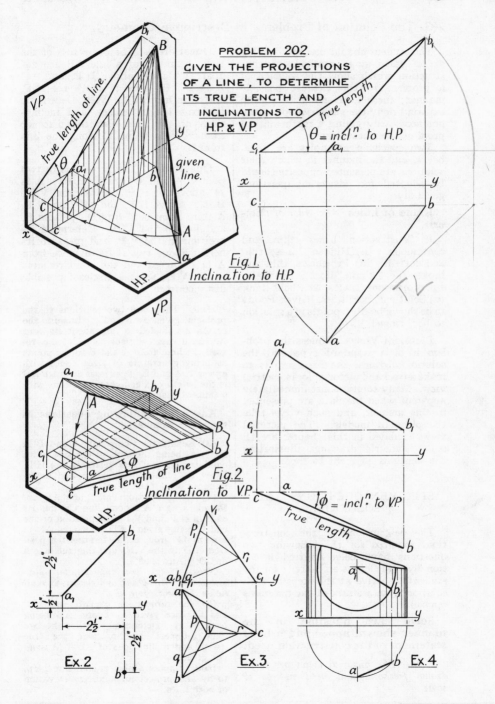

THE STRAIGHT LINE AND PLANE

PROBLEM 202.
GIVEN THE PROJECTIONS
OF A LINE, TO DETERMINE
ITS TRUE LENGTH AND
INCLINATIONS TO
H.P. & V.P.

true length of line

θ

given line.

true length

$\theta = incl^n$ to H.P.

Fig. 1.
Inclination to H.P.

TV

V.P.

true length of line

ϕ

Fig. 2.
Inclination to V.P.

H.P.

$\phi = incl^n$ to V.P.

true length

Ex. 2.

$2\frac{1}{2}''$

$\frac{1}{2}''$

$2\frac{1}{2}''$

$2\frac{1}{2}''$

Ex. 3.

Ex. 4.

203. The Solution of Problems in Descriptive Geometry.

The student should use and develop at an early stage methods of construction which show the solution of a problem in the clearest possible manner; further, he should be able to explain briefly the various steps taken, not necessarily giving a mathematical proof of their accuracy.

Two specimen examples are given below, and the manner in which their solutions are presented opposite should be adopted for answering questions generally.

Kinds of Lines.—*In order of thickness.*

(i) Construction Lines, thin and continuous. (ii) Hidden Lines, thin and dotted. (iii) Projectors, thin and broken (i.e. " chain " lines). (iv) Data, e.g. projections, traces, &c., bold lines. (v) Solutions, bold lines. Given Points to be shown boldly; points in a solution to be " ringed ".

Pictorial Views. Unless the problem is of a standard type and the solution obvious, the first step is to make a freehand pictorial (or isometric) view. Many solutions are immediately apparent when the data are presented in this manner, and such views take the place of models. The pictorial views included in this chapter are all scale isometric drawings illustrating the solutions adjacent to them.

SPECIMEN QUESTIONS AND SOLUTIONS

The descriptions of the constructions required for the following two questions have been omitted intentionally: it is left as an exercise for the student to arrive at the steps in each solution from a study of the diagrams opposite.

Solve each question in the manner shown, appending a brief statement of the construction used.

Note. The pictorial views are to be drawn freehand and need not be to scale.

Question 1. The projectors of the ends A and B of a line AB are $2\frac{1}{2}''$ apart; A is $\frac{1}{4}''$ above H.P. and $1\frac{1}{2}''$ behind V.P., B is $2''$ above the H.P. and $2''$ in front of the V.P. Determine the true length, traces, and inclinations of the line, also the distance from xy of a point $1''$ along the line from B.

Question 2. The projectors of two points, A and B, are $3''$ apart; A is $1\frac{1}{4}''$ above the H.P. and $1\frac{1}{2}''$ from V.P., B is $2''$ above H.P. and $1\frac{3}{4}''$ from V.P. A third point is $2''$ from A, $3''$ from B, and lies in the V.P. Determine its position in the V.P. and measure its height above xy. If its distance from A ($2''$) is fixed and the point remains in the V.P., what is its least possible distance from B?

Hint. There are two solutions to the problem, given by p and q. Imagine the rt. $\angle d$ $\triangle$s shaded to turn about Bb_1 and Aa_1 until their vertices coincide; i.e. regard the lines from A and B to the points as being generators of right cones with apices at A and B. (The least distance (D) is the length of a generator of a cone with a tangent base circle.)

Answer the following questions in the same manner.

(3) Regard the two points A and B in Ex. 2 as being the ends of the side of an equilateral $\triangle$ with its vertex in the H.P. Draw its plan and elevation and determine its side. (Ans. $3\cdot12''$).

(4) The projections of a line AB are shown in fig. A second line CD has its ends in H.P. and V.P., the position of one end C being given. CD intersects AB in a point $1\frac{1}{2}''$ from B. Determine the position of D, in the V.P., and the true length of CD. (Ans. $4\cdot68''$).

Note. If two lines intersect, the intersections of their plans and elevations must lie on the same projector.

(5) The projections of two lines AB and CD are given in figure, AB being perp. to xy. Determine whether the two lines intersect, and measure the true length and inclinations of AB. (Answer not given.)

Hint. Choose a new ground line perp. to xy and project an auxiliary elevation of both lines.

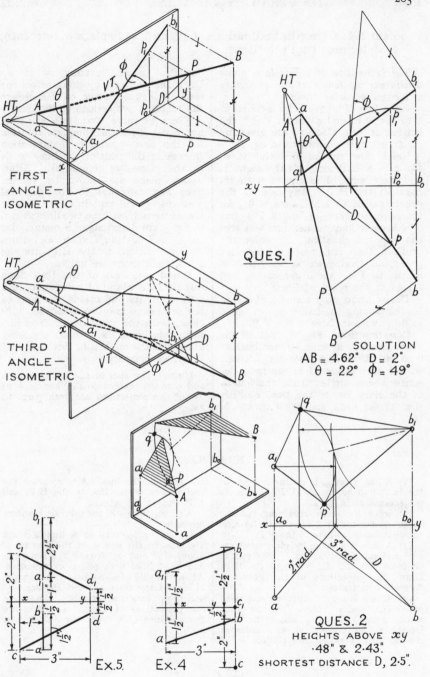

205

FIRST
ANGLE–
ISOMETRIC

THIRD
ANGLE–
ISOMETRIC

QUES. 1

SOLUTION
AB = 4·62" D = 2"
θ = 22° φ = 49°

Ex.5.

Ex. 4

QUES. 2
HEIGHTS ABOVE xy
·48" & 2·43"
SHORTEST DISTANCE D, 2·5".

Problem 204. Given the Inclinations of a Line to each plane of reference, to determine its Projections.

The inclination of a line to a plane determines the length of the projection of the line upon the plane.

Let a line of length L be inclined at θ to the H.P. and ϕ to the V.P.: the lengths of its projections are given by rt. $\angle$d triangles, drawn as in fig. 1.

Refer now to the pictorial view, fig. 2. A half-cone, slant height L and base angle θ is shown with its base on the H.P. Any one of the generators of this half-cone will give the required *plan*, of length PQ, but only *two* particular generators will give the required *elevation*. To solve the problem, then, it is necessary to select generators having elevations equal in length to PR; these generators will represent the required lines.

The solution may also be obtained by employing a half-cone of base angle ϕ, with its base in the V.P.

Construction. Fig. 3. Set off the given line Pb_1 of length L inclined at θ to xy. Draw the projector b_1b, and with b as centre and bP as rad., describe a semi-circle. Take the length of the elevation of the line, i.e. PR (fig. 1), as rad., and with centre b_1

describe an arc, cutting xy in a_1. Draw the projector a_1a to intersect the semi-circle in a. Join ab, a_1b_1: these are the required projections of the line.

There is more than one solution, and fig. 4 gives the complete solution for the first quadrant of projection (assuming the ends of the lines to lie in the planes of reference). Alternative plans are shown by broken lines and reference to the pictorial view, fig. 5, will explain the reason for these alternatives: the two lines shown in fig. 5 are diagonals of a rectangular solid, and although their elevations and inclinations to the H.P. are the same, their plans are unlike.

When one end of the line does not lie in a plane of reference, and is given by its projections, the apex of the semi-cone may be taken as coinciding with the given point. The pictorial solution to Ex. 4, shown opposite, illustrates the method in this case, and is self-explanatory.

Note. The sum of the inclinations θ and ϕ cannot exceed $90°$. When $(\theta + \phi) = 90°$ the projections are both perp. to xy.

EXAMPLES

(1) A line 3″ long, is inclined at 55° to the H.P. and 30° to the V.P. Draw its projections, as in fig. 4.

(2) Solve Ex. 1 by regarding the line as the generator of a cone with its axis horizontal and its base in the V.P.

(3) A line is 3″ long, its plan measures 1·9″ and its elevation 2·7″. One end is ·4″ from the V.P. and ·7″ above the H.P. Draw the projections of the line and determine its traces.

(4) A line is 2·75″ long; one end is 2″ above H.P. and 1·5″ from V.P., the other end is in the V.P. and ½″ above the H.P. Determine its projections and measure the distance between the projectors of

the ends of the line. Also measure the inclinations of the line to the H.P. and V.P. Ans. 1·72″ 33° each.

A pictorial view of the solution is shown opposite.

(5) The projectors of a line AB are 1·5″ apart; the end A of the line is ·5″ above H.P. and ·75″ from V.P.; the other end is 1″ from both planes of reference. AB is the side of a $\triangle$ABC; C lies in the H.P. and the sides AC and BC are 1·5″ and 1·75″ long respectively. Determine the projections of the $\triangle$, and measure the distance of C from the V.P. Ans. 2·1″.

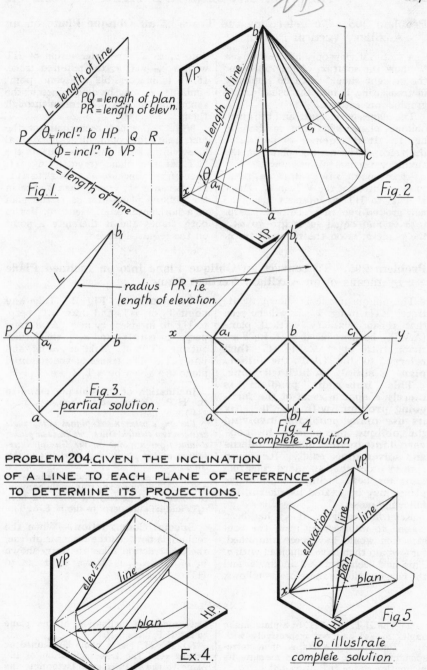

Fig. 1.

L = length of line

PQ = length of plan
PR = length of elevⁿ

θ = inclⁿ to HP.
ϕ = inclⁿ to VP.

P   Q   R

L = length of line

Fig. 2

VP.

L = length of line

HP.

Fig. 3.
partial solution.

L

radius PR, i.e.
length of elevation.

P θ

Fig. 4.
complete solution

(b)

PROBLEM 204. GIVEN THE INCLINATION
OF A LINE TO EACH PLANE OF REFERENCE,
TO DETERMINE ITS PROJECTIONS.

VP.

elevⁿ   line

plan

HP.

Ex. 4.

VP.

elevation   line   line

plan   plan

HP.

Fig 5

*to illustrate
complete solution.*

Problem 205. To determine the Trace of an Oblique Plane on an Auxiliary Vertical Plane.

A pictorial view only has been given to show the solution of this problem: the student should have no difficulty in translating it into the usual orthographic projections.

The planes of reference and the given oblique plane VTH are marked in fig. 1. It is required to determine the trace of the oblique plane on the auxiliary vertical plane shown.

At any point a in xy draw aa_1 perp. to xy to intersect the V.T. in a_1. Draw ab par[l] to HT to intersect x_1y_1, the new ground line, in b. Draw bb_1 perp. to x_1y_1 and equal in length to aa_1; the point b_1 is on the required trace.

Join b_1 to T_1, the intersection of HT with x_1y_1—b_1T_1 is the required trace. If T_1 is inaccessible a second point similar to b_1 may be determined in the same manner, and V_1T_1 drawn through them.

Note. Use has been made of the fact that the plan of a horizontal line lying in the plane is par[l] to the H.T. of the plane (refer to fig. 11, page 163). Because ab is par[l] to HT, and because $bb_1 = aa_1$, then b_1 lies in the oblique plane—for ab is the plan of a horizontal line. Hence b_1 lies in both planes and is therefore a point on the required trace.

Problem 206. To convert an Oblique Plane into an Inclined Plane by means of an Auxiliary Vertical Plane.

The oblique plane is shown by its traces VTH in fig. 2. It will be seen that if an auxiliary vertical plane (A.V.P.) be introduced with its horizontal trace at rt. ∠s to HT, then referred to this A.V.P. the oblique plane is simply an inclined plane.

This important problem is merely a special case of the foregoing problem, with $\varphi = 90°$. By its use many problems involving the oblique plane may be converted into inclined plane problems and solved more easily. It is given here in order that standard constructions for both inclined and oblique planes may be treated together in the following pages.

Refer first to fig. 1. If x_1y_1 had been chosen to intersect xy in a, the construction would have been simplified, for b would then have coincided with a. This has been arranged in fig. 2 and the construction is therefore as follows.

Construction. Fig. 3. Take any point T_1 on HT, and draw x_1y_1 perp. to HT to intersect xy in a. At a draw perps. to both xy and x_1y_1, the former cutting VT in a_1. Make $aa_2 = aa_1$ and join a_2T_1. The traces of the required plane are given by a_2T_1H, i.e. V_1T_1H.

Inclination of the oblique plane to the H.P. (Refer also to Prob. 233, page 234.)

The angle between two planes is the angle between two straight lines, one in each plane, drawn from a point on the line of intersection of the planes and at rt. ∠s to this line.

It will be seen that θ, the angle between a_2T_1 and x_1y_1, gives the true angle between the oblique plane and the H.P., for both a_2T_1 and aT_1 are perp. to the H.T. at T_1.

Special Construction.—When the real angle between the traces is obtuse, the construction takes the form shown in fig. 4; x_1y_1 is drawn at rt. ∠s to HT produced.

EXAMPLES

(1) The H.T. and V.T. of a plane make angles of 45° and 60° respectively with xy. Determine the traces of a corresponding inclined plane and measure its inclination to the H.P. Ans. 68°.

(2) The H.T. and V.T. of a plane make angles of 70° and 160° respectively with xy. Determine the inclination of the plane to the H.P. Ans. 21°.

(3) The H.T. of a plane is inclined at 65° to xy. The true inclination of the plane to the H.P. is 35°. Determine the V.T. and measure the angle between V.T. and xy. Ans. 32°.

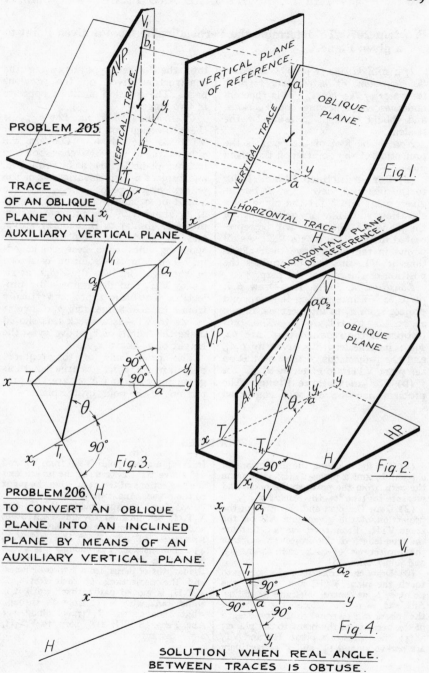

PROBLEM 205

TRACE
OF AN OBLIQUE
PLANE ON AN
AUXILIARY VERTICAL PLANE

Fig. 1.

Fig. 3.

Fig. 2.

PROBLEM 206
TO CONVERT AN OBLIQUE
PLANE INTO AN INCLINED
PLANE BY MEANS OF AN
AUXILIARY VERTICAL PLANE.

Fig. 4.

SOLUTION WHEN REAL ANGLE
BETWEEN TRACES IS OBTUSE.

Problem 207. To determine the Perpendicular from a given Point to a given Plane.

If a straight line is perp. to a plane, the projections of the line are perp. to the traces of the plane. This theorem underlies the following constructions, and should be kept in mind by the student.

Note. The foot of the perp. is the projection of the point upon the given plane.

(a) For an Inclined Plane. Refer to the pictorial view. Let P be the given point and VTH the plane. An elevation will show an edge view of the plane, i.e. the plane will be represented in elevation by its V.T. As all points in the plane have their elevations in VT, this line determines the position of the foot of the perp.

Construction. Fig. 1. Draw p_1a_1 perp. to VT intersecting it in a_1, and project from a_1 to intersect pa, drawn at rt. $\angle$s to HT, in a. The projections of the required perp. are pa, p_1a_1: its true length is given by p_1a_1, and the projections of the point P on the plane VTH are given by a, a_1.

(b) For an Oblique Plane. The problem may be at once converted into the former type by using the construction given on the previous page. The solution is shown opposite in two stages.

Construction. Fig. 2a. Let p, p_1 be the projections of the point and let VTH be the plane. Convert the oblique plane into the corresponding inclined plane V_1T_1H, as in Prob. 206, and project a new elevation p_2 of the point: pp_2 is perp. to x_1y_1 and the height of p_2 above x_1y_1 = the height of p_1 above xy. The problem is now precisely similar to that in (a) above, and to show the similarity corresponding lines have been thickened-in. The complete solution is shown on the right, fig. 2b; draw p_2a_2 at rt. $\angle$s to V_1T_1 and determine a by projection. To obtain a_1 project vertically from a to meet a line from p_1 drawn at rt. $\angle$s to VT—as a check this should make the height of a_1 above xy = the height of a_2 above x_1y_1.

The projections of the required perp. are pa p_1a_1: its true length is given by p_2a_2, and a, a_1 are the projections of the point on the plane.

EXAMPLES

(1) The figure shows the projections of a point P, and a plane VTH. Determine the perp. from the point to the plane and measure its true length. Ans. 1·7″.

(2) Draw the plan and elevation of the projection of the given line AB on the plane VTH. From these views determine the true length of the projection. (Obtain the projections of each end, as in (1), and join.) Ans. 2·1″.

(3) Draw the projections of a sphere to touch the plane VTH, and to have the point p, p_1 as centre. Measure its radius. Ans. rad. = 1·05″. (Note: the radius of the sphere will be equal to the true length of the perp. from the point to the plane.)

(4) The traces of a plane, HT and VT, are par¹ to xy and 1¼″ and 3″ respectively from it; a given point is 2¼″ from VP and 2¾″ above HP. Determine the length and the projections of the perp. from the point to the plane. Ans. 1·92″.

(5) Two oblique planes VTH and $V_1T_1H_1$ are inclined towards each other and their traces make the following angles with xy:—HT 45°, VT 60°, H_1T_1 55°, V_1T_1 45°. The distance TT_1, along xy, is 5″. A point P in xy is 2″ from T. Determine the lengths of perps. from P to the planes, and the projections of their feet. If $V_1T_1H_1$ is moved par¹ to itself until it is equidistant with VTH from P, through what distance will T_1 travel along xy? Ans. Perp. to VTH 1·3″, perp. to $V_1T_1H_1$ 1·9″; distance ·93″.

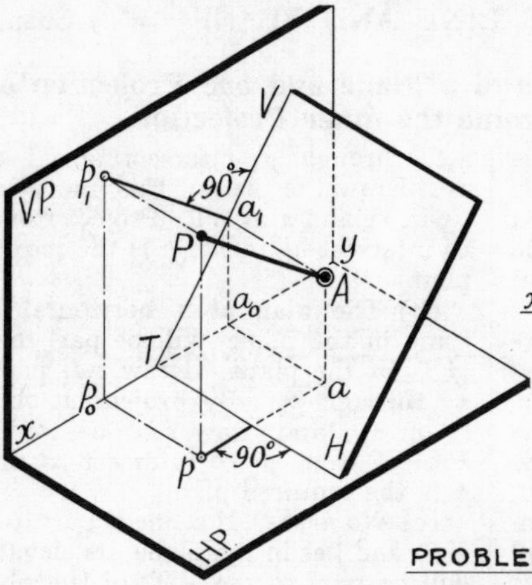

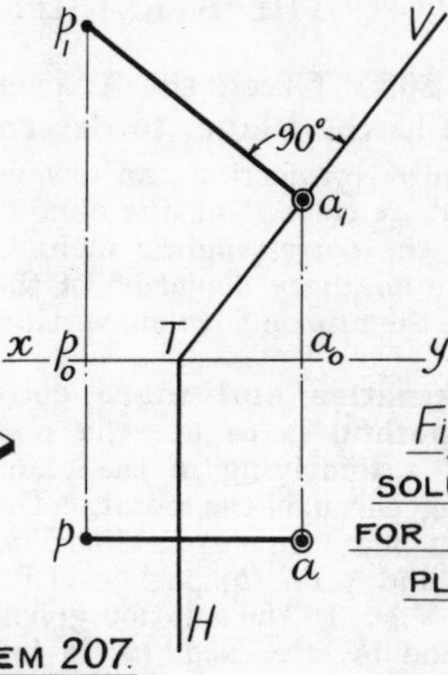

Fig.1.
SOLUTION FOR INCLINED PLANE.

PROBLEM 207.
PERPENDICULAR FROM A POINT TO A PLANE

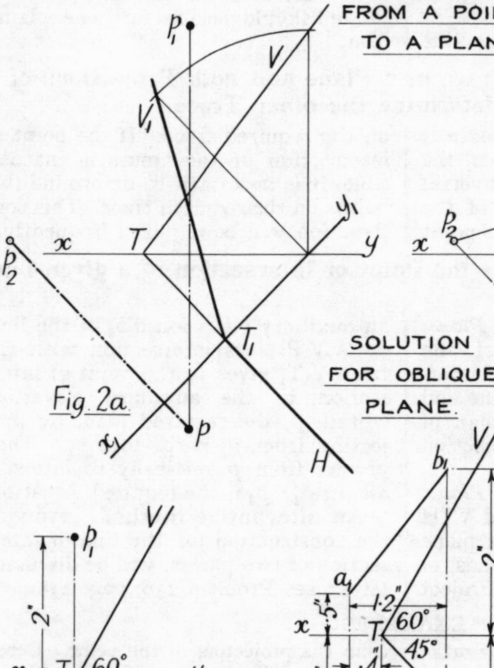

Fig. 2a.

SOLUTION FOR OBLIQUE PLANE.

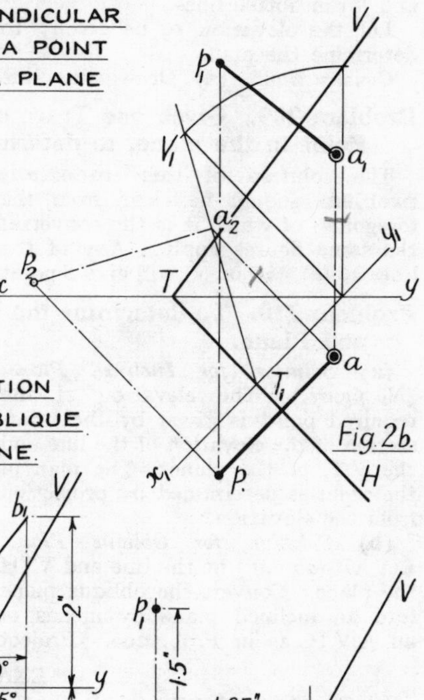

Fig. 2b.

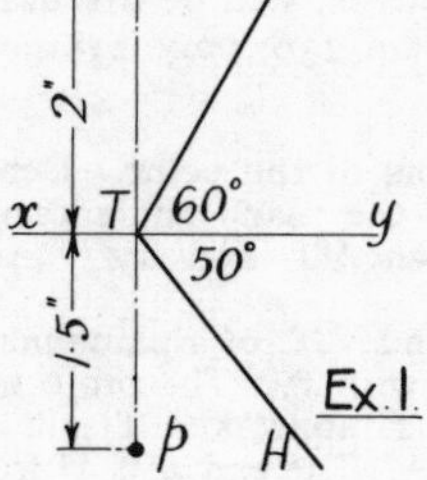

Ex.1.

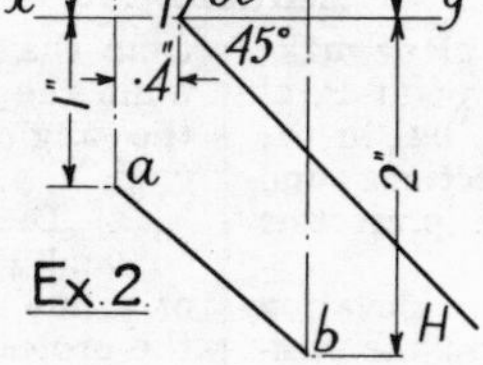

Ex.2.

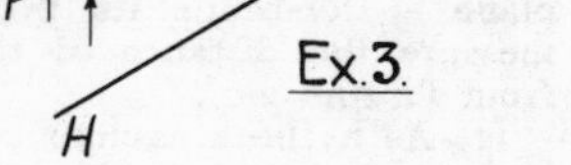

Ex.3.

Problem 208. Given the Traces of a Plane and one Projection of a Point in the Plane, to determine the other Projection.

The required projection can always be obtained by converting the oblique plane into the corresponding inclined plane. The auxiliary elevation of the point must then lie on the new vertical trace.

An alternative and more convenient method is to use the projections of a line lying in the plane and passing through the point. The line chosen may be (a) inclined to both H.P. and V.P., (b) par^l to H.P., (c) par^l to V.P. In the solution given, both (a) and (b) are used, (a) in full and (b) in dotted lines.

Let the elevation p_1 be given; to determine the plan.

Construction. (a) Draw any line a_1b_1 through p_1 intersecting VΓ and xy. Draw the plan of this line: b will lie in xy and a in HT. Project from p_1 to intersect ab in p; p is the required plan.

(b) The plan of a horizontal line lying in the plane, will be par^l to the H.T. of the plane. Draw c_1d_1 par^l to xy through p_1. By projection obtain c, on xy, and draw cd par^l to H.T. Project from p_1 to intersect cd in p: p is the required plan.

(c) *No figure.* If a line is par^l to the V.P. and lies in the plane, its elevation will be par^l to the V.T. of the plane. The construction is similar to that of (b) and should need no further explanation.

Problem 209. Given one Trace of a Plane and both Projections of a Point in the plane, to determine the other Trace.

The solution of this **important problem** should be clear from the foregoing, of which it is the converse: the same figures apply. Any of the lines at (a), (b), or (c) will give a point on the required trace. If the point of intersection of the traces is inaccessible, it is necessary to determine two points on the required trace. This construction will be required frequently.

Problem 210. To determine the Point of Intersection of a given Line and Plane.

(a) *Solution for Inclined Plane.* (*No figure.*) The elevation of the required point is given by the intersection of the elevation of the line and the V.T. of the plane. The plan of the point is determined by projection from the elevation.

(b) *Solution for Oblique Plane.* Let AB (ab a_1b_1) be the line and VTH the plane. Convert the oblique plane into an inclined plane by means of an A.V.P. as in Prob. 206. Project an auxiliary elevation a_2b_2 of the line, on A.V.P.; its intersection with the trace V_1T_1 gives p_2 the point of intersection in the auxiliary elevation. Obtain p, the required plan, by projecting from p_2 perp. to x_1y_1. Then project from p vertically to intersect a_1b_1 in p_1; p_1 is the required elevation.

An alternative method, involving the construction for the line of intersection of two planes, will be discussed later; see Problem 236, page 238.

EXAMPLES

(1) The H.T. and V.T. of a plane make 35° and 60° resp. with xy. A point P, 2" above H.P. and 1" from V.P., lies in the plane. Determine its projections and measure the distance of the projectors from T. Ans. 2·6".

(2) As a check assume your elevation in (1) to be correct and confirm the position of the plan by the three methods given above.

(3) A point is $1\frac{3}{4}$" above H.P. and $1\frac{1}{4}$" from V.P. The H.T. of a plane containing it makes 40° with xy and meets xy $2\frac{1}{4}$" from the projectors of the point. Determine the V.T. of the plane and measure the angle between VT and xy. Ans. $67\frac{1}{2}$°.

(4) The H.T. and V.T. of a plane make 60° and 45° resp. with xy. The projectors of a line AB are 2" apart and T falls on the projectors of A. The end A is 2" from H.P. and $2\frac{1}{2}$" from V.P.; B is $\frac{1}{4}$" from H.P. and $1\frac{1}{2}$" from V.P. Determine the point of intersection and measure its real distance from A. Ans. 2·25".

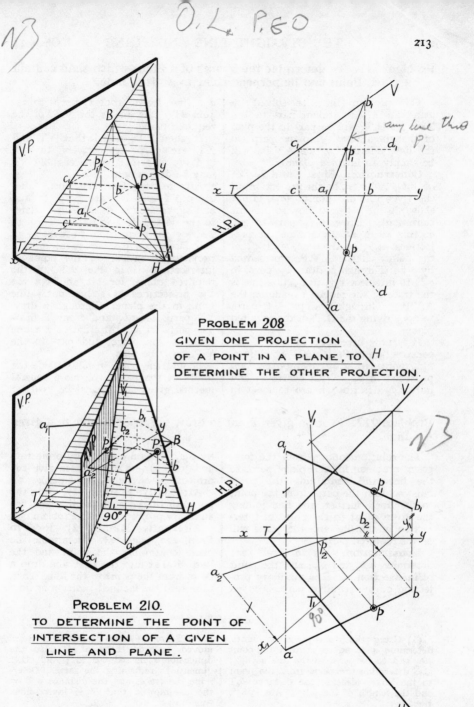

PROBLEM 208
GIVEN ONE PROJECTION
OF A POINT IN A PLANE, TO
DETERMINE THE OTHER PROJECTION.

PROBLEM 210.
TO DETERMINE THE POINT OF
INTERSECTION OF A GIVEN
LINE AND PLANE.

Problem 211. To determine the Traces of a Plane which shall contain a given Point and be perpendicular to a given Line.*

This problem may be solved by using an A.V.P., taken par[1] to the line; its H.T. will be par[1] to the plan of the given line. A plane perp. to the line, referred to this A.V.P., will be simply an inclined plane.

Construction. Figs. 1 and 2. Let ab, a_1b_1, and p, p_1 be the projections of the given line and point: (a_1b_1 is not shown in the pictorial view to prevent confusion). Draw x_1y_1 par[1] to ab, cutting xy in x_1 Determine the projections, a_2b_2 of the line, and p_2 of the point, on the A.V.P. represented by x_1y_1. Through p_2 draw p_2e perp. to a_2b_2 to intersect x_1y_1 in e; this line is the trace of the required plane on the A.V.P. Through e draw HT perp. to x_1y_1, giving the H.T. of the required plane.

It is now necessary to reverse the construction of Prob. 206. At x_1 draw x_1d_2 perp. to x_1y_1 to intersect ep_2 produced in d_2. With centre x_1 and rad. x_1d_2 describe an arc to meet in

d a line drawn from x_1 perp. to xy. Join dT: this line is the V.T. of the required plane.

It should be noted that VT and HT are perp. respectively to a_1b_1 and ab, and if T is inaccessible VT may be drawn perp. to a_1b_1.

Alternative Construction. Because VT and HT are perp. to a_1b_1 and ab, an *easy solution* results. Refer to the dotted lines in fig. 2. Through p draw pm perp. to ab to meet xy in m. Draw p_1n par[1] to xy to meet a projector from m; n, the point of intersection, is in the V.T. of the required plane, for pm and p_1n are the projections of a horizontal line lying in the plane. Through n draw VT perp. to a_1b_1 and from T draw TH perp. to ab. Alternatively a line lying in the plane, and par[1] to the V.P., may be used.

Should the point m fall outside the limits of the paper the more general method given above must be used.

Problem 212. From a given Point to draw a Perpendicular to a given Line.

The solution follows from the foregoing problem, for the plane perp. to the line and containing the point must contain the perp. from the point to the line: further, the line joining the given point to the point of intersection of the line and the plane will be the required perp.

Construction. As in Prob. 211 determine a_2b_2 and p_2e, and the point of intersection c_2. The auxiliary projection of the required perp. is given

by p_2c_2; obtain c and c_1 by projection and join pc and p_1c_1—these give the projections of the required perp.

Alternative Method (1). Join the point to the ends of the line and use auxiliary projections as in Problem 152.

Alternative Method (2). Refer to Prob. 220, page 224. Determine the plane containing the point and the line. Rabat into the H.P. and drop a perp. from the point to the line. Then raise into the inclined position again.

EXAMPLES

(1) Using the data given in figure, determine a plane to contain the point pp_1 and to be perp. to the line ab a_1b_1; also determine the perp. from the point to the line. Measure the distance p_0T and the length of the perp. Ans. 5·85″, 1·07″.

(2) Solve (1) if the projectors of B are moved towards those of A until the dimension 1″ is reduced to $\frac{1}{4}$″, all other dimensions remaining the same. Determine the traces and the distance p_0T on the assumption that T is inaccessible. Ans. 23·75″, ·69″.

* This problem, and Problem 212, may be found difficult at a first reading. If so, the study of them may be deferred.

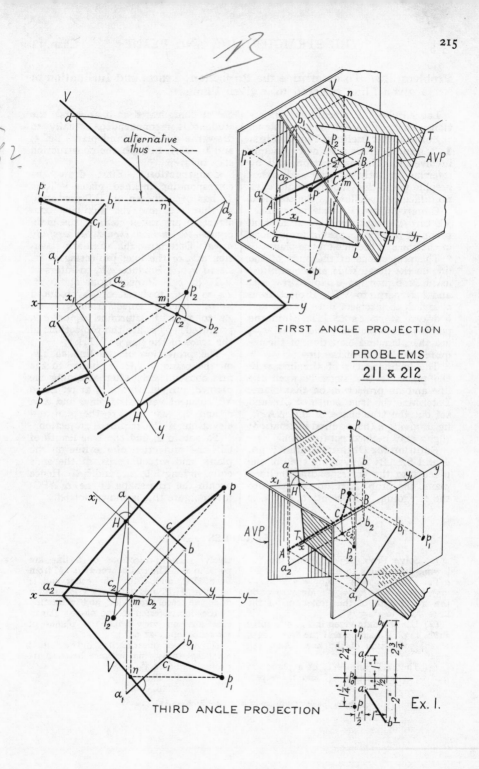

alternative thus

FIRST ANGLE PROJECTION

PROBLEMS
211 & 212

THIRD ANGLE PROJECTION

Ex. I.

Problem 213. To determine the Projection, Trace, and Inclination of a given Line, referred to a given Plane.

Let AB be the given line, projections ab a_1b_1, and VTH the given plane.

Solution for Inclined Plane. Figs. 1, 2, and 3. There is no new principle involved in this construction, and by referring to both pictorial and orthographic views the student should have no difficulty in following the solution.

Construction. Produce a_1b_1 to meet VT in p_1; project from p_1 to meet ab produced in p: p, p_1 are the projections of the trace of the line on the plane.

The projections of the ends of the line on the plane VTH are determined as in Prob. 207, page 210. Draw a_1c_1 and b_1d_1 perp. to VT. Project from c_1 and d_1 to intersect lines from a and b drawn at rt. ∠s to HT. Join the points of intersection c and d: cd, c_1d_1 are the plan and elevation of the required projection of the line.

The inclination α of the line with the plane, is the angle between the line and its projection on that plane. Determine the true length of CD, and set out the true shape of the $\triangle APC$, fig. 2, giving α the required inclination: (fig. 2 may be incorporated in fig. 3).

Solution for Oblique Plane. Figs. 4 and 5.* By converting the oblique plane into the corresponding inclined plane, the problem resolves itself into the foregoing. The solution, fig. 4, is not as complicated as it appears; the student is recommended actually to draw the given line and plane, fig. 5, and to work through the construction step by step.

Construction. First draw the corresponding inclined plane V_1T_1H; T_1 has been chosen on HT produced, so that x_1y_1 and the auxiliary construction are taken clear of the main figure—*the method should be carefully noted.* Determine the auxiliary elevation a_2b_2 of the line, projecting at rt. ∠s to x_1y_1. Produce a_2b_2 to intersect V_1T_1 in p_2. Project from p_2, at rt. ∠s to x_1y_1, to intersect ab produced in p. Thence project from p at rt. ∠s to xy and determine p_1 on a_1b_1 produced: p, p_1 are the projections of the trace of the line on VTH.

The projection of the line ab a_2b_2 on the plane V_1T_1H is similar to the first construction; after obtaining the plan cd, project from a_1b_1 at rt. ∠s to VT to meet verticals from c and d in c_1 and d_1. cd, c_1d_1 are the plan and elevation of the required projection.

To obtain α, find the true length of CD, the projection of the line on the plane, and set off perps. at the ends equal in length to a_2c_2 and b_2d_2. Hence obtain the true shape of the $\triangle$ APC; in the figure this is shown dotted.

EXAMPLES

(1) Solve Prob. 213 applying the following dimensions to fig. 3. $VTy = 30°$, $a_0T = \cdot2''$, $a_0b_0 = 3''$, $a_0a_1 = 3''$, $a_0a = \cdot75''$, $b_0b_1 = 2\cdot5''$, $b_0b = 3''$. Measure α and the true length of the projection of the line. Ans. $30\cdot5°$, $3\cdot25''$.

(2) Use the data given in fig. 5 to solve Prob. 213. Measure α and the true length of the projection of the line. Ans. $43°$, $1\cdot82''$.

(3) The H.T. and V.T. of a plane are par^l to xy and distant $3\frac{1}{2}''$ and $2\frac{1}{4}''$ respectively. The projections of a line are perp. to xy. The end nearer xy is $1\frac{3}{4}''$ from both H.P. and V.P.; the other end is $2\frac{1}{4}''$ from H.P. and $3\frac{1}{2}''$ from V.P. Determine the trace, projection, and inclination of the line, referred to the plane. (Draw first an end view with the planes of reference open at 90°.)

(4) Answer question (3) if the line is *not* perp. to xy, the distance between its projectors being $3''$.

* For the determination of α only see Problem 228, p. 228

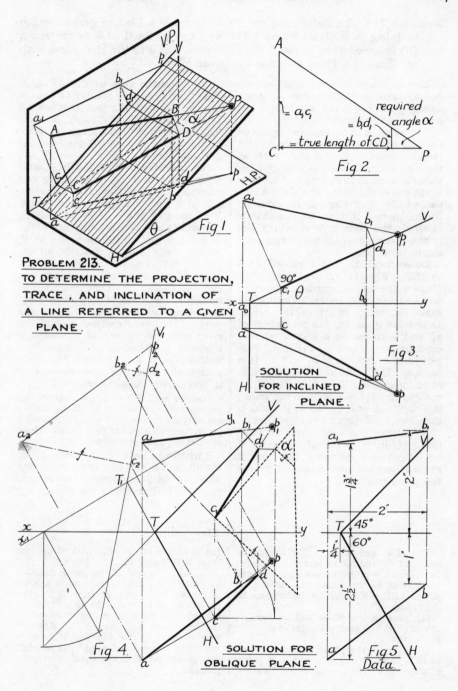

Fig 2.

Fig 3.

SOLUTION FOR INCLINED PLANE.

PROBLEM 213.

TO DETERMINE THE PROJECTION, TRACE, AND INCLINATION OF A LINE REFERRED TO A GIVEN PLANE.

Fig 1

Fig 4

SOLUTION FOR OBLIQUE PLANE.

Fig 5
Data.

Problem 214. To determine the Projections of a Line of given Length L, lying in a given Plane VTH, and (a) inclined at θ to the H.P., (b) inclined at φ to the V.P., (c) inclined at θ to the H.P., and with one End at a given Point in the given Plane.

This problem is used when determining the projections of a plane figure, given the plane in which it lies and the inclination of one edge.

The solution is as follows. Suppose the line to be the generator of a right circular cone, having a base angle = the inclination of the line, and situated with its apex in the plane and its base (a) in the H.P.: (b) in the V.P. A line lies wholly in a plane if its ends lie in the plane; if then a generator is selected, the base end of which lies in the plane, this generator represents the required line.

Construction. (a) Inclined at θ to H.P. Fig. 1. (i) *Full lines only.* Draw a line a_1c, of length L, above xy and inclined at θ to xy. Through a_1 draw the traces VTH of the plane. Determine a on xy, the projection of a_1, and with centre a and ac as rad. describe an arc cutting TH in b. Project from b and obtain b_1 on xy. Join ba, b_1a_1. The point a_1, a is in the V.T. of the plane, and b_1, b is in the H.T.; the line ab a_1b_1 lies wholly in the plane; it is inclined at θ to the H.P., and is therefore a solution to the problem.

(ii) *Dotted lines only.* Another solution, of little practical value, will be given by continuing the conical surface into the 2nd quadrant to cut the plane produced.

(b) Inclined at φ to V.P. Fig. 2. This construction is similar to that of (a), except that the line L is set off *below xy* at an angle φ to xy, and the base of the cone drawn *above xy*. Here also another solution obtains, in the 4th quadrant.

(c) Inclined at θ to the H.P. and with one end at a, a_1 in the given plane. Fig. 3. Draw the elevation a_1B of a line of length L par[l] to the V.P., inclined at θ to H.P., and having one end at a_1. Produce this line to the H.P., and project its plan ac. With centre a and rad. ac describe an arc cutting HT in c_2 and c_3. Join ac_2 and ac_3. These are the plans of two suitable lines; it only remains to mark off the given length along them and determine the projections of the part. The construction should be clear from the figure; ab a_1b_1 are the required projections, and the alternative plan is shown dotted (the alternative elevation has been omitted as it practically coincides with a_1B).

Another method is as follows. Draw the projections of a line of length L, exactly as in (a) above, and draw equal and parallel projections through the given point. These are the required projections.

Limiting Value. It should be noted that if the line is to lie in the plane, its inclination cannot be greater than the inclination of the plane.

EXAMPLES

The H.T. and V.T. of a plane make angles of 45° and 60° respectively with xy. Determine the projections of a line 3″ long:

(1) lying in the plane and inclined at 30° to H.P. Measure its inclination to V.P. Ans. 47°.

(2) lying in the plane and inclined at 40° to V.P. Measure its inclination to H.P. Ans. 47°.

(3) lying in the plane, inclined at 30° to H.P., and having one end in a point 2″ above H.P. and 1″ from V.P. Measure

the distances of the other end of the line from H.P. and V.P. Ans. ½″; 3¼″.

(4) which is par[l] to the plane, inclined at 35° to H.P., and has one end in a point 2″ above H.P. and ¼″ from V.P., the projectors of the point being 1″ from T. Measure the distances of the other end of the line from H.P. and V.P. Ans. ¼″. 2·67″. (*Hint.* Draw any line 3″ long, in the plane and inclined at 35° and through the given point draw equal par[l] projections.)

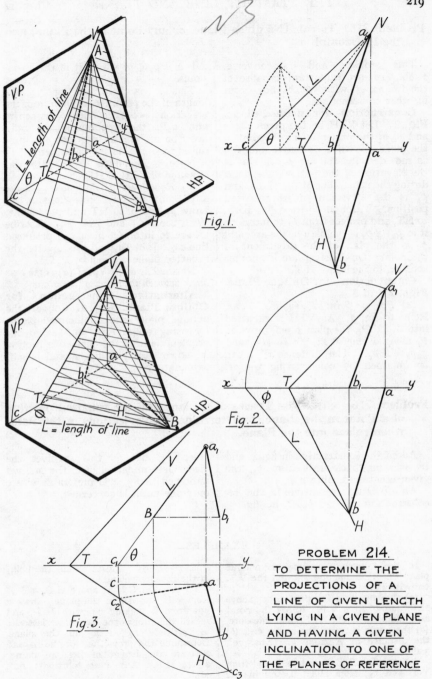

Fig. 1.

Fig. 2.

Fig. 3.

PROBLEM 214.
TO DETERMINE THE
PROJECTIONS OF A
LINE OF GIVEN LENGTH
LYING IN A GIVEN PLANE
AND HAVING A GIVEN
INCLINATION TO ONE OF
THE PLANES OF REFERENCE

Problem 215. To rabat* a given Plane, or any Point in the Plane, into the Horizontal.

This problem and its converse, Prob. 216, are important because of the frequency with which they occur in other constructions.

Construction—for Inclined Plane. Fig. 1. Let VTH be the given plane and p, p_1 the projections of a point in the plane. With T as centre and Tp_1 as rad. describe the arc p_1p_2; this is the elevation of the path of the point during the rabatment of the plane, and p_2 is the elevation of its rabatted position. Through p draw pp_3 (perp.) to HT, and project from p_2 to intersect it in p_3; pp_3 is the plan of the arc, and p_3 is the plan of the rabatment of the point. Obviously there is another solution to the left of H.T.

Construction—for Oblique Plane. Figs. 2 and 3.

(a) *For a point P in the plane.* Refer to fig. 2. As VTH is rabatted into the H.P., the plan p will move to p_3 along a line at rt. $\angle$s to HT, and $p_2p_3 = Pp_2$. The distance Pp_2 can be obtained by constructing the rt.

$\angle$d $\triangle$ Pp_2p, of which it is the hypotenuse.

This $\triangle$ is drawn in fig. 3: the length of the perp. Pp is taken from the elevation ($=p_1p_0$). With p_2 as centre and p_2P, the hypotenuse, as rad., describe an arc to intersect pp_2 produced in p_3; p_3 is the rabatment of P.

Note. *The angle Pp_2p (θ) = inclination of VTH to HP.*

(b) *Rabatment of the plane.* Take any point in VT, projections a, a_1. Draw aa_0 perp. to HT and produce it. With centre T and rad. Ta_1, describe an arc to intersect in a_2 the produced line aa_0. Join Ta_2 and produce it: the rabatted plane is given by V_2TH.

Note. *The angle V_2TH (α) is the true angle between the traces of the plane.*

Alternative Construction for Oblique Plane. Fig. 4. Convert the oblique plane into an inclined plane, by using an A.V.P., and solve the problem as in fig. 1. The construction is shown clearly in fig. 4, and is self-explanatory.

Problem 216. Given the Traces of a Plane and the rabatted position of a Point in the Plane, to determine the Projections of the Point when raised into the Plane.

Adopt the construction in fig. 4, and by working backwards from p_3, the given rabatment, obtain p, p_1.

An obvious alternative is the reconstruction of $\triangle$ p_2pP in fig. 3, given p_3, but as this involves the direct determination of θ, the method indicated in fig. 4 is preferable where only one point is concerned.

EXAMPLES

In the following, assume an oblique plane, the H.T. making 45° and the V.T. 60° with xy.

(1) A point A in the plane is 2" above H.P. and 1" from V.P. Rabat the point and plane into the H.P. and measure, (a) the angle between the traces and (b) the distance of A from T. Ans. 70°, 3·1".

(2) A point B in the plane is 1¼" from V.T. and 1½" from H.T. Determine its projections and measure (a) its height above H.P., (b) its distance from V.P.

Ans. 1·38", ·95". (First obtain the point in the rabatted plane.)

(3) A point C is 3" above H.P. and 2" from V.P. and lies in the plane. Draw a line from C making 60° with H.T., and regard this line as the base of an isosceles $\triangle$, altitude 1½", lying in the plane. Determine the projections of the $\triangle$ and measure the height of the 3rd corner above H.P. Ans. (two solutions) 2·2", ·8".

(4) Solve (1) by rabatting into the V.P. Check the answers.

* A term taken from Mongean geometry.

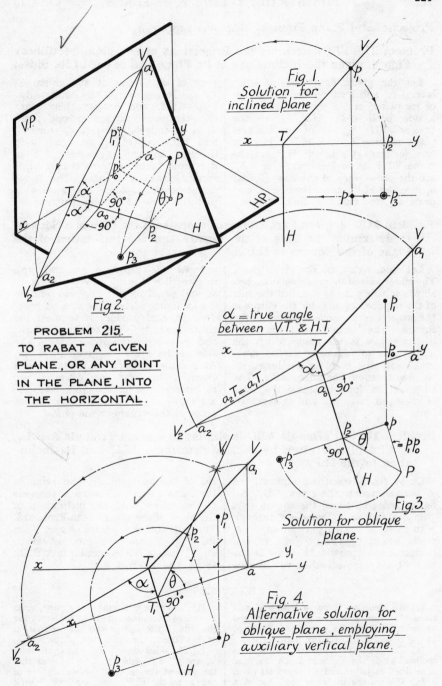

Fig. 1.
Solution for
inclined plane

Fig 2.

PROBLEM 215.
TO RABAT A GIVEN
PLANE, OR ANY POINT
IN THE PLANE, INTO
THE HORIZONTAL.

α = true angle
between V.T. & H.T.

$a_2T = a_1T$

Fig.3.
Solution for oblique
plane.

Fig. 4.
Alternative solution for
oblique plane, employing
auxiliary vertical plane.

Projections of Plane Figures. (*See also page 226.*)

Problem 217. To determine the Projections of any plane rectilinear Figure, given the Inclinations of its Plane and of one of its Sides.

Let the given figure be an equilateral $\triangle$, and let θ be the inclination of its plane, and α the inclination of a side, both to the H.P. Draw the traces VTH, fig. 1, of an inclined plane making θ with H.P. Place a line, inclined at α to the H.P., in the plane VTH, as in Prob. 214; ab a_1b_1, are the projections of the line. Rabat b_1 into the H.P., as in Prob. 215, and draw ab_2, the rabatted line. On any part of ab_2 construct the given $\triangle$, $c_2d_2e_2$. Raise each of the points $c_2d_2e_2$ into VTH, and project their plans cde—the construction for one point d_2 is as follows: project d_0 from d_2, and with centre a_1 and rad. a_1d_0, describe an arc intersecting VT in d_1; project from d_1 to meet in d a line from d_2 drawn par[l] to xy. The $\triangle$ cde is the required plan. Fig. 2 shows a pictorial view of the construction.

Problem 218. A given Parallelogram represents the Plan of a Square; to determine the Side of the Square and the Inclination of the Plane of the Square to H.P.

Let $abcd$, fig. 3, be the given par[m]. The major axis of the principal inscribed ellipse to this par[m] is equal to the side of the required square, for the ellipse is the projection of a circle inscribed to the square. The inclination of the plane of the ellipse is the angle which the major axis (or diam. of circle) must make with the H.P. for the length of its plan to equal the minor axis. The construction for obtaining the axes is given on page 32, and is repeated here.

Draw the bisectors gg_1 and pp_1, intersecting in n. Draw pm perp. to ad and equal to pd. Join mn, and on mn as diam., describe a circle, centre o. Join po, and produce it to meet the circle in r: pr = the semi-major axis, and pq = the semi-minor axis. The side of the square is given by $2 \cdot pr$. At q draw qk perp. to pr, and with centre p and rad. pr, draw an arc to intersect qk in k. Join pk: the angle qpk (θ) is the inclination of the plane of the square to the H.P.

Problem 219. A Triangle ABC is similar to a given Triangle $A_1B_1C_1$, and its Plan is given by abc: to determine ABC, and its Inclination to the H.P.

On bc, fig. 6, construct (dotted) the $\triangle$ bca_1, similar to the given $\triangle B_1C_1A_1$, fig. 5, and (full lines) the square $bcfe$. Join a_1f and aa_1. Draw agh through g to meet fh drawn par[l] to aa_1: i.e. make $hg : ga :: fg : ga_1$. Join hc and complete the par[m] $bchk$: this is the plan of a square attached to the side BC of the required $\triangle$, and lying in the same plane. Determine the side of the square, and the inclination of the plane of the square as in Prob. 218. The side of the square = $2pr$. On BC (fig. 4) as base (and equal to $2 \cdot pr$) draw a $\triangle ABC$ similar to $A_1B_1C_1$: this is the required $\triangle$.

EXAMPLES

(1) An isosceles $\triangle$ has a base 2·5″ and sides 3″. The $\triangle$ lies in a plane inclined at 50° to H.P., and a long side is inclined at 30° to H.P. Project the plan.

(2) The plane of a square, 2½″ side, is inclined at 60° to H.P., and one diagonal is inclined at 30° to H.P. Project its plan.

(3) The square in (2) is the base of a cube. Project the plan of the cube.

(4) The plan of a square is a par[m], sides 3″ and 4½″, included angle 45°. Determine the side of the square and the inclination of its plane to the H.P. Ans. 5″, 68°.

(5) The sides of the plan of an equilateral $\triangle$ are 3″, 2½″, and 1¾″ long. Determine the side of the $\triangle$ and the inclination of the $\triangle$ to the H.P. Ans. 3·1″, 58° (check the answer by applying Prob. 217).

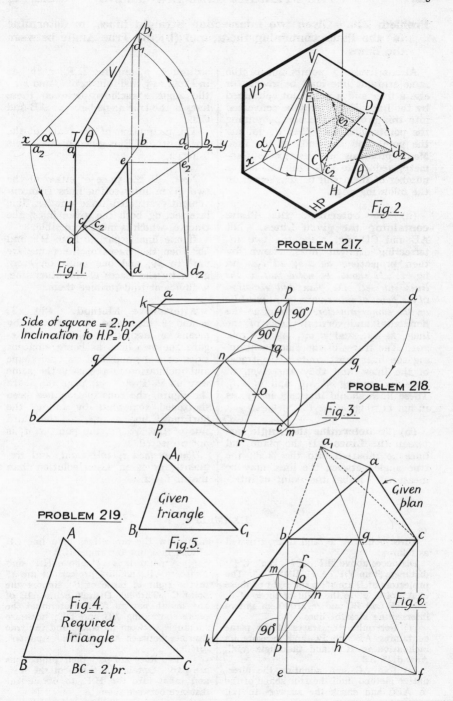

Fig. 1.

PROBLEM 217

Fig. 2.

Side of square = 2.br.
Inclination to H.P. = θ

PROBLEM 218.

Fig. 3.

PROBLEM 219.

Given triangle

Fig. 5.

Fig. 4.
Required triangle

BC = 2.br.

Given plan

Fig. 6.

Problem 220. Given two intersecting Straight Lines, to determine (a) the Plane containing them, and (b) the True Angle between the Lines.

Alternatively, 3 points not in the same straight line may be given; or else a line, and a point not contained by the line. Either may be converted into this problem, the one by joining the points, and the other by joining the point and the ends of the line. Many problems in Descriptive Geometry reduce to this problem, and a number of them will be considered on the following page.

(a) To determine the Plane containing the given Lines. Let AB and CD, fig. 1, be the two intersecting straight lines, shown by their projections ab a_1b_1, cd c_1d_1, in fig. 2. *It should be noted that as the lines intersect, the plan and elevation of the point of intersection p, p_1 must lie on the same projector.* Determine the horizontal and vertical traces of the lines, ht h_1t_1, and vt v_1t_1, as in Prob. 201. The traces of the plane containing the lines must contain the traces of the lines, and they are given at once by joining ht h_1t_1 and vt v_1t_1. These lines should intersect in xy, as in fig. 11, page 163.

(b) To determine the Angle between the Lines. If the plane and lines be rabatted into the H.P., the true angle between the lines may be measured. Rabat the point of inter-

section p, p_1 into the H.P., at p_2, as in Prob. 215, and join p_2 to ht and h_1t_1; the angle α included between these lines is the true angle between AB and CD.

The inclination of the plane of the lines to the H.P. is given by the angle θ, fig. 2.

Note. If the trace of either of the two given intersecting lines is inconveniently situated, take another line intersecting both the given lines, the trace of which is more accessible.

If one line is par[l] to both HP and VP then the traces of the plane are par[l] to xy, and one point on each, given by the traces of the other line, is sufficient to determine them.

Alternative Method. Fig. 3. α and θ may also be obtained by means of auxiliary projections. Regard the two lines as being adjacent sides of a $\triangle$, and find its true shape and inclination in precisely the same way as in Prob. 152, page 166. In the figure, the construction has been shortened somewhat by taking the horizontal line ae a_1e_1 as the third side of the $\triangle$: i.e. the point d, d_1 is not projected.

This method is important, and frequently gives an easier solution than that in fig. 2.

EXAMPLES

Three points A, B, and C are situated as follows:

Distances above H.P., A $\frac{1}{2}''$, B 2$''$, C $\frac{3}{4}''$; distances from V.P., A 1$''$, B $\frac{1}{2}''$, C 2$''$. The projectors of B are 1$''$ to the right of those of A; of C, 2$''$ to the right of those of B. Join AB and BC and regard them as two intersecting straight lines.

(1) Determine the traces of the plane containing AB and BC and measure its inclination to H.P. and the angle ABC. Ans. 61$\frac{1}{2}$°, 83°.

(2) Join AC, and adopting the alternative method, find the true shape of the $\triangle$ ABC and check the answers to (1).

Also draw the projections of a line CD 2$''$ long bisecting the angle ACB.

(3) A point D is 2$\frac{1}{2}''$ above H.P. and 1$''$ from V.P., and its projections are 1$''$ to the right of those of B. Ignore the point C. Through D draw a line DE of any length par[l] to AB. Determine the plane containing AB and DE; measure the angle between its traces and the true distance between AB and DE. Ans. 60°, ·91$''$.

(If two lines are par[l] their projections are par[l]. Obtain the traces of the lines and rabat into the H.P. to obtain the distance between them.)

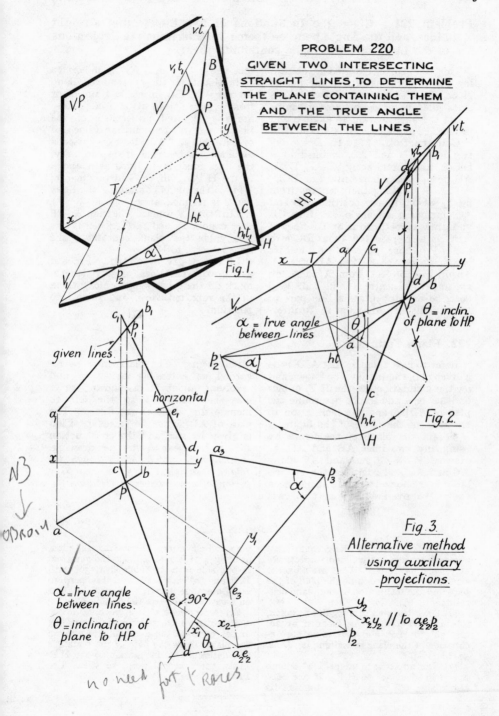

PROBLEM 220.
GIVEN TWO INTERSECTING
STRAIGHT LINES, TO DETERMINE
THE PLANE CONTAINING THEM
AND THE TRUE ANGLE
BETWEEN THE LINES.

V.P.

Fig. 1.

α = true angle
between lines

θ = inclin.
of plane to H.P.

Fig. 2.

given lines

horizontal

Fig. 3.
Alternative method
using auxiliary
projections.

α = true angle
between lines.

θ = inclination of
plane to H.P.

$x_2 y_2$ // to $ae_2 b_2$

90°

N3

no need for V rotes

Problem 221.   Given the Inclinations of two intersecting straight Lines, and the Angle between them; to determine the Projections of the Lines and the Plane containing them.

This problem is required in obtaining the projections of a plane figure, given the inclinations of two of its sides and the angle between them.

Let the lines contain an angle φ and be inclined at angles α and β to the H.P.

Construction. Fig. 1. Draw *any* two lines AB and AC, inclined at α and β, and intersecting at A. Draw AD perp. to *any* horizontal line BCD.

Fig. 3. Taking their lengths from fig. 1, set off, in any position, AB and AC, including an angle φ. Join CB and draw *xy* perp. to it at T. Regard ABC as a △ rabatted about BC into the H.P. Raise the vertex A until its height above the H.P. = AD (fig. 1). To do this project from A to A_0 on *xy*, and with centre T and rad. TA_0, describe an arc; draw a line parl to *xy*, and distant AD from it, to inter-

sect the arc in a_1. As A_0 moves to a_1 in elevation, the plan moves along Aa perp. to BC, meeting a projector from a_1 in a. Join aB and aC; these are the required projections of the lines, for it will be seen that if one end of both AB and AC (fig. 1) is raised a distance = AD, measured vertically, the lines must be inclined at α and β to the H.P. Join Ta_1 and produce it; VTH is the plane containing the lines and θ is its inclination.

Limiting Values. The sum α + β + φ cannot exceed 180°: when it has this value the plane containing the lines is perp. to H.P.

Lines of given length. Proceed as above with lines of any length, mark off the given lengths along them in the rabatted view, and raise into position.

222. Plane Figures.

Refer to figs. 2 and 4. Let ADEF be a given plane figure; let φ be known and let the inclinations to the H.P. of AD and AF be α and β. To determine the plan of ADEF and the inclination to the H.P. of the plane of the figure.

Obtain the plane of the figure by using any two lines AB and AC, as above. Mark off AD and AF along AB and AC and complete the quadrilateral, as shown in fig. 4; it has been chosen to overlap BC, as this case

often occurs. The plane may be regarded as extending below *xy* and revolving about TC as shown; or it may be translated to the left to include the whole of the figure. The plan of ADEF in the raised position is given by *adef* and the construction should be clear from the drawing. *It should be noted that although the data are sufficient to fix a plan they permit of any number of elevations.*

EXAMPLES

(1) Lines 2″ and $1\frac{1}{2}$″ long contain an angle of 60° and are inclined respectively at 30° and 45° to H.P. They are adjacent sides of a parm; determine the plan of the parm and the inclination of the plane containing it. (Ans. 47°.)

(2) Draw the plan of a regular hexagon, side $1\frac{1}{2}$″, adjacent edges inclined at 20° and 30° to H.P. and determine the inclination of the plane containing it. (Ans. $57\frac{1}{2}$°.)

(3) The adjacent sides of a 2″ square are inclined at 30° to H.P. If one edge also makes 60° with V.P., determine its

plan and elevation. (First draw the plan, as in fig. 4; this plan satisfies the question. If one side is to be inclined at 30° with H.P. and 60° with V.P., it must be perp. to *xy*. Determine then an auxiliary elevation on a new *xy* line perp. to one side.)

(4) A rectangle 3″ × 4″ revolves about a diagonal, which is horizontal, until the angles opposite the diagonal are 120° in the plan. Draw the plan of the rectangle and determine its inclination to the H.P. (Take the diagonal perp. to V.P. Use Prob. 5, page 10) (Ans. 55°.)

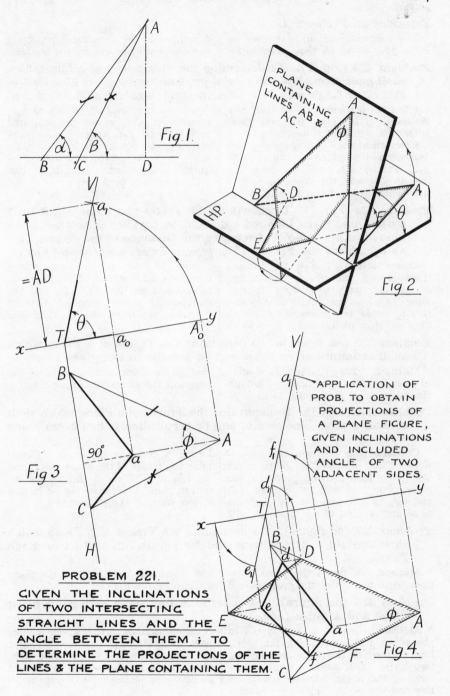

Fig. 1.

Fig. 2.

= AD

Fig 3

90°

APPLICATION OF
PROB. TO OBTAIN
PROJECTIONS OF
A PLANE FIGURE,
GIVEN INCLINATIONS
AND INCLUDED
ANGLE OF TWO
ADJACENT SIDES.

Fig. 4.

PROBLEM 221.

GIVEN THE INCLINATIONS
OF TWO INTERSECTING
STRAIGHT LINES AND THE
ANGLE BETWEEN THEM ; TO
DETERMINE THE PROJECTIONS OF THE
LINES & THE PLANE CONTAINING THEM.

Exercises on Problem 220.

The following are six exercises on Prob. 220, to which they may be re- | duced. Those illustrated opposite should be worked out by the student.

Problem 223 (Fig. 1). To determine the Projections of a Line which shall pass through a given Point pp_1 and meet a given Line ab a_1b_1 at a given Angle, say 30°. (When the angle is 90° refer to Problem 212.)

(*Measure the length of the line, also the length of the perp. from the point to the line, produced if necessary.*)

Construction. Join pa p_1a_1, and determine the plane containing pa p_1a_1 and ab a_1b_1. Rabat both the line ab a_1b_1, and the point p, p_1 into the H.P., given by a_2b_2 and p_2 in

figure. Draw p_2c_2 at rt. $\angle$s to b_2a_2 produced; this gives the required perp. Draw p_2d_2 at 30° to b_2a_2 and raise d_2 into position. Join pd p_1d_1; these give the projections of the required line. Ans. Length of line 4·4″; length of perp., 2·2″.

Problem 224 (Fig. 2). To determine the Traces of a Plane which shall contain one given Line, cd c_1d_1, and be parallel to another given Line ab a_1b_1. Also to determine the Distance of the second Line ab a_1b_1 from the Plane. (*This problem is used on the following page.*)

Construction. Draw the projections of any line ef e_1f_1 intersecting cd c_1d_1 and par¹ to ab a_1b_1. Determine the plane containing cd c_1d_1 and ef e_1f_1; this is the required plane. Convert this plane into an inclined

plane, V_1T_1, and project an auxiliary elevation a_2b_2 of ab. This elevation a_2b_2 should be par¹ to V_1T_1, and the distance D is the required distance between the line ab a_1b_1 and the plane. Ans. D = ·59″.

Problem 225 (no figure). To determine the Traces of a Plane which shall contain a given Point and be parallel to two given Lines.

Method. This problem is similar to the previous problem. Through the point draw lines par¹ to each of | the given lines and determine the traces of the plane containing them.

Problem 226 (Fig. 3). To determine the Traces of a Plane which shall contain a given Line ab a_1b_1 and be perpendicular to a given Plane VTH.

Construction. Draw the projections cd c_1d_1 of any line to intersect the given line and be perp. to the given plane: cd will be perp. to HT, and c_1d_1 to VT, and the intersections of ab and cd, and of a_1b_1 and c_1d_1 will

lie on a common projector. Determine the traces of the plane containing ab a_1b_1 and cd c_1d_1; this is the required plane. (*Measure the inclination of this plane to HP.*) Ans. 62°.

Problem 227 (no figure). To determine the Traces of a Plane which shall contain a given Point and be perpendicular to two given Planes.

Method. Through the point draw lines perp. to each of the given planes | and determine the traces of the plane containing the lines.

Problem 228 (no figure). To determine the Angle between a given Plane and a given Line.

Method. Draw a line at rt. $\angle$s to the plane and intersecting the given line; determine the angle between the two lines. This angle is the *complement* of the angle between the given line and the plane. This construction

is an alternative to that given for one part of Prob. 213, page 216. Solve by this method Ex. 2, page 216, obtaining α only—use the method of auxiliary projections, fig. 3, page 225.

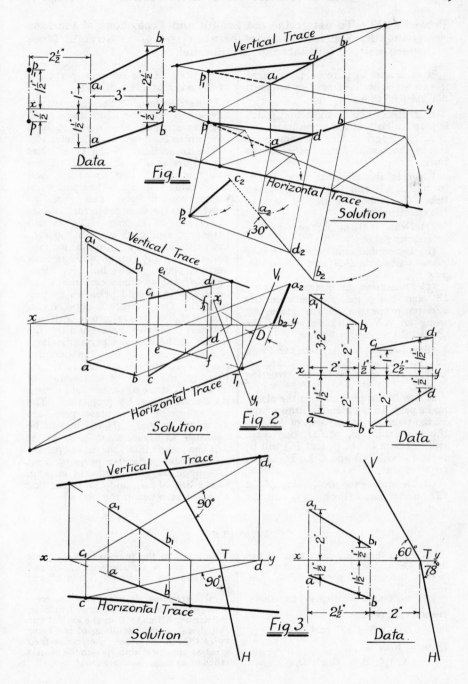

Fig 1

Data

Vertical Trace

Horizontal Trace

Solution

Fig 2

Vertical Trace

Horizontal Trace

Solution

Data

Fig 3

Vertical Trace

Horizontal Trace

Solution

Data

Skew lines

Problem 229. To determine the Length and Projections of the Line giving the Shortest Distance between two given Straight Lines which neither intersect nor are parallel.

It is proved in pure geometry that if two straight lines neither intersect nor are par[l] then:

1. There is one straight line which is perp. to both of the given lines.
2. The length of this common perp. is the shortest distance between the lines.

Refer to the pictorial view, fig. 1, in which AB and CD are the given lines. The following are the steps in the construction:

(a) Draw the line EF par[l] to CD, intersecting AB.

(b) Determine the traces of the plane containing AB and EF—Prob. 220.

(c) Determine the distance between CD and the plane, i.e. between CD and its projection on the plane, by using an A.V.P. and converting the plane into an inclined plane.

(Steps (a) (b) and (c) occur in Prob. 224 on previous page; they are repeated here to make the construction clear.)

If now CD be projected on the plane, and a perp. to the plane be drawn from P, the point of intersection of AB and the projection c_3d_3 of CD, this perp. will intersect CD in Q, and PQ will be perp. to both AB and CD. PQ is the required line. Hence·

(d) Determine the projection, c_3d_3, of CD on the plane (Prob. 213) and the projections of the common perp. PQ at the point of intersection P.

Construction. (The student is advised to draw the figure as he reads the text using the lines shown in fig. 2 —to which both figs. 1 and 3 refer.)

The projections of the given lines (fig. 3) are $ab\ a_1b_1$, and $cd\ c_1d_1$. Draw any line $ef\ e_1f_1$ par[l] to $cd\ c_1d_1$ and intersecting $ab\ a_1b_1$—the points of intersection in plan and elevation must lie on the same projector. Obtain the traces, V.T. and H.T., of the plane containing $ab\ a_1b_1$, and $ef\ e_1f_1$. Convert this plane into an inclined plane V_1T_1, by using an auxiliary vertical plane, ground line x_1y_1. Project c_2d_2, the auxiliary elevation of cd on x_1y_1—c_2d_2 should prove to be par[l] to V_1T_1. The distance L between c_2d_2 and V_1T_1 is the shortest distance between the two lines, and is the length of the common perpendicular.

Project $cd\ c_2d_2$ on the inclined plane V_1T_1 and obtain the plan c_3d_3. From p, the point of intersection of ab and c_3d_3, draw pq par[l] to d_3d; obtain its elevation p_1q_1 by projection. The projections of the common perp. are given by $pq\ p_1q_1$. (If the line cd be regarded as a succession of points it will be seen that we have merely selected the particular projector that falls upon ab; this projector is perp. to the line $cd\ c_1d_1$, and also to the line $ab\ a_1b_1$ which lies in the plane.)

EXAMPLES

(1) Fig. 2 shows the projections of two lines. Determine the length and projections of the common perpendicular. Ans. L = 1·05″.

(2) Four points are disposed as follows:

Distance above H.P.:

A, 1″; B, 4″; C, o; D, 1¼″.

Distance from V.P.:

A, 3¼″; B, o; C, 4″; D, o.

Referred to the projectors of C, those of B, D, and A are displaced 1″, 4″, and 7″ to the right, respectively. Join AB and CD.

(a) Determine the length of the common perp. to AB and CD. Ans. 1·82″; or,

(b) Regard AB and CD as the axes of two cylindrical shafts, the diam. of one being twice that of the other. Determine their greatest possible diameters—to the nearest tenth of an inch. Ans. 1·2″, 2·4″.

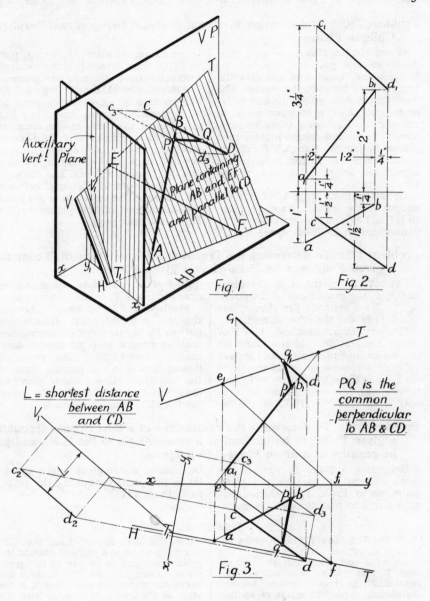

L = shortest distance between AB and CD.

PQ is the common perpendicular to AB & CD.

Fig. 1

Fig. 2

Fig. 3

PROBLEM 229.

TO DETERMINE THE LENGTH AND THE PROJECTIONS OF THE LINE GIVING THE SHORTEST DISTANCE BETWEEN TWO GIVEN STRAIGHT LINES.

Problem 230. To determine the Least Distance between two Parallel Oblique Planes.

If two oblique planes are par^1, their traces are also par^1, but the distance between the traces does not give the true distance between the planes. The least distance is given by the inter-cepted length of a common perp. to the planes. To draw a common perp. convert the oblique planes into inclined planes, by means of an A.V.P., shown pictorially in fig. 2, and measure the intercepted length of a perp. to the vertical traces.

Construction. Fig. 1. Let VTH and *vth* be the traces. Draw x_1y_1 perp. to HT and *ht*, and convert *either* of the planes into an inclined plane; V_1T_1H

is the converted plane. Draw v_1t_1 par^1 to V_1T_1; the distance D is the required distance between the two par^1 planes.

Alternative Method. Fig. 3. At any point *a* in *xy*, draw *afg* perp. to *xy* to meet the vertical traces in *f* and *g*. Regard this line as the common axis of two right circular semi-cones. With centre *a* draw tangent arcs to the horizontal traces, meeting *xy* in *e* and *d*. Join *ge* and *fd*. The distance D is the required distance between the planes. (The angle *fda* or *gea* gives the inclination of the planes to the H.P.—see next page.)

Problem 231. To determine the Traces of a Plane which shall contain a given Point and be Parallel to a given Plane.

Let *vth* be the traces of the plane and *p*, p_1 the projections of the point, fig. 4. Two methods are shown, one in full lines and the other dotted.

Construction. Method 1. *Full lines.* Convert the oblique plane *vth* into an inclined plane v_1t_1h, and project an auxiliary elevation p_2 of the point on x_1y_1. Through p_2 draw V_1T_1 par^1 to v_1t_1; at the intersection of V_1T_1 with x_1y_1 draw the H.T. of the re-

quired plane par^1 to *ht*. Complete by drawing the V.T. from T par^1 to *vt*.

Method 2. *Dotted lines.* Through the plan *p* of the point, draw a line par^1 to *ht*. At its point of intersection with *xy* erect a perp. to meet a hori-zontal through p_1. The point of intersection is in the vertical trace of the required plane, which may be drawn par^1 to *vt*. Complete by draw-ing the H.T. from T par^1 to *ht*.

Problem 232. To determine the Projections of a Line to pass through a given Point, to be inclined at a given Angle to the H.P., and to be parallel to a given Plane. (No figure.)

Determine a plane to contain the point and to be par^1 to the given plane, as in Prob. 231; then deter-mine a line to pass through the point,

be inclined at the given angle, and to lie in the plane containing the point, as in Prob. 214.

EXAMPLES

(1) The H.T. and V.T. of a plane are inclined at 45° and 60° resp. to *xy*. The traces of a par^1 plane intersect *xy* $1\frac{1}{2}''$ from T and outside the angle VTH. Determine the least distance between the planes. Draw the traces of another par^1 plane on the opposite side of the given plane and distant $1\frac{1}{2}''$ from it. Ans. ·98″.

(2) Solve Prob. 231 using the data given in the figure.

(3) The H.T. and V.T. of a plane make ∠s of 120° and 30° resp. with *xy*. The projectors of a point are $\frac{1}{4}''$ from T and outside the angle VTH; the point is $1\frac{1}{2}''$

above the H.P. and 2″ from the V.P. Draw the traces of a plane to contain the given point and to be par^1 to the given plane. Determine the true distance be-tween the planes and show that it is the same as the length of a perp. from the point to the given plane. Ans. ·8″.

(4) The H.T. and V.T. of a plane are par^1 to *xy* and distant 3″ and $2\frac{1}{2}''$ resp. A point is 2″ from the H.P. and V.P. Determine the projections of a line, par^1 to the plane, passing through the point, and inclined at 30° to H.P. Measure the inclination of the line to the V.P. Ans. $35\frac{1}{2}°$.

233

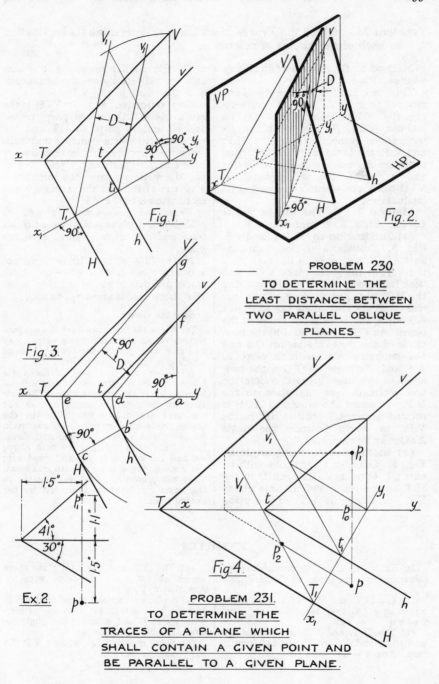

$Fig.1$

$Fig.2$

PROBLEM 230.

TO DETERMINE THE
LEAST DISTANCE BETWEEN
TWO PARALLEL OBLIQUE
PLANES

$Fig.3$.

Ex.2.

$Fig.4$.

PROBLEM 231.

TO DETERMINE THE
TRACES OF A PLANE WHICH
SHALL CONTAIN A GIVEN POINT AND
BE PARALLEL TO A GIVEN PLANE.

Problem 233. Given the Traces of a Plane, to determine its Inclination to each of the planes of reference.

Method 1. By the use of auxiliary planes. Figs. 1 and 2.

This method has already been referred to in Prob. 206. By converting the oblique plane into (a) an inclined plane, (b) a perp. plane, its inclinations respectively to the H.P. and V.P. are obtained. The constructions are shown separately in figs. 1 and 2, which are self-explanatory, VTH being the plane, and θ and φ its inclinations.

Method 2. By the use of semi-cones. Figs. 3, 4, and 5.

(a) Inclination to H.P.; angle θ.
Fig. 3. Consider a vertical semi-cone axis AO, arranged with its apex A in the V.T. of the given plane VTH, and with its base in the H.P. and touching the H.T. of the plane at B. The plane will touch the cone along the generator AB, for A and B lie in the traces of the plane. As TH is tangential to the base circle of the cone, the radius OB and the generator AB are each perp. to TH, and the angle ABO (i.e. the base angle of the cone) gives therefore the true inclination of the plane to the H.P. In effect the rt. ∠d △ AOB is rotated about AO until it lies in the V.P., in which position the angle AEO may be measured.

(b) Inclination to V.P., angle φ.
Fig. 4. Consider a semi-cone with its axis OD horizontal, its apex D in the H.T. of the plane and its base in the V.P., touching the V.T. of the plane

at C. By similar reasoning, the base angle of this cone gives the inclination φ of the plane to the V.P.

Construction. Fig. 5. VTH is the given plane. Draw aod perp. to xy, and ob and oc perp. to HT and VT respectively. With centre o and radii ob and oc describe arcs intersecting xy in e and f respectively. Join ae and fd; the angle aeo gives the inclination θ to the H.P., and the angle dfo the inclination φ to the V.P.

From the construction it will be evident that a plane is determined if one trace and one angle of inclination are given.

Note. (θ + φ) must lie between 90° and 180°; when (θ + φ) = 90°, both traces are par[l] to xy; when (θ + φ) = 180°, both traces are perp. to xy.

Special Cases.

(1) *When the traces are par[l] to xy,* a projection on an auxiliary plane perp. to xy will give a rt. ∠d △, the acute angles of which give the required inclinations.

(2) *When the real angle between the traces exceeds* 90°, the semi-cones will *not* be situated in the 1st quadrant. The construction follows the usual method but the arcs are drawn tangential to the traces *produced.* Fig. 6 shows the solution for the plane VTH; the plan and elevation of the cone giving θ are both above xy, and the cone is in the 2nd quadrant; the cone giving φ is in the 4th quadrant. They may, however, both be situated in the same quadrant as indicated by the dotted lines.

EXAMPLES

In Ex. 1, 2, and 3 determine the inclinations of each plane to the H.P. and V.P.

(1) The H.T. and V.T. of a plane make angles of 45° and 60° resp. with xy. Ans. θ = 67½°, φ = 49°.

(2) The H.T. and V.T. of a plane are par[l] to xy and distant 3″ and 2¼″ from it resp. Ans. θ = 40°, φ = 50°.

(3) The H.T. and V.T. of a plane make angles of 45° and 120° resp. with xy. Ans. θ = 67½°, φ = 49°.

(4) A plane is inclined at 50° to the H.P. and its V.T. makes 40° with xy. Determine its H.T. and measure the angle between xy and H.T. Ans. 45°.

(5) Solve question (4) reading V.P. for H.P. Ans. 37·5°.

PROBLEM 233. GIVEN THE TRACES OF A PLANE, TO DETERMINE ITS INCLINATION TO EACH OF THE PLANES OF REFERENCE.

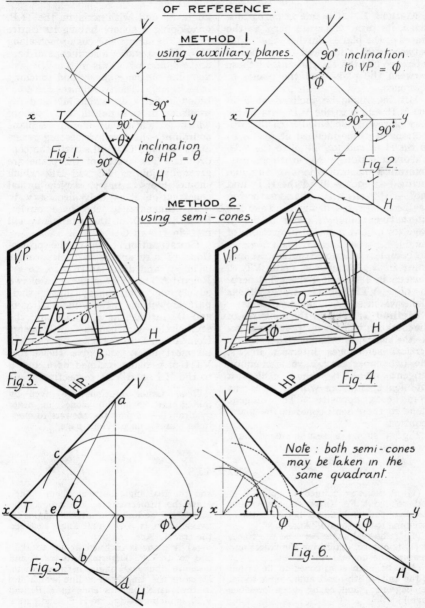

METHOD 1.
using auxiliary planes.

90°

inclination to VP = φ

90°

inclination to HP = θ

Fig. 1.

Fig. 2.

METHOD 2.
using semi-cones.

Fig. 3.

Fig. 4.

Fig. 5.

Note: both semi-cones may be taken in the same quadrant.

Fig. 6.

Problem 234. Given the Inclinations of a Plane to the Planes of reference, to determine its Traces.

Method 1. If a line is perp. to a plane, its projections are perp. to the traces of the plane; further, the angles between such a line and the planes of reference are the *complements* of those between the plane and the planes of reference.

Let the plane be inclined at θ to the H.P. and φ to the V.P.; its traces may be readily obtained from the projections of a line inclined at (90° − θ) to the H.P. and (90° − φ) to the V.P.

Construction. As in Prob. 204, determine the projections of any line inclined at (90° − θ) to the H.P. and (90° − φ) to the V.P.: the construction has been repeated in fig. 1. Take the alternative plan a_1b_2 (refer to fig. 5, page 207). The required projections of the line have been drawn again in fig. 2. To complete the construction, at any point in a_1b_1 draw the perp. VT to meet xy in T, and at T draw TH perp. to a_1b_2. VTH are the traces of *one* plane which satisfies the conditions

Method 2. This is the converse of the method given on the previous page. If the axes of the horizontal and vertical semi-cones intersect in xy, the two cones may have a common tangent plane if they envelop a common sphere, as will be apparent from fig. 5 opposite. The tangent plane to these semi-cones is the plane required.

Fig. 3 shows a semi-cone of base angle φ, and with axis in the H.P. enveloping a sphere having its centre in xy; the sphere and cone touch along MN. Fig. 4 shows a semi-cone of base angle θ, and with axis in the V.P., enveloping *the same sphere* and touching it along PQ. The two figures are combined in fig. 5, in which MN and PQ intersect in two points, only one of which, E, will be considered. A plane containing the two intersecting generators AB and CD will satisfy the conditions. The traces of the plane are given by joining AC and DB, which should intersect in xy. In the actual construction it is only necessary to draw tangents to the base circles, from A and D, as these tangents will include B and C.

Construction. Draw the projections of any sphere with its centre O in xy, and draw AOD perp. to xy. Regard AO and OD as the axes of two enveloping semi-cones; draw their part projections, i.e. Ab inclined at θ, and Dc inclined at φ, to xy, and the semi-circles with centre O. Tangents from A and D to these semi-circles will intersect in xy and give the traces VTH of a plane inclined at θ and φ to the H.P. and V.P. respectively.

Note. Other solutions are given by taking the enveloping cones in other quadrants—compare the answers to questions 1 and 3 on previous page.

EXAMPLES

(1) A plane is inclined at 35° to H.P. and 70° to V.P. Draw its traces and measure the real angle between them, assuming it to be acute. Ans. 59·5°.

(2) If the real angle between the traces in (1) is obtuse, determine the traces and measure the angle. Ans. 120·5°.

(Assume both semi-cones to be either in the 2nd or 4th quadrants; refer to fig. 11, page 163, and to fig. 6 on previous page.)

(3) A plane is inclined at 60° to the H.P. and 45° to the V.P. Determine its traces and the real angle between them. Then draw the projections of the intersecting generators and show that the angle between them is = the real angle between the traces. Ans. 55°.

(4) A plane is inclined at 40° to H.P. and 70° to V.P. Determine its traces and measure the real angle between them. Measure the length of a line which lies in the plane, has its ends in H.P. and V.P., is inclined at 35° to H.P., and passes through a point which is 1″ from each plane of reference. Ans. 64°, 3·05″.

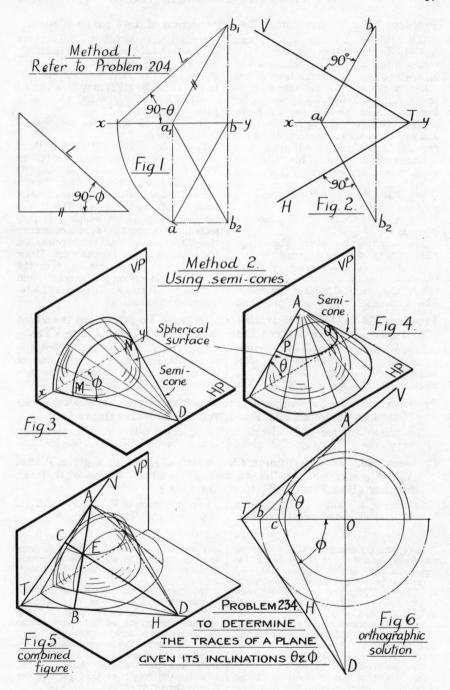

Method 1.
Refer to Problem 204.

$90-\theta$

$90-\phi$

Fig 1

Fig 2.

Method 2.
Using semi-cones.

VP

Spherical surface

Semi-cone

Semi-cone

Fig 3

VP

Semi-cone

Fig 4.

ϕ

θ

P

HP

V

A

VP

N

Fig 5
combined figure.

PROBLEM 234.
TO DETERMINE
THE TRACES OF A PLANE
GIVEN ITS INCLINATIONS θ & ϕ

θ

ϕ

Fig 6.
orthographic solution

Problem 235. To determine the Intersection of Two given Planes.

Refer to fig. 2. Let $V_1T_1H_1$ and $V_2T_2H_2$ be the given planes. The traces of the line of intersection, a and b_1, are given by the intersections of the traces of the planes. Because b_1 is in the V.P., b will lie in xy; similarly, because a is in the H.P., a_1 will lie in xy also. Obtain b and a_1 and join a_1b_1 and ab; these are the projections of the required line—given by AB in fig. 1.

Other Examples. In each, the given planes are lettered $V_1T_1H_1$ and $V_2T_2H_2$.

(a) *When the traces of one plane are parl to xy. Fig. 3.*

(b) *When the traces of one plane are perp. to xy. Fig. 4.*

The construction given above applies to both of these figures.

(c) *When only one pair of traces intersect. Fig. 5.*

Let the vertical traces intersect. The horizontal traces are parl and therefore meet at infinity. The plan ab, drawn from b on xy to this infinitely distant point, must be parl to the traces, and its elevation a_1b_1 is parl to xy. (The elevation may be a point.)

(d) *When all the traces meet at a point in xy. Fig. 6.*

(e) *When all the traces are parallel to xy. Fig. 7.*

Draw the traces LMN of a third plane which is perp. either to HP or VP; in the figs. a vertical plane is shown, LM being perp. to xy. Determine the lines of intersection of LMN with $V_1T_1H_1$ and $V_2T_2H_2$: these lines intersect in p_1, from which p is projected. This point p, p_1 is common to the three planes and is therefore on the required line of intersection. Draw lines through p, p_1, for fig. 6 to the point of intersection of the traces, and for fig. 7 parl to xy. These are the required projections.

Problem 236 (*alternative to Problem 210, p. 212*). To determine the Point of Intersection of a given Line ab a_1b_1 and a given Plane VTH.

Fig. 8. Draw the traces of a vertical plane LMN to contain the line: MN will coincide with ab. Determine the intersection of LMN and VTH. The points p and p_1 in which ab and a_1b_1 meet the projections of the line of intersection of the planes are the projections of the required point.

Problem 237. To determine a Line which shall pass through a given Point and be parallel to Two given Planes. (No figure.)

Determine the line of intersection of the planes. Through the given point draw a line parl to this line; this is the required line. See Ex. 5.

Problem 238. To determine a Line which shall lie in a given Plane, $V_2T_2H_2$, and be parallel to and at a given distance D from another given Plane $V_1T_1H_1$. (No figure.)

Determine a plane $V_3T_3H_3$ parl to $V_1T_1H_1$ and at a distance D from it. The intersection of $V_3T_3H_3$ and $V_2T_2H_2$ is the required line. See Ex. 6.

EXAMPLES

In (1) and (2) determine the projections and length of that part of the line of intersection lying in the 1st quadrant:—

(1) For two planes arranged as in fig. 1. Their V.T.'s make 30° and 60° with xy and corresponding H.T.'s 50° and 30°. $T_1T_2 = 4''$. Ans. 2·87''.

(2) For two planes arranged as in fig. 4. V_2T_2 makes 60°, and H_2T_2 20°, with xy. T_2 is 1¼'' from T_1. Ans. 2·65''.

(3) Solve Ex. 4, page 212, using the method illustrated in fig. 8.

(4) Determine the intersection of the given planes assuming that the traces do not intersect within the paper. Measure the distances along xy of the projectors of the ends of the line from T_1. Ans. 2·05'', 2·55''.

(5) Determine the projections of a line to pass through the point p, p_1 and be parl to the planes $V_1T_1H_1$, $V_2T_2H_2$. Measure the length of that part of the line in the 1st quadrant. Ans. 4·02''.

(6) Determine the projections of a line to lie in the plane $V_2T_2H_2$ and to be ½'' from and parl to the plane $V_1T_1H_1$. Measure the length of that part of the line in the 1st quadrant. Ans. 3·68''.

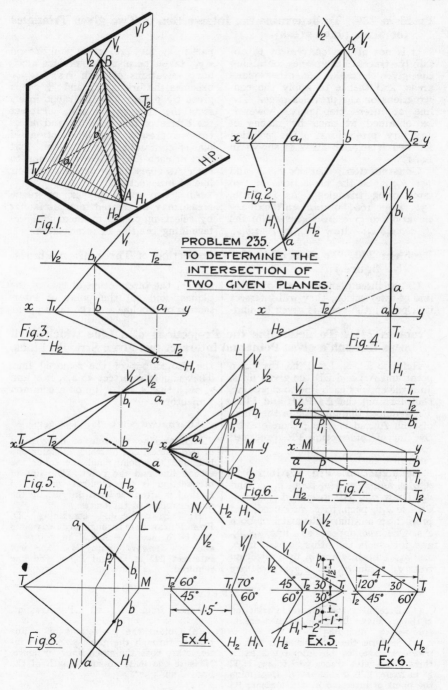

PROBLEM 235.
TO DETERMINE THE
INTERSECTION OF
TWO GIVEN PLANES.

Fig.1.

Fig.2.

Fig.3.

Fig.4.

Fig.5.

Fig.6.

Fig.7.

Fig.8.

Ex.4.

Ex.5.

Ex.6.

Problem 239. To determine the Intersection of Two given Triangles (or other plane areas).

It is not always convenient to obtain the traces of the planes containing the given triangles, or other plane areas, and thence to apply the constructions on the previous page. The line of intersection may, however, be obtained at once by means of auxiliary projections. The solution for two triangular areas is shown in fig. 1.

Construction. Let abc $a_1b_1c_1$ and pqr $p_1q_1r_1$ be the projections of two intersecting triangles. As in Prob. 152, page 166, obtain an auxiliary elevation to give an edge view of the $\triangle$ abc $a_1b_1c_1$:—draw b_1d_1, in $\triangle$ $a_1b_1c_1$,

par[l] to xy, and project its plan bd; on x_1y_1, taken perp. to bd, project auxiliary elevations of both $\triangle$s; $\triangle$ abc becomes the line $a_2b_2c_2$, and $\triangle$ pqr is given by $p_2q_2r_2$. The line $a_2b_2c_2$ intersects the $\triangle$ $p_2q_2r_2$ in m_2n_2. Project the corresponding lines mn and m_1n_1; these represent the intersection of planes containing the $\triangle$s, and that part of each line which is common to both $\triangle$s gives the projections of their line of intersection.

Any two given planes, or plane areas, may be treated in this manner by selecting points in them and determining the line of intersection.

Problem 240. To determine the Intersection of Three given Planes. (No figure.)

Unless three planes have a common line of intersection, they will intersect in a point; the point is given by finding (a) the intersection of two of the planes, and (b) the point of intersection of this line and the third plane.

Problem 241. To determine the Projections of a Line which shall pass through a given Point and intersect two given Straight Lines.

Refer to fig. 2. Let P be the given point and AB and CD the given lines. Join the ends of each line to the point, thus forming the $\triangle$s PAB and PCD. The required line must lie in the plane of each $\triangle$ and is given, therefore, by the line of intersection PR, produced to Q.

Construction. Fig. 3. Join p to a, b, c, and d, and p_1 to a_1, b_1, c_1, and d_1. Draw p_1e_1 par[l] to xy and project e. Choose x_1y_1 perp. to pe produced, and project an auxiliary elevation of both $\triangle$s. One, apb, appears as a line, $a_2p_2b_2$, and intersects the other, $c_2p_2d_2$, in p_2 and q_2. Obtain q and q_1 by projection from q_2. Join pq and p_1q_1; these are

the projections of the required line. They should intersect ab a_1b_1 in r and r_1 such that r and r_1 lie on a common projector.

Alternative Methods. (No figure.)

I. Determine the line of intersection of two planes, one of which contains one line and the point, and the other the second line and the point. The line of intersection is the required line. The method is often rendered difficult by the inaccessibility of the traces.

II. Take an A.V.P. containing CD. From P draw lines through any two points in AB to meet the A.V.P. in points M and N. The line joining M and N will intersect CD in Q, and PQ is the line required.

EXAMPLES

(1) Determine the line of intersection of the $\triangle$s shown in figure and measure its true length. Ans. 2·85".

(2) Suppose the two planes in Ex. 1, previous page, to be intersected by a third plane with traces par[l] to xy, H.T. $\frac{3}{4}$" in front, V.T. 2" below xy. Determine the point of intersection and measure its

distances from H.P. and V.P. Ans. ·9", ·73".

(3) Obtain the projections of a line passing through the given point and intersecting both the given lines, in figure. Measure the least possible length of the line. Ans. 3·57".

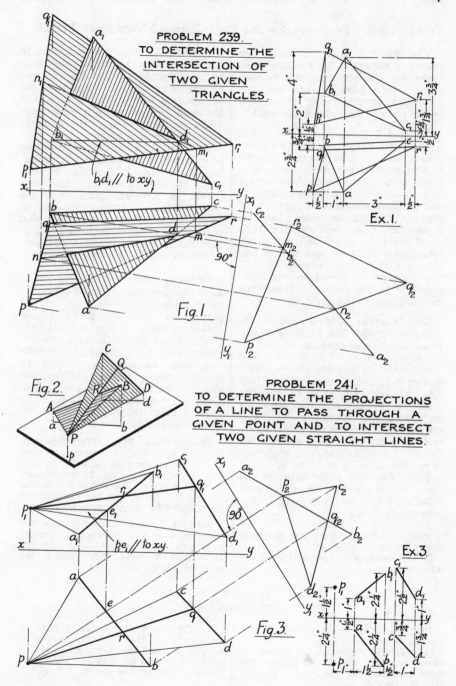

PROBLEM 239.
TO DETERMINE THE
INTERSECTION OF
TWO GIVEN
TRIANGLES.

$b_1 d_1$ // to xy.

Fig.1.

90°

Ex.1.

Fig.2.

PROBLEM 241.
TO DETERMINE THE PROJECTIONS
OF A LINE TO PASS THROUGH A
GIVEN POINT AND TO INTERSECT
TWO GIVEN STRAIGHT LINES.

pe_1 // to xy.

90°

Fig.3.

Ex.3.

Problem 242. To determine the Angle between Two given Planes.

The dihedral angle between two planes is measured by the angle between two straight lines, one in each plane, which meet in, and are perpendicular to, the line of intersection of the planes. (See also Prob. 206.)

Method·1. Refer to fig. 1. Let $V_1T_1H_1$ and $V_2T_2H_2$ be the two given planes, and let ab be the plan of their line of intersection AB. Suppose a plane, perp. to the line of intersection AB, to cut the given planes·along cf and df; the angle cfd (α) is the required angle. It should be noted that this intersecting plane will cut the H.P. along cd; that cd will be perp. to ab, intersecting it at e; and that a line joining e and f will be perp. to AB. The problem is solved by the construction of the real $\triangle$ cdf and the measurement of the angle cfd.

Construction. Fig. 2. Obtain ab, the plan of the line of intersection. Draw any line cd perp. to ab and intersecting it in e. With centre b and radii bc and ba describe arcs to intersect xy in e_2 and a_2 respectively. Join a_2b_1 and draw e_2f_2 perp. to it. The $\triangle cdf$ (fig. 1) may now be drawn, for its base, cd, and its altitude at e, e_2f_2 are known. Complete the construction *in the plan* by making $ef_1 = e_2f_2$; join f_1c and f_1d. The required angle between the planes is given by α.

When the real angle between the traces of one plane is obtuse, the construction takes the form given in fig. 2 on the following page.

Note. θ = the inclination to the H.P. of the line of intersection.

β = the inclination to the H.P. of the intersecting plane.

Method 2. Conceive the planes in fig. 1 to be so arranged that they may be viewed along the line AB, this line appearing as a point. Each plane will appear as a line, not as an area, and the angle between these lines, or edge views of the planes, is the required angle between the planes.

If, therefore, an auxiliary projection be obtained of one point in each plane, and of the line AB, such that AB is a point, then lines joining these selected points to AB will include α, the required angle. It is convenient to choose both points in the horizontal traces of the planes, such as c and d; then the $\triangle$ formed is the $\triangle$ cdf, projected on a plane par¹ to itself.

Construction. Fig. 3. Obtain ab a_1b_1 the projections of the line of intersection, and take any line cd perp. to ab. Draw x_1y_1 par¹ to ab and obtain, in the usual way, the auxiliary elevations of points a, b, c, and d, given by a_2, b_2, c_2, and d_2. Now take x_2y_2 perp. to b_2a_2 and project an auxiliary plan; the line AB appears as the point a_3b_3, and the positions of c and d are given by c_3 and d_3. The angle α gives the required angle between the planes.

Special Case. Fig. 4. This shows the application of Method 2 for the solution of the type of example discussed on page 239 fig. 6, in which all the traces meet at a point in xy. The solution is similar to that for fig. 3, and the construction given above for fig. 3 applies here also.

Method 3. (No figure.) Draw the projections of two intersecting straight lines perp. to the planes. Determine the angle between these lines: this is the required angle.

EXAMPLES

Note. Give the acute angles between the planes, not the obtuse.

(1), (2), (3) Three pairs of planes are shown by their traces. Determine the angle between each pair. Ans. (1) 78°, (2) 22½°, (3) 49°.

(4) The traces of two planes meet in xy as in fig. 4. The V.T.'s make angles of 60° and 30°, and the corresponding H.T.'s angles of 30° and 60° with xy. Determine the angle between the planes. Ans. 46°.

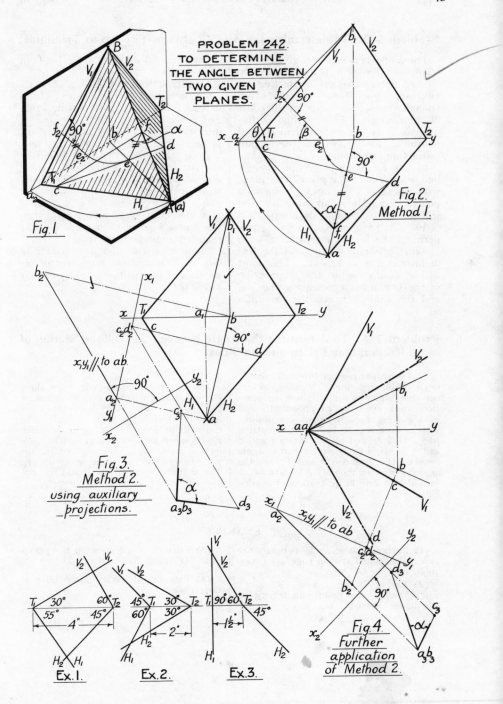

PROBLEM 242.
TO DETERMINE
THE ANGLE BETWEEN
TWO GIVEN
PLANES.

Fig.1

Fig.2.
Method 1.

Fig.3.
Method 2.
using auxiliary
projections.

x,y_1 // to ab

Fig.4
Further
application
of Method 2.

x,y_1 // to ab

Ex.1.

Ex.2.

Ex.3.

Problem 243. To determine the Angle between Two given Triangles.

The angle between two given plane triangular (or other) areas may be readily determined by projecting an auxiliary view of each of the given triangles on a plane perp. to their line of intersection. This method is frequently more convenient than that of obtaining the traces of the planes containing the triangles.

In any auxiliary projection along a line which lies in a plane, the plane will appear as a line, and the line as a point. Similarly in an auxiliary projection along the line of intersection of two △s, i.e. a projection on a plane perp. to the line of intersection, each △ will appear as a line, and the line of intersection as a point. The point will of course lie on the intersection of the two lines representing the △s, and the angle between the two lines

is the required dihedral angle between the two △s.

Construction. Fig. 1. Let abc $a_1b_1c_1$, and pqr $p_1q_1r_1$ be the projections of two intersecting △s. As in Prob. 230 determine mn m_1n_1, the line of intersection. Obtain an auxiliary projection of each △ along this line. To do this, first project auxiliary elevations on x_1v_1 parl to mn; these give $a_2b_2c_2$ and $p_2q_2r_2$ for the △s and m_2n_2 for the line. Then project auxiliary plans on x_2v_2 perp. to m_2n_2; these give the lines $a_3b_3c_3$ and $p_3q_3r_3$, m_3n_3 being now a point. The angle α is the required angle between the △s.

Accurate construction is essential in this problem and the alignment of the points and the position of m_3n_3 provide excellent tests of this.

Problem 244. To determine the Angle between an Oblique Section of a Rectangular Prism and its Sides.

This is an exercise on Prob. 242 and is dealt with separately because of its common occurrence in engineering practice. The prism in fig. 2 represents the end of a cutting tool in which an oblique section O gives "rake" to the cutting edge. O is defined by the angles α and β and it is required to find the true angle between the surfaces O and P: i.e. the angle γ_1 between the dotted lines in fig. 2, each dotted line being perp. to the edge EF.

Construction.

It will be clear that an edge view along EF will show EF as a point and will show the planes O and P as lines the angle between which is the required angle. The construction is set out in fig. 3, in which x_1v_1 is taken perp. to e_1f_1, so that EF projects as a point e_2f_2. The required angle is γ_1. Obviously the angle γ is $90-\gamma_1$. The angle between surface O and the side opposite to P is given by $\gamma_2 = 90° + \gamma$.

EXAMPLES

(1) Determine the angle between the two triangles in Ex. 1 on page 241. Ans. 79·5°.

(2) Taking AB, fig. 2, as 2″, BC as $2\frac{1}{2}$″, $\alpha = 30°$ and $\beta = 20°$, find the angle between the planes O and P. Ans. 72·5°.

(3) Find the angle between the planes O and Q, fig. 2. Ans. 118·5°.

(4) Establish the relationship $\tan \gamma = \tan \beta \cos \alpha$.

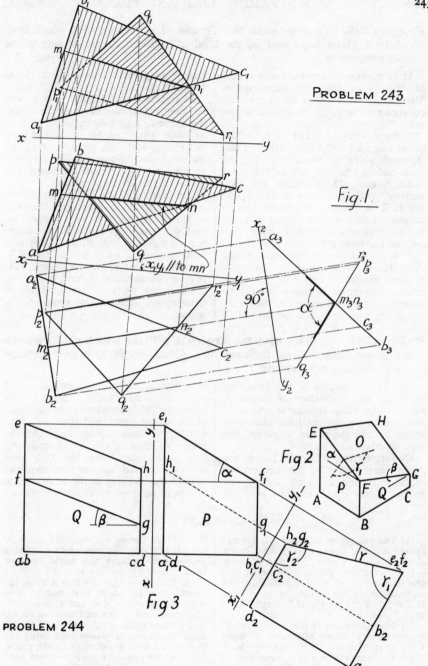

PROBLEM 243.

Fig.1.

$x_1 y_1$ // to mn

$90°$

α

Fig 2

Fig 3

PROBLEM 244

Problem 245. To determine the Traces of a Plane which shall contain a given Line and be inclined at a given Angle to one plane of projection.

It is assumed here that the traces of the given line are inaccessible. Should they fall within the paper the construction may be simplified somewhat.*

Refer to Fig. 1. Let AB be the given line and θ the inclination of the required plane to the H.P. Regard each end of the line AB as the apex of a right cone of base angle θ, with axis vertical, and with base in the H.P.; the H.T. of the required plane will be tangential to the base circles of these cones. Determine the V.T. by applying the methods given in Prob. 209.

Two planes satisfy the conditions, their H.T.s being given by the two *outside* tangents common to both circles—the cross tangents cannot be taken. If the inclination to the V.P. be given, instead of to the H.P.

choose cones with axes horizontal and bases in the V.P.

Construction. Fig. 2. Let $ab\ a_1b_1$ be the projections of the given line. Set off a_1c_0 and b_1d_0 inclined at θ to xy, and with centres a and b and radii a_0c_0 and b_0d_0 respectively, describe circles. Draw outside tangents to the circles; these represent the H.T.s of the required plane.

To determine the V.T.s: draw any line mn through a, m being in the H.T. of the plane and n in xy. Project m_1 and draw m_1a_1 produced; the intersection of this line with a projector from n gives n_1, a point on the V.T. A similar point q_1 on the second V.T. is given by the lines pq and p_1q_1. One other point on each trace, similarly determined, enables the traces to be drawn.

Problem 246. To determine the Traces of a Plane which shall contain a given Line and be inclined at a given Angle to a given Plane which does not contain the Line.

This is an exercise on the foregoing problem, and a solution is shown in fig. 3. In this, the given plane is assumed to be an inclined plane—to which an oblique plane, if given, can be converted.

Construction. Let VTH be the given plane, $ab\ a_1b_1$ the projections of the given line, and let β be the given angle. Draw a_1m_1 and b_1n_1 perp. to

V.T. and regard these lines as axes of right cones, base angle β, having their bases on VTH. Determine the rabatted plans of the bases and draw cd_2 tangential to them. Raise cd_2 into the plane VTH, as given by cd and c_1d_1. The plane containing the two lines $ab\ a_1b_1$ and $cd\ c_1d_1$ is the required plane: the last step in the construction has been omitted to avoid confusion.

EXAMPLES

(1) The projectors of a line AB are $2\frac{1}{2}''$ apart; A is $1\cdot3''$ above H.P. and $1\cdot5''$ from V.P., B is $1\cdot6''$ above H.P. and $3''$ from V.P. Determine planes containing AB and inclined at 45° to H.P. Measure the inclinations of the traces to xy. Ans. H.T.1 31°, V.T.1 36·5°; H.T.2 25°, V.T.2 $22\frac{1}{2}°$ (below xy).

(2) The projectors of a line AB are $1\frac{1}{2}''$ apart; A is $3''$ above H.P. and $\cdot5''$ from V.P., B is $\frac{3}{4}''$ above H.P. and $2\frac{1}{4}''$ from V.P. AB is in the line of intersection of two planes inclined at 60° and 45° to H.P.,

their H.T.s being on opposite sides of the plan of AB. Determine two suitable planes and measure the angle between them. Ans. 73° (or 107°).

(3) The projectors of a line AB are $1\frac{1}{2}''$ apart; A is $1\frac{1}{2}''$ above H.P. and $1''$ from V.P., B is $1''$ above H.P. and $2''$ from V.P. The lower end of the line touches an inclined plane making 30° with H.P. Determine the traces of planes containing the line and inclined at 60° to the given plane. Measure the apparent angle between each pair of traces. Ans. 84°, 148·5°.

---

* The traces of the plane contain the traces of the line, and when the latter are accessible the projections of *one* cone only are required: tangents may then be drawn to its base circle from the H.T. of the line.

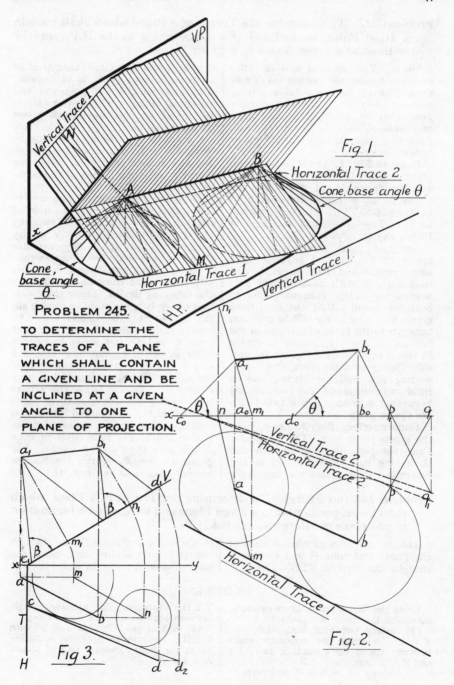

Fig. 1.

Vertical Trace 1

V.P.

W

Horizontal Trace 2.

Cone, base angle θ.

B

A

x

M

Cone, base angle θ.

Horizontal Trace 1.

Vertical Trace 1.

H.P.

PROBLEM 245.

TO DETERMINE THE TRACES OF A PLANE WHICH SHALL CONTAIN A GIVEN LINE AND BE INCLINED AT A GIVEN ANGLE TO ONE PLANE OF PROJECTION.

Vertical Trace 2.

Horizontal Trace 2.

Horizontal Trace 1

Fig. 2.

Fig. 3.

Problem 247. To determine the Traces of a Plane which shall contain a given Point, be inclined at a given angle to the H.P., and be inclined at a given Angle to a given Plane.

Note. *This problem will be better understood after the chapter on* **Tangent Planes** *has been taken; it is included here in order that further projections of solids, involving these constructions, may be proceeded with. Prob. 265, page 264 may with advantage be taken now.*

Refer to Fig. 1. Let P be the given point and VTH the given plane, assumed inclined; to determine the traces of all planes inclined at θ to H.P., and β to VTH. (P may, of course, lie in VTH—it is thus required in Probs. 257 and 258.)

Let two right cones have their apices at P: A, base angle θ, with axis vertical and base in H.P.; B, base angle β, with base in, and axis normal to, VTH. Tangent planes to both cones will satisfy the conditions given, and their H.T.s will be common tangents to the horizontal traces of the two cones. Four H.T.s may be drawn in the example shown,* given by outside and crossed tangents, and representing four suitable planes, one in front and one behind, and two crossing between the cones. Only two H.T.s are shown in fig. 1.

Construction. Fig. 2. Let p, p_1 be the given point and VTH the plane. Draw the projections of cones A and B, having base angles θ and β, and determine their traces on the H.P.:

that of A will be a circle, that of B an ellipse. Several methods of drawing the ellipse are given on page 264 and are shown in fig. 2. The major axis $= ab$; to obtain the minor axis bisect ab at rt. $\angle$s, and at c, on xy, draw dce perp. to dp_1, the axis of cone B. With centre d and rad. de describe an arc, and draw cf perp. to de to cut the arc in f; cf is the length of the semi-minor axis.

Draw the four common tangents to the ellipse and to the base circle of cone A, these giving the required H.T.s. Determine the V.T.s by means of Prob. 209; only two are shown, the other two being inaccessible.

Note. The construction of the elliptical trace may be avoided by the following device. Draw the elevation of a sphere inscribed in cone B (in the figure the focal sphere has been used) and draw the outlines of two right cones C and D, one upright and the other inverted, having base angles θ and with axes vertical, circumscribing the sphere. In the pictorial view cone C only is shown. Draw the circles representing the traces of these cones on the H.P., as marked in figure. Common tangents to either of these circles and to the base circle of cone A will give the required traces. *The principles underlying this construction are fully discussed in Chap. 18.*

Problem 248 (no figure). To determine the Traces of a Plane which shall be perpendicular to a given Plane and have a given Inclination to one plane of reference, say the H.P.

This is merely a particular case of the above and cone B will become a straight line perp. to VTH. The H.T.

of the required plane will pass through the H.T. of this line and touch the base of cone A (refer to p. 256).

EXAMPLES

Using the data given in figure opposite, determine the traces of a plane:

(1) Making 70° with both H.P. and V.T.H.; measure the apparent angle between traces corresponding to V.T.3 and H.T.3. Ans. 131°.

(2) Making 70° with H.P. and perp. to

V.T.H.; measure the apparent angle between both traces. Ans. 65·5°, 156·5°.

(3) Making 70° with H.P. and 60° with V.T.H.; measure the apparent angle between the more accessible pair of traces. Ans. 156·5°.

* The number possible obviously depends upon the relative positions of the ellipse and· the circle.

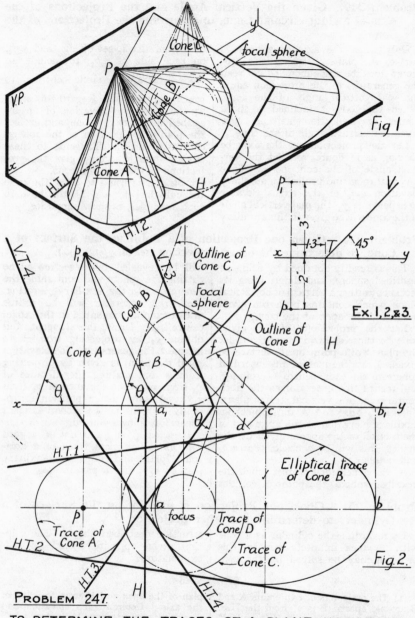

Fig 1

focal sphere

V.P.

Cone C

P

Cone B

T

Cone A

H.T.1

H.T.2.

Ex. 1, 2, & 3.

P₁

3″

1·3″

x ———— T ———— y

2″

p

V

45°

H

Fig. 2.

VT4

P₁

Cone A

Cone B

V.T.3

Outline of Cone C.

Focal sphere

V

Outline of Cone D

θ

θ

B

f

f

x ———— θ ———— y

T a₁   θ   d   c   b₁

e

H.T.1

p

a   focus

Trace of Cone A

Trace of Cone D

Elliptical trace of Cone B.

b

Trace of Cone C.

H.T.2.

H.T.3.

H

H.T.4

PROBLEM 247.

TO DETERMINE THE TRACES OF A PLANE WHICH SHALL
CONTAIN A GIVEN POINT, BE INCLINED AT A GIVEN ANGLE TO HP,
AND BE INCLINED AT A GIVEN ANGLE TO A GIVEN PLANE.

Problem 249.   Given the Vertical Angle and the Projections of the Axis of a right circular Cone, to determine the Projections of the Cone.

Only when the axis of a cone is par[1] to the plane on which it is projected does the angle included between the generators in the projection equal the real vertical angle of the cone. When the axis is inclined to both planes of projection neither view gives the vertical angle of the cone.

Let the projections of the axis be ab a_1b_1, as in figure, and let the vertical angle of the cone be α. Take x_1y_1 par[1] to ab and project an auxiliary elevation a_2b_2 of the axis. Because ab is par[1] to x_1y_1, the real vertical angle of the cone will be given in this auxiliary

elevation. At a_2 set off a_2c_2 and a_2d_2, on each side of a_2b_2, to include an angle $\dfrac{\alpha}{2}$. Draw any circle, centre o_2, to touch a_2c_2 and a_2d_2. Regard this circle as the auxiliary elevation of an inscribed sphere to the cone, and project the plan and elevation of the sphere, centres o and o_1. Tangents to these circles, from a and a_1, give the projections of the required cone.

Note. It is usual to show the cone (or cylinder) broken at the base in such examples, as shown opposite.

Problem 250.   Given one Projection of a Point on the Surface of a Cone, to determine the other Projection (*see also page 190*).

This is readily obtained by using an inscribed sphere, and determining its section by either a vertical or horizontal plane which contains both the given point and the apex of the cone.

Let the projections of the given cone be those shown in figure, and let the plan p of a point on its surface be given. Draw the plan of any inscribed sphere to the cone, centre o. Join ap, and regard this generator as the horizontal trace of a vertical section plane. Take x_1y_1 *par[1] to ap* and project an auxiliary elevation of the apex and of the section of the sphere given by the plane; this will be a circle, centre o_2 and radius r.

As the generator ap must touch the inscribed sphere, the point of contact

will be revealed by a section plane cutting the sphere and containing the generator. In the auxiliary elevation, then, the generator will be a line drawn from a_2 tangential to the circle, centre o_2. Draw this tangent and obtain p_2 by projecting from p to intersect the generator. The elevation p_1 of the point is given by projecting from p and making the distance of p_1 from xy = the distance of p_2 from x_1y_1. An alternative position for p_1 is given by using the second tangent generator, shown dotted.

If p_1 is given, and not p, regard a_1p_1 as the vertical trace of a horizontal plane and project an auxiliary plan on x_1y_1 drawn par[1] to a_1p_1.

Problem 251.   Given one Projection of a Point on the Surface of a Cylinder, to determine the other projection.

By regarding the cylinder as a cone with its apex infinitely remote, this problem may be solved by the construction described above. The solution given opposite should be self-explanatory.

EXAMPLES

(1) The projectors of two points A and B are $2\frac{1}{2}''$ apart. B is $2''$ from the H.P. and $1\frac{1}{4}''$ from V.P.; A is in H.P. and $3\frac{1}{2}''$ from V.P. AB is the axis of a cone, apex A, having a vertical angle of $30°$. Determine its projections.

(2) The plan of a point on the upper surface of the cone in (1) is $2''$ from the

plan of the apex and $\frac{1}{4}''$ from the plan of the axis. Determine its elevation and measure its height above H.P. Ans. $1.65''$.

(3) Regard AB in (1) as the axis of a cylinder $1\frac{1}{4}''$ diam. The elevation of a point on its upper surface is on a generator $\cdot4''$ from the elevation of the axis. Determine the plan of the point.

251

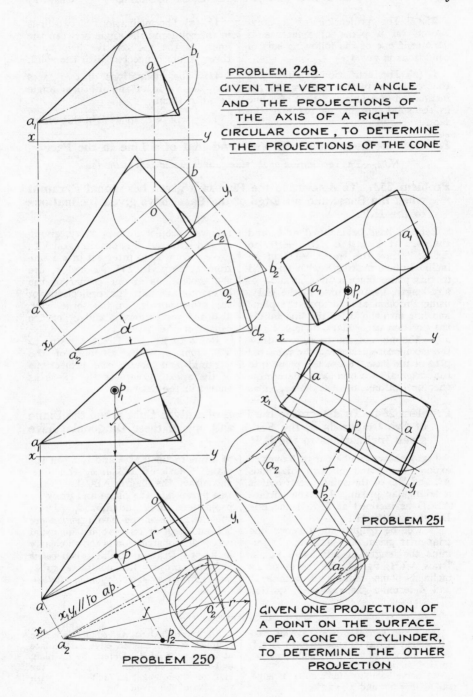

PROBLEM 249.
GIVEN THE VERTICAL ANGLE
AND THE PROJECTIONS OF
THE AXIS OF A RIGHT
CIRCULAR CONE, TO DETERMINE
THE PROJECTIONS OF THE CONE

PROBLEM 251

PROBLEM 250

GIVEN ONE PROJECTION OF
A POINT ON THE SURFACE
OF A CONE OR CYLINDER,
TO DETERMINE THE OTHER
PROJECTION

252. The projection of a poly-hedron on a plane of reference is definite if one of the following sets of conditions is given:—

I. (a) The inclination of a face and that of a line in the face, or **(b)** the inclination of a face and that of a line in the solid, not in the plane of the face.

II. (a) The inclinations of two lines in the solid and the angle between the lines, or **(b)** the relative heights of three points connected with the solid.

III. The inclinations of two faces and the value of the dihedral angle between them.

They will be dealt with in this order.

I. Given the Inclination of a Face and that of a Line in the Face.

Note.—The inclination of the line cannot exceed that of the face.

Problem 253. To determine the Plan of a given hexagonal Pyramid when the Base, and an Edge of the Base, have given Inclinations to the H.P.

Let the base be inclined at θ, and the edge at α, to the H.P. Refer to fig. 1. Draw the traces VTH of an inclined plane making θ with HP, and in this plane place a line, ab a_1b_1, of any length, inclined at α to the H.P., using Problem 214. Rabat the plane and line into the H.P., the line taking the position ab_2. Mark off c_2d_2 along ab_2 = the edge of the base, and draw the hexagon representing the rabatted plan of the base. Raise the plane into the original position and determine the projections of the hexagon as

follows: project from c_2 to xy, giving c_0, and with rad. a_1c_0 obtain c_1 on VT; project from c_1 to intersect in c a line from c_2 perp. to TH, c and c_1 are the projections of one corner of the hexagon. Obtain the projections of the other corners in the same way, also the projections o and o_1 of the centre of the base o_2.

Refer to fig. 2. Erect a perp. to VT from o_1 = the altitude of the pyramid and obtain the projections of the apex. Complete the views as shown in the drawing.

Problem 254. To determine the Plan of a given Cube when the Plane of two Diagonals of the Solid, and one of these Diagonals, have given Inclinations to the H.P.

Fig. 3 shows a cube cut in halves to expose the plane of two diagonals AC and BD of the solid: each half is a triangular prism. Let the surface ABCD be inclined at θ, and the line DB at α, to the H.P.

It will be helpful to consider one triangular prism only at first. Determine the length of DB as in fig. 4. Draw VTH, fig. 5, the traces of an inclined plane making θ with HP, and determine the rabatted position

of a line inclined at α and lying in the plane. Mark off DB along this line, and draw the figure ABCD. Raise this figure into the plane and draw its projections, $abcd$ and $a_1b_1c_1d_1$. Draw the elevation of the half prism *above* VT and project its plan in the usual way: the construction is shown clearly in fig. 5. Draw a similar prism *below* VT and project the complete cube; *this is left as an exercise for the student.*

EXAMPLES

(1) A pentagonal prism, edge of base $1\frac{1}{4}''$, height $3''$, has its base inclined at $60°$, and a diagonal of the base inclined at $50°$, to H.P. Draw its plan.

(2) Solve Prob. 254 for a cube, $2''$ edge, taking $\theta = 60°$ and $\alpha = 42°$.

(3) A tetrahedron, $3\frac{1}{2}''$ edge, has a face inclined at $60°$, and an edge in the face at $40°$, to the H.P. Determine its plan, and draw an elevation on an A.V.P. the H.T. of which makes an angle of $35°$ with the plan of the given edge.

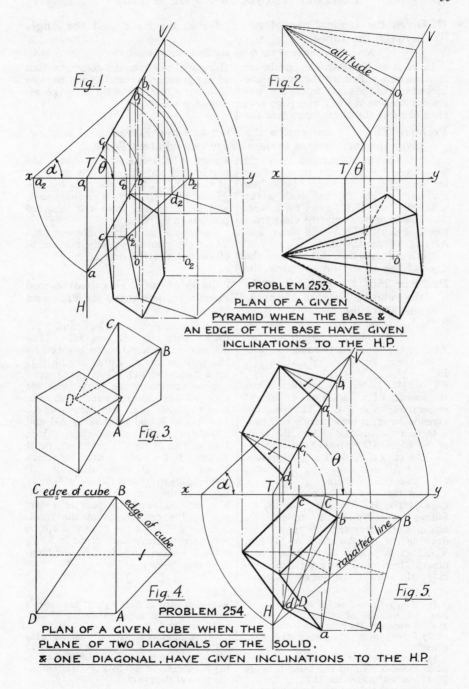

Fig. 1.

Fig. 2.

altitude

PROBLEM 253.

PLAN OF A GIVEN
PYRAMID WHEN THE BASE &
AN EDGE OF THE BASE HAVE GIVEN
INCLINATIONS TO THE H.P.

Fig. 3.

C *edge of cube* B

edge of cube

Fig. 4.

PROBLEM 254.

PLAN OF A GIVEN CUBE WHEN THE
PLANE OF TWO DIAGONALS OF THE SOLID,
& ONE DIAGONAL, HAVE GIVEN INCLINATIONS TO THE H.P.

rabatted line

Fig. 5.

II. Given the Inclinations of two Lines in the Solid and the Angle between the Lines.

Note.—The sum of the three angles cannot exceed 180°.

In each of the following problems it is necessary first to determine the projections of the plane figure containing the lines, as in Prob. 221, page 226; then to rabat the plane figure into the H.P. and to draw the complete plan of the given solid; and finally to raise the figure and the solid into the required position.

Problem 255. To determine the Plan and one Elevation of a Cube, given the Inclinations to the H.P. of two adjacent Edges.

Let the given inclinations be α and β. Determine first the traces of an inclined plane containing two lines mutually at rt. ∠s, and inclined at α and β to the H.P., as in Prob. 221. The construction should be clear from figs. 1 and 2: VTH is the plane, and AF and AE are the lines.

Mark off AB and AD each = the edge of the cube, and complete the square ABCD. Raise this into the plane VTH and complete the elevation of the cube. Obtain the plan by projecting perp. to *xy* from the elevation, and perp. to TH from the rabatted plan, as in fig. 2.

Note: the data given are sufficient to fix the plan only; any number of elevations may be drawn.

Problem 256. Given the Lengths of the Edges of a Tetrahedron and the relative Heights of three Corners: to determine its Plan and one Elevation.

The solution is best explained by means of a numerical example. Let ABC be the base of a tetrahedron, and V the apex; let AB = $1\frac{3}{4}$″, AC = $2\frac{1}{4}$″, BC = $2\frac{1}{4}$″, AV = $2\frac{1}{2}$″, BV = $2\frac{3}{4}$″, CV = $2\frac{1}{4}$″. Determine its plan when the heights of A and B are 2″ and 1″ respectively above C. The solution given is drawn approximately half size.

Obtain the inclinations to the H.P. of the edges AB and AC by constructing the rt. ∠d △s in fig. 3. Produce AB and AC and draw any horizontal DEF to intersect them.

Draw the plan of the solid with its base in the H.P. *in any position*, as follows: first draw the base ABC, fig. 4, and determine the apex V by drawing the *rabatted* plans V_1AB and V_2AC of adjacent slant faces, and raising them until their vertices coincide; the plan of the apex V is given by the intersection of V_1V and V_2V, drawn perp. to AB and AC. (The *altitude* of the tetrahedron is given by the perp. of the rt. ∠d △ having V_1G for hypotenuse and VG as base.)

Now determine the projections *abc* $a_1b_1c_1$ of the base of the tetrahedron in the required position: produce AC and AB to E and D, making AE and AD (fig. 4) = AE and AD (fig. 3). Join DE, and at M on DE produced, draw *xy* perp. to it. Raise A until its height above the H.P. = the distance AF from fig. 3 (its elevation is given by a_1), and determine the plane LMN. The problem now consists in determining the projections of the tetrahedron VABC on the base *abc* $a_1b_1c_1$; reference to the construction in fig. 4 should render any further explanation unnecessary.

EXAMPLES

(1) Two adjacent edges of a cube, 2″ long, are inclined at 20° and 60° to the H.P. Determine its plan.

(2) Solve Prob. 256 using the data given.

(3) Draw the plan of a cube, $2\frac{1}{4}$″ edge, when three corners of the same face are $\frac{3}{4}$″, $1\frac{1}{2}$″, and 2″ above the H.P.

(4) Draw the elevation of a pentagonal pyramid, edge of base $1\frac{1}{4}$″, 3″ high, when three corners of the base are 0″, 1″, and $1\frac{1}{4}$″ from the V.P.

(5) A hexagonal prism, edge of base 1″, 3″ long, has a long edge inclined at 50° to H.P. and an edge of the base meeting it inclined at 35° to H.P. Determine the plan of the prism.

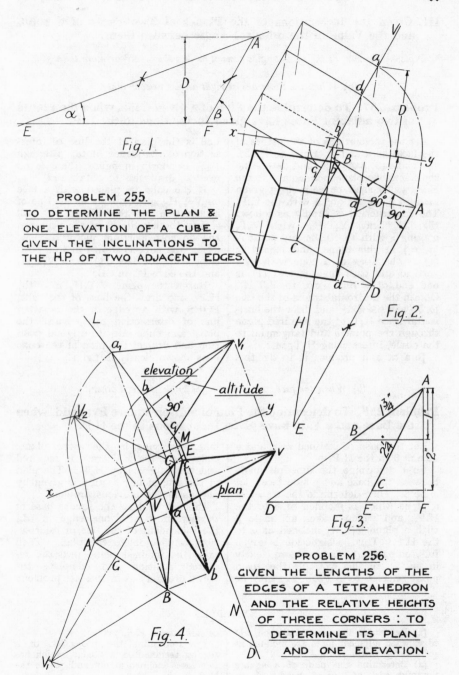

Fig. 1.

PROBLEM 255.
TO DETERMINE THE PLAN &
ONE ELEVATION OF A CUBE,
GIVEN THE INCLINATIONS TO
THE H.P. OF TWO ADJACENT EDGES.

Fig. 2.

Fig. 4.

elevation

altitude

90°

plan

Fig. 3.

PROBLEM 256.
GIVEN THE LENGTHS OF THE
EDGES OF A TETRAHEDRON
AND THE RELATIVE HEIGHTS
OF THREE CORNERS : TO
DETERMINE ITS PLAN
AND ONE ELEVATION.

III. Given the Inclinations of the Planes of Two Faces of a solid, and the Value of the dihedral Angle between them.

Note.—The sum of the three angles cannot be less than 180° *or more than* 360°.

(a) When the faces are at right angles to each other.

Problem 257. To determine the Plan of a given Cube, when the Planes of two adjacent Faces have given Inclinations to the H.P.

Let the inclinations of the planes of the faces be α and θ. Refer to Probs. 247 and 248, page 248, and determine the traces of a plane inclined at θ to the H.P. which shall be perp. to a given *inclined* plane making α with the H.P. The construction is briefly as follow: (fig. 1), draw $V_1T_1H_1$, with V_1T_1 making α with xy. Take any point A (a, a_1) in this plane, and regard it (1) as the apex of a cone with axis vertical and base angle θ, and (2) as one end of a line perp. to $V_1T_1H_1$. Obtain the horizontal traces of the line (b, b_1) and the cone, and draw the horizontal trace H_2T_2 of the required plane through the point b and tangential to the circle, intersecting H_1T_1 in c.

Join ca and produce it to d: this line is the plan of the line of intersection of the planes (note: although V_2T_2 is shown in figure, there is no need to draw this to obtain cad).

If the cube be placed with a face on $V_1T_1H_1$, and an edge in the line of intersection of the planes, it will satisfy the conditions of the problem. In the corresponding pictorial view, fig. 2, the cube is shown in this position, with the face PQRS in $V_1T_1H_1$ and the edge PQ in CD.

Rabat the plane $V_1T_1H_1$ into the H.P., and draw the base of the cube, PQRS, with an edge in the rabatted line of intersection cd_2. Raise the plane and cube into its original position and project the plan of the cube —as shown clearly in fig. 1.

(b) When the faces are not at right angles to each other.

Problem 258. To determine the Plan of a given square Pyramid, when the Base and a Face have given Inclinations to the H.P.

Let the base be inclined at α and a face at θ to the H.P.

First determine the dihedral angle between the base and a side face: let it be β. Then determine the traces of a plane which is inclined at θ to the H.P., and which makes an angle β with a plane $V_1T_1H_1$ inclined at α to the H.P. This construction is given fully on page 248, and is shown clearly in fig. 3: in this example the given point referred to in Prob. 247 may be taken as lying in the given plane. Only the horizontal trace is required and this is given by H_2T_2. The plan of the line of intersection is given by cad, as in the previous problem.

Proceed to draw the square base of the pyramid with one edge in cd_2, the rabatment of the line of intersection, and complete the plan. Then raise the inclined plane into the required position and complete the views precisely as in the last problem

EXAMPLES

(1) Determine the plan of a cube, $2\frac{1}{4}''$ edge, when adjacent faces are inclined at 40° and 65° to the H.P.

(2) Determine the plan of a square pyramid, edge of base 2″, height 3″, inclination of base to H.P. 50°, inclination of side face fo H.P. 75°.

(3) Determine the projection of a regular tetrahedron, $2\frac{1}{2}''$ edge, which has two faces inclined at 60° and 70° to the H.P.

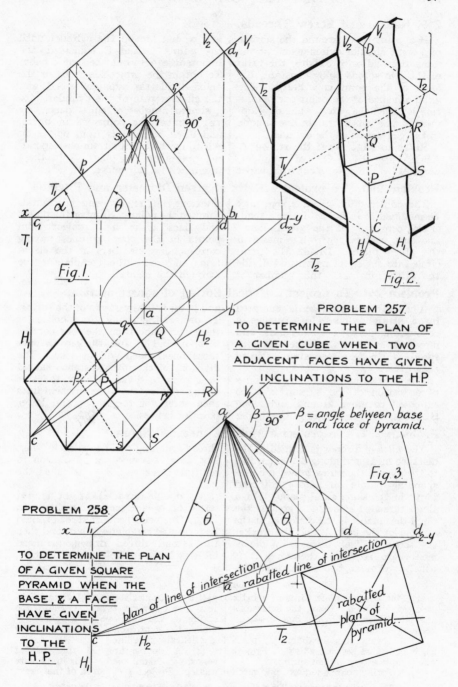

Fig.1.

Fig.2.

PROBLEM 257.

TO DETERMINE THE PLAN OF
A GIVEN CUBE WHEN TWO
ADJACENT FACES HAVE GIVEN
INCLINATIONS TO THE H.P.

β = angle between base
and face of pyramid.

Fig.3.

PROBLEM 258.

TO DETERMINE THE PLAN
OF A GIVEN SQUARE
PYRAMID WHEN THE
BASE, & A FACE
HAVE GIVEN
INCLINATIONS
TO THE
H.P.

plan of line of intersection

rabatted line of intersection

rabatted
plan of
pyramid.

259. Helices and Screw Threads.

If a point move around the surface of a right circular cylinder and at the same time advance axially, the ratio of the two speeds being constant, the locus of the point is a **helix**. The axial advance of the point per revolution is called the **pitch**. The development of a helix is a straight line oblique to the generators of the cylinder.

Refer to Fig. 1. If the rt. ∠d △ ABC, base BC = O° of cylinder, perp. AC = pitch, were cut from paper and wound around the cylinder until the points B and C coincided, the hypotenuse AB would become a helix. To obtain the projection A_1B_1 of the helix, divide the hypotenuse AB, and the circular plan of the cylinder, into a similar number of equal parts, say 12 ; number the points as in figure, and draw horizontals from points on AB to meet verticals from corresponding points on the plan. Join these points with a fair curve.

Problem 260. To project a Helix of given Diameter and Pitch.

The construction given above may be modified as in fig. 2. Draw a semicircle, centre o, of the given diam., and on a vertical centre line mark off the given pitch. Divide the semicircle into six equal parts, and divide the pitch into *twice* this number of divisions, i.e. 12, drawing horizontals through the points. Number the points and lines as in fig., project from points in the plan to lines having corresponding numbers in the elevation, and draw the complete curve through the points.

Problem 261. To project a Helical Spring of square section.

A helical or screw surface is generated by a straight line which makes a constant angle with a fixed axis, and moves around and along the axis with speeds which bear a constant ratio to each other.

. The lines ab and cd, fig. 3, the sides of a section of the spring, may be supposed to move around and along the axis, thus generating the upper and lower helical surfaces of the spring. To project the spring, draw helices of the given pitch from the four corners a, b, c, and d, and thicken in only those lines which will be seen. The divisions of the pitch distance should be chosen to divide equally the side of the square; the same division lines then serve for the upper and lower pairs of helices.

Problem 262. To project a Multiple-threaded Screw.*

The *pitch* of a screw is defined as the distance measured along a line parallel to the axis of the screw between corresponding points on adjacent thread forms in the same axial plane. For a single-threaded screw, the greater the pitch the smaller is the diameter at the bottom of the thread, and the weaker the screw. When large axial movements are required, therefore, it is usual to provide two or more similar threads running parallel to each other. The axial advance per revolution in multiple-threaded screws is called the *lead*.

A double-threaded right-handed screw of square section is shown in fig. 4. One helical square thread should be projected completely, first, and then the second thread drawn, arranged midway between the turns of the first.

EXAMPLES

(1) Plot two complete turns of a helix 4″ diam., 2″ pitch. Using the same axis and pitch, plot, within the first, a second helix 2½″ diam.

(2) A spring, outside diam. 5″, pitch 4″, is of square section, 1″ side. Project one complete turn of the spring.

(3) A spring, outside diam. 4″, pitch 3½″, is of circular section, ¾″ diam. Project one complete turn. (Regard the spring as the envelope of a sphere, ¾″ diam., the centre of which moves along the helical centre line.)

(4) A double-threaded right-handed screw, 4″ diam., lead 3″, is of square section. Project a 4″ length of the screw.

* The screwed or " threaded " part of an ordinary bolt should be examined by the student.

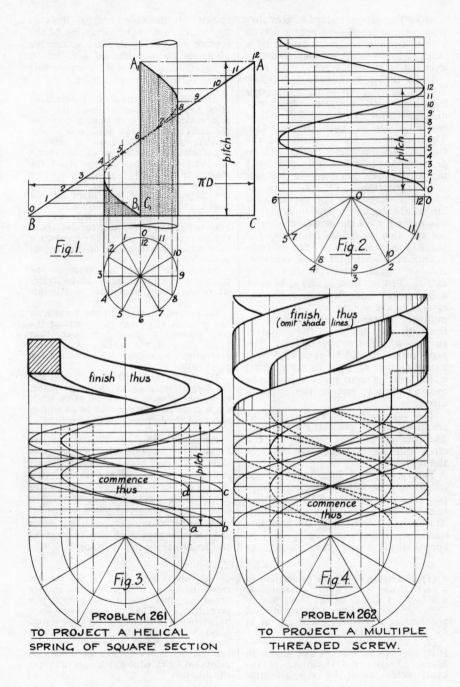

Fig. 1.

pitch

πD

Fig. 2.

Fig. 3.

finish thus

commence thus

pitch

Fig. 4.

finish thus
(omit shade lines.)

commence thus

PROBLEM 261

TO PROJECT A HELICAL
SPRING OF SQUARE SECTION

PROBLEM 262

TO PROJECT A MULTIPLE
THREADED SCREW.

263. The Hyperboloid of Revolution (of one sheet—see next col.*) is a ruled surface, and is generated by a straight line which revolves about an axis not in the same plane with it. **Refer to fig. 1.** Let the axis OO be vertical, and let AB be the generating line in its initial position. The upper and lower limits to the surface will be horizontal circles, radii OA and OB. The outline of the elevation of the surface between these circles will be given by the projection of the meridian† parl to the V.P., determined as follows:

Construction. Fig. 2. Let ab a_1b_1 be the generator and oo_1 the axis. Draw MN through o parl to xy. Take any point c, c_1 on the generator ab a_1b_1. As the line revolves about oo_1 the point c moves around the circumference of a horizontal circle, centre o, rad. oc, which intersects MN in c_2 and c_4. Project from c_2 and c_4 to meet a horizontal through c_1 in c_3 and c_5; c_3 and c_5 are points on the required outline. Determine other points in the same way and draw the complete curves—which are hyperbolas. The same surface would be generated by a straight line ae a_1e_1 (broken lines) which has the same inclination to the axis as ab a_1b_1 but in the opposite direction.

The asymptotes to the hyperbolas are given by the projections a_1e_1 a_1b_1. The smallest circular section, i.e. the circle rad. oa, is called the *throat* of the surface.

The generating line may extend indefinitely in either direction. Suppose the line to extend in the direction BA, fig. 1, until A is a mid-point. The surface generated by the complete line—b_1d_1 in elevation, fig. 2—will be symmetrical about the throat, as shown by the added dotted lines in fig. 2. An idea of the formation of the surface is given by drawing successive positions for the line bd b_1d_1, as in fig. 3, in which 12 positions are shown; the envelope is a hyperboloid.

Alternative Method of Generation. If a hyperbola revolve about its conjugate axis, fig. 4, each branch describes the same surface; this surface is similar to that discussed above and is called a hyperboloid of revolution of *one sheet*.* (If the two branches revolve about the transverse axis, fig. 5, they describe separate similar surfaces, the two forming the hyperboloid of revolution of *two sheets*; this is not a ruled surface.)

If the asymptotes, as well as the branches of the hyperbola, revolve about the conjugate axis, fig. 4, they generate a cone which is asymptotic to the hyperboloid.

Notes.—I. Through any point on the surface, two straight lines can be drawn which are wholly contained by the surface.

II. A tangent plane to the surface at any point is the plane that contains the two generators passing through that point; it cuts the surface along these generators and touches the surface at the point only.

III. All sections of the hyperboloid and its asymptotic cone by the same plane, or by parl planes, are of the same kind; e.g. a plane cutting the cone in an ellipse, also cuts the hyperboloid in an ellipse.

Applications. A second hyperboloid may be generated *by the same line* revolving about a second axis inclined to OO and not intersecting it. Two such surfaces may rotate about their axes and touch along a line. Skew bevel wheels are formed with hyperboloids as pitch surfaces; the teeth engage with line contact at the imaginary pitch surface.

EXAMPLES

(1) The figure shows the projections of a vertical axis o, o_1, and a generator ab a_1b_1 inclined at 60° to H.P. and parl to V.P. Project the line in 12 equidistant positions as it revolves about oo_1, as in fig. 3, and sketch in the envelope.

(2) A second axis q, q_1, inclined at 30° to H.P. and parl to V.P., is also shown in figure. Project the elevations of the hyperboloids traced by the generator ab a_1b_1 about *both* axes, in the manner shown in fig. 2. (*Hint.* For the surface about q_1q_1 first take a new ground line perp. to q_1q_1.) The two projections show hyperboloids in contact along a generator.

(3) A cube $2\frac{1}{2}''$ edge revolves about a long diagonal, which is vertical. Project the outline of the surface of revolution generated by an edge which does not meet the diagonal.

† Meridian sections are sections by planes containing the axis.

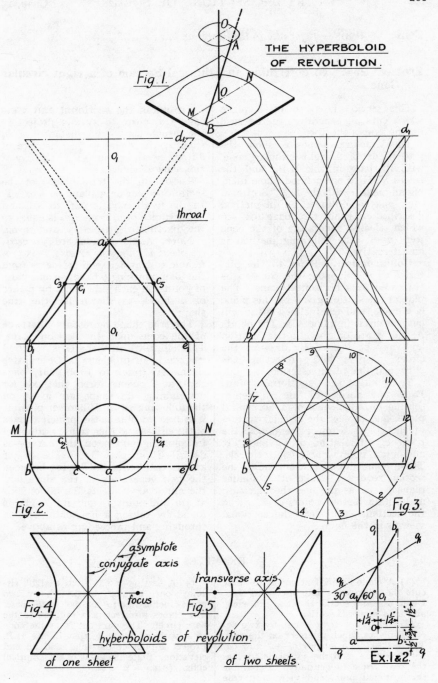

THE HYPERBOLOID
OF REVOLUTION.

Fig. 1.

throat

Fig. 2.

Fig. 3.

asymptote
conjugate axis

focus

Fig. 4.

transverse axis

Fig. 5

Ex. 1 & 2.

hyperboloids of revolution.

of one sheet

of two sheets.

Conic Sections.—*Refer also to the figures on p. 23.*

Problem 264. To determine an Elliptical Section of a right circular Cone.

This section is given by a plane which cuts all generators of the cone on one side of the apex.

The drawings opposite show the projections of a right circular cone, with its base in the H.P., and the traces VTH of an inclined plane making an angle θ with the H.P. Sectional views may be obtained by the method described in Prob. 189, page 190—in which selected generators of the cone were used: the following method is an alternative.

Construction. Refer to the pictorial view. Suppose the cone to be truncated by a horizontal plane. The plan of the section given by this plane is a circle, and the inclined plane will cut this circle in a line AB. The points A and B are on the elliptical section and may be located by drawing the circle in plan and projecting the particular line of intersection.

To obtain the **sectional plan.** Draw any horizontal line hp, representing an edge view of a horizontal plane, and project the circular section of the cone. Project from a_1b_1, the point of intersection of hp and VT, and obtain points a and b on the circle. These points lie on the outline of the section required. Take other similar planes, such as the one through c_1d_1, and obtain further points c, d . . . on the section. Complete the sectional view as in the figure.

To obtain the **sectional end view** take x_1y_1 perp. to xy and project in the usual way: the distance of b_2 from x_1y_1 is equal to the distance of b from xy, and so on. One half only of this view is shown.

To obtain the **true shape** of the section project on a plane par¹ to VT. In the figure x_2y_2 is par¹ to VT and the distance of b_3 from x_2y_2 is equal to the distance of b from xy, and so on.

Note. As the sections are *symmetrical about a centre-line or axis*, it is preferable to transfer semi-ordinates from the plan, marking off the various pairs of points on each side of CL by means of dividers—as shown for the true shape.

The true shape represents the **trace** of the cone upon the section plane. This elliptical true shape is required frequently in problems in Descriptive Geometry (refer to Prob. 281, page 281), and various direct methods for determining its shape are given on the following page. Wherever possible it is best to obtain the principal axes of the ellipse and then to use a trammel for plotting points on the curve—as described on page 30. The major axis is that part of VT lying between the two generators in the elevation; the minor axis bisects the major axis at its mid-point e, and its length is found by taking a horizontal section through e and proceeding as above.

EXAMPLES

(1) A plane inclined at 30° to the base cuts the given cone and bisects its axis. Determine the sectional plan, end view, and true shape of section.

(2) The given cone is cut by two inclined planes which intersect on the axis. The planes make angles of 30° and 60° with the base, and the 30° plane touches the base circle tangentially. Determine the sectional plan, end view, and true shape of section.

(3) A section is cut from a right circular cone, base 4″, height 5″, by two planes inclined at 10° and 60° to the base. The planes intersect in a tangent to the base circle. The part cut from the cone rests with its lower section in the H.P. Project a plan of the solid and an end elevation. State the areas of the elliptical ends. (Area = $\pi \cdot a \cdot b$.)

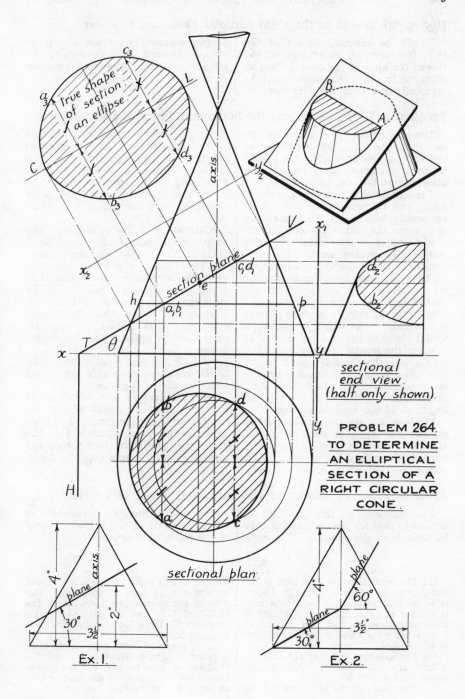

true shape of section: an ellipse

L

C

axis

section blane

sectional end view. (half only shown)

PROBLEM 264.
TO DETERMINE
AN ELLIPTICAL
SECTION OF A
RIGHT CIRCULAR
CONE.

sectional plan.

4" plane axis 2" 30° 3½"

Ex.1.

4" plane 60° plane 30° 3½"

Ex.2.

Horizontal Traces of the right Circular Cone and Cylinder.

It will be assumed here that the H.P. intersects all generators on one side of the apex of the cone, giving an elliptical trace. The cylinder may be regarded as a cone with its apex in-finitely distant; its trace also is an ellipse. Several ways of drawing the ellipses are given below in their order of usefulness.

Problem 265.   To determine the horizontal Trace of a given Cone.

The axis of the cone is assumed to be par[1] to the V.P.; if it is inclined to the V.P. an auxiliary elevation should be used, projected on a ground line taken par[1] to the plan of the axis.

The outline of the cone in elevation, $a_1b_1c_1$, and the projections of its axis, are readily determined. Three methods of drawing the ellipse are as follows:

Method 1.   By obtaining the major and minor axes. The projections of the major axis are given by bc, b_1c_1. The axes bisect each other at rt. ∠s at d, the mid-point of bc. To obtain the minor axis, suppose the cone to be cut at d_1 by a plane normal to the axis. The width of the horizontal chord of the circular section at d_1 gives the length of the minor axis. At d_1, then, draw a diam. perp. to the axis and describe as much of the circle as is necessary; the half-chord d_1e_1, drawn perp. to the diam., gives the length de of the semi-minor axis.

Method 2.   By obtaining points on the curve. The construction of Method 1 may be applied to determine other points. For example, a normal section at g_1 gives the half-chord g_1h_1 which is the half-ordinate gh of the ellipse. Any number of points may be determined by this method, which is merely that of Prob. 264 on the previous page: it is of value when the ellipse is so elongated that the end c_1 is inaccessible.

Method 3.   By obtaining the major axis and foci. Inscribe a circle to the △ $a_1b_1c_1$ and regard it as the elevation of the focal sphere to the cone, with reference to the H.P. Its point of contact f_1 with xy is the elevation of one focus and f is its plan. The distance of the second focus from c is equal to fb. Prob. 41, page 28, may then be applied and the ellipse drawn.

Note. Another focal sphere may be drawn below xy to give the second focus, as shown opposite.

The **directrices** are obtained as follows: draw the traces of the two planes containing the circles of contact between the focal spheres and the cone: they will be inclined planes and their horizontal traces give the directrices.

Problem 266.   To determine the horizontal Trace of a Cylinder.

The solution to this problem is similar to that for the cone and should be clear from the drawing. The construction is simplified somewhat as the minor axis of the ellipse is equal to the diam. of the cylinder.

EXAMPLES

(1) The vertical angle of a cone is 40° and its apex is $2\frac{1}{2}''$ above the H.P. Determine its horizontal trace when the axis is inclined at 35° to the H.P. Use one of the methods given and test the other two. What is the value of the eccentricity of the ellipse?

(2) The axis of a cylinder $2\frac{1}{4}''$ diam. is inclined at 30° to the H.P. Determine the horizontal trace of the cylinder.

(3) The traces on the H.P. and V.P. of the axis of a cylinder $2\frac{1}{2}''$ diam. are distant respectively $4\frac{1}{2}''$ and $4''$ from xy. The ends of the cylinder are bounded by the H.P. and V.P., and the plan of its axis measures $6''$. Draw the plan and elevation of the cylinder.

(4) A sphere $2''$ diam. rests on the H.P. A point of light is situated $3''$ above the H.P. and $2\frac{1}{2}''$ from the centre of the sphere. Determine the outline of the shadow cast by the sphere on the H.P.

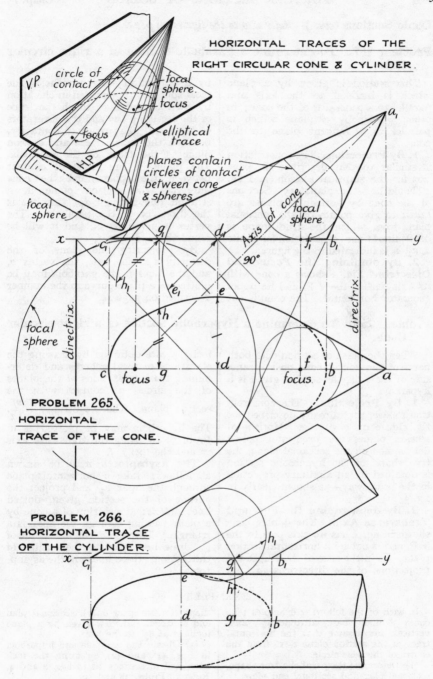

HORIZONTAL TRACES OF THE
RIGHT CIRCULAR CONE & CYLINDER.

PROBLEM 265.
HORIZONTAL
TRACE OF THE CONE.

PROBLEM 266.
HORIZONTAL TRACE
OF THE CYLINDER.

Conic Sections (*cont.*)—*Refer also to the figures on p. 23.*

Problem 267. To determine a Parabolic Section of a right circular Cone.

This section is given by a plane which is inclined to the axis and parallel to a generator of the cone; in other words, by a plane which is parallel to a tangent plane to the cone.

1. By Projection. The construction is similar to that described on page 262 for the elliptical section, and fig. 1 should be self-explanatory. Sections of the cone by horizontal planes are taken to give points on the sectional plan, from which the true shape is determined: the construction for points a, a_1, a_2 is indicated in the figure.

2. By obtaining the Focus and Directrix. Fig. 2 shows a cone with its axis par[l] to the V.P. and its upper generator horizontal. The cone is cut by a horizontal section plane. The focal sphere is represented in elevation by a circle drawn to touch the trace of the section plane and the generators of the cone. The point of contact f_1 between the focal sphere and section plane is the focus of the parabola. The plane of the circle of contact between the sphere and the cone intersects the section plane in a straight line through d_1; this line is the directrix of the parabola. The vertex is given by v_1, and it will be found that $v_1 d_1 = v_1 f_1$.

By projecting the plans of the directrix $d_1 d$, the axis, the vertex v, and the focus f, the parabola may be plotted as a plane curve in the manner described on page 24.

Problem 268. To determine a Hyperbolic Section of a right circular Cone.

When the plane of section cuts both parts of the double cone on the same side of the axis, the section given is a hyperbola.

1. By Projection. The construction follows that already described for the ellipse on page 262. Horizontal section planes are used and points determined on the sectional plan; the true shape of the hyperbolic section is obtained by an auxiliary projection in the usual way—as shown clearly in fig. 3.

2. By determining the Foci and Transverse Axis. The double cone shown in fig. 4 has its axis par[l] to the V.P. and is cut by a horizontal section plane. As in the case of the ellipse the projections of the directrices $d d_1$ and foci f, f_1 are obtained by drawing the elevations of focal spheres and determining the intersections of the planes of the circles of contact with the section plane. The eccentricity $= \dfrac{vf}{vd}$.

The hyperbola may be plotted as in Prob. 50 by using the transverse axis vv and the foci f, f.

The asymptotes may be drawn as follows: take a horizontal plane through the apex a_1, and project the outline of the section given, dotted lines. (*Note*: the section of a cone by a plane passing through the apex is a triangle.) Through c, the mid-point of vv, draw lines par[l] to the outline of the section: these lines are the asymptotes.

EXAMPLES

In each of the following examples take cones 5″ diam. base, altitude 4½″, axes vertical, and assume that the horizontal trace of the section plane is 1″ from the centre of the base circle.

(1) Determine the parabolic true shape, sectional plan and sectional end view.

(2) Determine the hyperbolic true shape (for one part only), sectional plan and sectional end view given by a plane inclined at 85° to the base.

(3) Obtain the parabola and hyperbola in Exs. (1) and (2) by using the foci, axes, and directrices, as in figs. 2 and 4. Refer to Probs. 32 and 50.

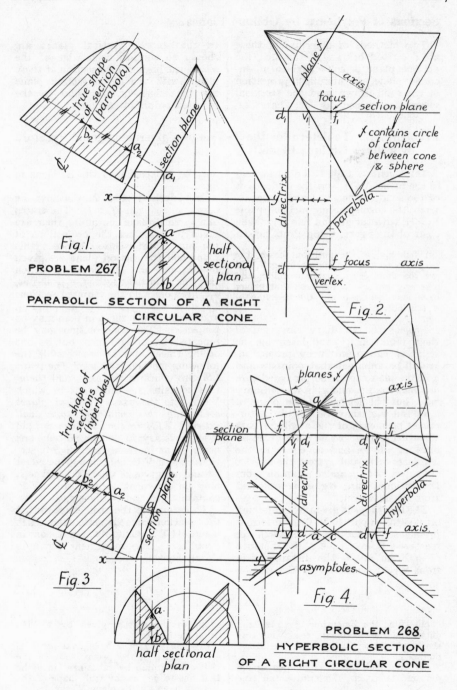

true shape
of section (parabola)

b_2

a_2

℄

section plane

a_1

x ———— y

a

half
sectional
plan.

b

Fig.1.

PROBLEM 267

PARABOLIC SECTION OF A RIGHT
CIRCULAR CONE

plane f

axis

focus

section plane

d_1 v_1 f_1

f contains circle
of contact
between cone
& sphere

directrix

parabola

d   v   f focus   axis.

vertex.

Fig. 2.

true shape of
sections (hyperbolas)

b_2

a_2

℄

section plane

a

section plane

x ———— y

Fig 3

a

b

half sectional
plan

planes x

axis

a

section
plane

f_1 f_1

v_1 d_1 d_1 v_1

directrix   directrix

hyperbola

f v d a c   d v f   axis.

asymptotes

Fig. 4.

PROBLEM 268.

HYPERBOLIC SECTION
OF A RIGHT CIRCULAR CONE

Sections of Polyhedra by Oblique Planes.

Two methods of determining these are described below. In the first, the oblique plane is converted to an inclined plane, an auxiliary projection of the solid drawn, and the sectional views determined as in Chap. 12.

In the second, vertical planes are chosen to contain each edge of the solid; the lines of intersection of these planes with the given section plane cut the edges of the solid in the required points of section.

Problem 269. To determine the Section of a given hexagonal Pyramid by a given Oblique Plane.

1st Method. Fig. 1. Let the traces of the plane, VTH, and the projections of the solid, be those shown opposite. Draw the corresponding inclined plane V_1T_1H, with the ground line x_1y_1 taken clear of the figure, using Prob. 206.

(With x_1y_1 in the position shown, the perps. to xy and x_1y_1 determining V_1T_1H are too short for accurate working, and a second line x_2y_2 is taken so that longer lines may be used and the *direction* of V_1T_1H obtained.)

Project the auxiliary elevation of the pyramid on x_1y_1 and determine the *sectional plan* from it by projection. It will be apparent that points n and q in the plan cannot be projected *with accuracy* from n_2 and q_2 in the ordinary way, but the lengths of the *plans* of v_2n_2 and v_2q_2 may be readily determined by means of the auxiliary part elevation shown: n and q are given by marking off vn and vq equal to the distances of n_3 and q_3 from the axis.

Project the *sectional elevation* from the sectional plan; e.g. m_1 is projected from m, and so on.

The *true shape* is given by projecting on x_3y_3 taken par^l to V_1T_1 as in former problems. In the figure distances are measured from x_2y_2, e.g. the distance of m_3 from x_3y_3 = the distance of m from x_2y_2. Alternatively the section

may be rabatted into the H.P., as in fig. 2.

2nd Method. Fig. 2. Consider the edges vc, v_1c_1 and vf, v_1f_1. The traces of a vertical plane containing them are given by t_1th, and the projections of the line of intersection between this plane and the section plane are given by t_1h_1 and th. The points in which this line cuts the edges vf, v_1f_1 and vc, v_1c_1 are points on the required section: t_1h_1 intersects v_1f_1 and v_1c_1 in r_1 and o_1, and from these r and o in plan may be projected. The construction may be repeated for other edges, but as one of the traces may be inaccessible the following method is of use. The plane t_1th also contains the axis, and therefore contains s_1—the point in which VTH cuts the axis. *Vertical planes containing the remaining edges will intersect VTH in lines through s_1*, and it is necessary to determine only one other point in these lines of intersection: e.g. a V.P. containing the edges va and vd gives w on H.T., and w_1s_1 may be drawn at once; also, a V.P. containing vb and ve gives z_1s_1.

To determine the true shape, rabat the section $mnopqr$ into the H.P. about H.T., point by point, as in Prob. 215; the rabatment for n is clearly shown.

EXAMPLES

(1) Using the dimensions given in fig. 2 determine the sectional views and true shape of section, using both the above methods.

(2) The plan of a right pentagonal pyramid is given. Determine the true

shape of the section given by V.T.H., using Method 1.

(3) The plan of a hexagonal prism is given, a face of the prism being in the H.P. Determine by rabatment into the H.P. (as in fig. 2) the true shape of the section given by the plane VTH.

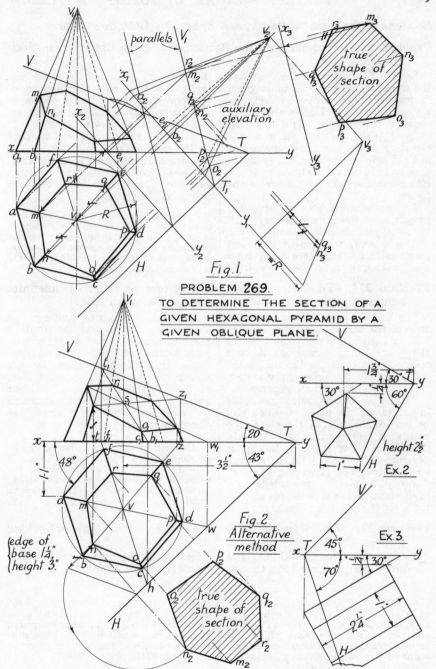

Fig.1

PROBLEM **269**.

TO DETERMINE THE SECTION OF A
GIVEN HEXAGONAL PYRAMID BY A
GIVEN OBLIQUE PLANE.

true shape of section

auxiliary elevation.

parallels

Fig.2
Alternative method

true shape of section

(edge of base 1¼"
height 3".

Ex.2

height 2½"

Ex.3

Sections of the Cone, Cylinder, and Sphere by Oblique Planes.

Problem 270. To determine the Section of a given Cone * by a given Oblique Plane.

Let the projections of the cone and the traces of the plane be those given in figure. Take a new ground line x_1y_1 perp. to the H.T. of the plane, convert the oblique plane into an inclined plane, project an auxiliary elevation of the cone, and proceed as in Prob. 264 to determine the sectional plan and true shape. This part of the construction should be clear from the figure.

To determine the sectional elevation, project first the circumscribing par^m to the ellipse. In plan this will be the rectangle $abcd$, with edges par^l and perp. to H.T. Project $a_1b_1c_1d_1$ from $abcd$, taking heights above xy from

the auxiliary elevation: e.g. the distance of a_1 from xy = the distance of a_2 from x_1y_1. Inscribe an ellipse in the par^m $a_1b_1c_1d_1$, using Prob. 47, and complete the view as shown.

As an alternative, the 2nd method on the previous page may be applied if the cone is regarded as a pyramid. Vertical planes containing selected generators will give the required points on the section.

If the cone is given with a generator in the H.P., obtain the section curves by plotting the points in which various generators are cut by the oblique section plane, first converting the oblique plane into an inclined plane.

Problem 271. To determine the Section of a Cylinder of indefinite Length, in a given Position, by a given Oblique Plane.

The section is an ellipse, and the ends of the conjugate diams. in each view are given by the projections of the points of intersection of four selected generators with the plane VTH, namely the highest, the lowest, and the two intermediate generators.

By the use of horizontal lines lying in the plane, the plans of which will be par^l to the H.T. of the plane, the points of intersection are quickly determined. For example, the upper generator intersects VTH in a horizontal through e_1, given in plan by ea drawn par^l to H.T.; the intersection of ea with the plan of the upper generator gives the point a. Similarly fdc is

a horizontal through f_1, giving points d and c. The ellipses inscribed in par^{ms} having sides par^l to these conjugate diams. are the required sections.

The true shape is given by the rabatment into the H.P., about H.T. of the sectional plan $abcd$. Rabat points a, b, c, and d (the construction for a is indicated) and draw the ellipse in the par^m $a_2b_2c_2d_2$.

Alternatively, the cylinder may be enveloped by a square prism having a face in the H.P. The sections of the prism by the plane are the par^{ms} shown, in which the elliptical sections may be inscribed.

Problem 272. To determine the Section of a Sphere by a given Oblique Plane. (No figure.)

By changing the ground line and converting the problem into the in-

clined plane type, the sections may be obtained as in Prob. 181.

EXAMPLES

(1), (2), (3) The figure shows the plan of (1) a cone 3″ high, (2) a cylinder 3″ high, (3) a sphere, each resting on the H.P., and also the traces of a plane. Obtain the sectional plan, sectional elevation, and true shape of section for each. Use the 2nd Method of Prob. 269 for (1).

(4), (5) The figure shows the plan of the axis of (4) a cylinder 2″ diam. with a generator in the H.P., (5) a hexagonal prism, 1″ edge, with a face in the H.P. Determine the true shapes of the sections given by the plane VTH.

* A cylinder standing with an end in the H.P. may be treated in the same way by regarding it as a cone with its apex at infinity.

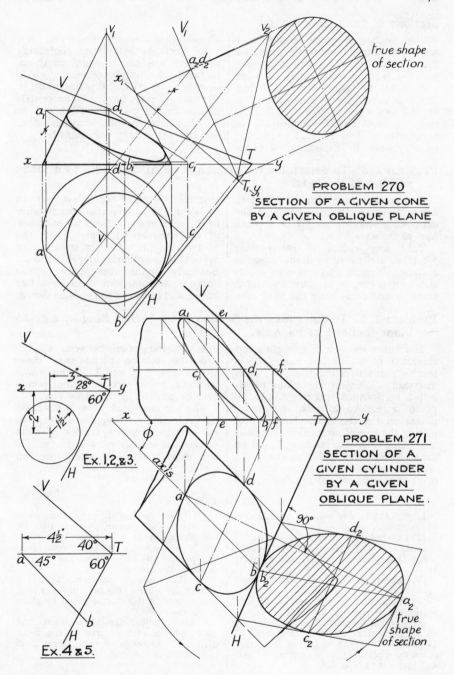

PROBLEM 270
SECTION OF A GIVEN CONE
BY A GIVEN OBLIQUE PLANE

true shape of section.

Ex. 1, 2, & 3.

PROBLEM 271
SECTION OF A
GIVEN CYLINDER
BY A GIVEN
OBLIQUE PLANE.

true shape of section.

Ex. 4 & 5.

Sections of an Anchor Ring.

The Anchor Ring or Tore * is a solid or surface of revolution, generated by a circle revolving about an external axis in its own plane; consider a rectangle to rotate about an edge, which remains fixed—a circle drawn on the rectangle generates an anchor ring as it revolves. It is helpful also to regard the surface as that swept out by a sphere in traversing a plane circular path—refer to page 171.

All sections by planes containing the axis are circles equal in diam. to the diam. d of the generating circle or sphere. Sections by planes perp. to the axis are rings formed by concentric circles, and varying in width from o to d. From these considerations other sections are readily determined.

Problem 273. To determine the Section of an Anchor Ring by a Plane parallel to its Axis.

A side view of a ring is shown in figure, D being its mean diam., d the diam. of a section: let VT be a plane par¹ to the axis.

Take any radius or intersecting VT in a, and on rc as diam. describe a circle. Regard this circle as a section of the ring by a plane cutting it along ro and containing the axis: the length of the chord through a, perp. to ro, is the width, measured horizontally, of the section at a. Draw an end view of the complete ring and insert the centre line CL. Project from a and mark off the ordinates a_1b_1 equal to ab. Similarly obtain other points on the curve of section, join them by a fair curve, and complete the view as shown.

Problem 274. To determine the Section of an Anchor Ring by a given Plane inclined to its Axis.

The figure shows a simple plan and elevation of an anchor ring, and a section plane dividing the solid symmetrically. Project the plans of selected sections of the ring given by planes normal to its axis, and obtain points on the sectional plan by projecting the line of intersection between the plane and the section. As an example consider the plane $a_1c_1d_1b_1$, which cuts the ring in an annulus and gives the circles ab and cd in plan; the line of intersection between $a_1c_1d_1b_1$ and the section plane cuts these circles in e, f, g, and h, giving points on the curve of section. Other points may be plotted in the same way.

The sectional end view is obtained by projecting on x_1y_1 drawn perp. to xy: the distance of e_2 from $x_1y_1 =$ the distance of e from xy, and so on. To determine the true shape, project on a plane par¹ to the section plane.

EXAMPLES

In Ex. (1) to (4) take an anchor ring of the dimensions given in figure.

(1) Determine a section by a plane par¹ to the axis and (a) 1·2″ from it, (b) 1″ from it.

(2) Determine a sectional plan by plane (1).

(3) Determine a sectional plan by plane (2).

(4) Determine a sectional plan and sectional end view by plane (3).

(5) A prolate spheroid is generated by an ellipse, major axis 3½″, minor axis 2½″. Determine the true shape of a section given by a plane bisecting the major axis and inclined to it at 60°.

(6) Refer to Ex. 1, page 260. The hyperboloid of revolution generated by the given line ab a_1b_1 is cut by a plane which is tangential to the throat circle and par¹ to the axis. Determine the true shape of the section.

(7) More difficult question. Project a plan of the anchor ring used in Ex. 1 to 4, arranged with the section given by plane (2) in the H.P.

Hint. Project the mean diam. and draw spheres with centres on the mean diam. The envelope of the spheres gives the outline of the ring.

* In mathematics a tore is a surface or solid generated by the revolution of *any conic* about an axis in its plane.

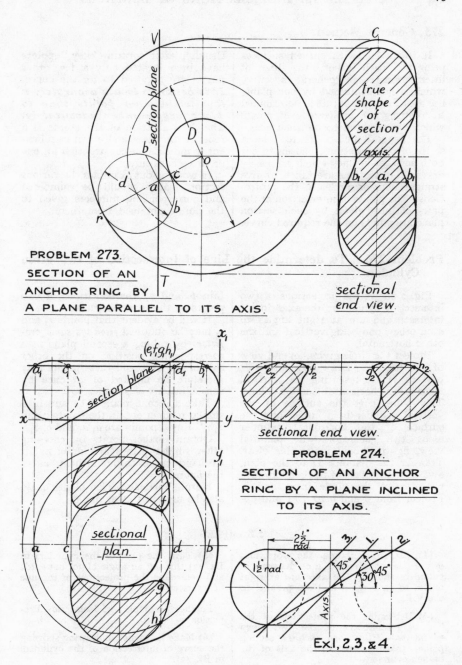

PROBLEM 273.
SECTION OF AN
ANCHOR RING BY
A PLANE PARALLEL TO ITS AXIS.

true shape of section

sectional end view.

section plane

sectional plan.

section plane

sectional end view.

PROBLEM 274.
SECTION OF AN ANCHOR
RING BY A PLANE INCLINED
TO ITS AXIS.

Ex.1, 2,3, & 4.

275. General Method.

If two solids with curved surfaces penetrate each other, the line of intersection is in general a curve which is not contained by one plane, i.e. a tortuous curve: it is determined by plotting the projections of points which are common to both surfaces.

The general method is to take a series of plane sections, chosen to cut both surfaces in lines which are easily drawn; i.e. sections are taken to give straight lines or circles in the projections. The points of intersection of the pairs of lines given by each section plane are points on the required curve.

Usually, after certain "key" points have been located only a few other points are required to fix the curve. *It is better to locate accurately a few well-chosen points than to plot a large number inaccurately.* The determination of the points is a repetition process, and it will be sufficient to indicate the construction for one or two points only.

In working out examples the various section planes should be numbered and corresponding numbers given to the points obtained from them.

Problem 276.   To determine the Line of Intersection of two given Cylinders.

Fig. 2 shows the projections of two intersecting cylinders; the axes do not intersect and are at right angles to each other, one being vertical and the other horizontal.

Project the circle giving an end view of the branch. Draw any line VT1 par[l] to *xy*, representing the vertical trace of a horizontal section plane. This plane cuts the surface of the vertical cylinder in a circle, and the surface of the horizontal cylinder along two generators—the pictorial view, fig. 1, should make this clear. Transfer the distance D to the plan, and draw the plans of the generators— to intersect the circle in p and q. Project from p and q to VT1 and ob-

tain p_1 and q_1, the elevations of two points on the required curve.

It will be evident that points r_1 and s_1 may be obtained *from the same projectors* by taking a second plane in a corresponding position on the other side of the axis, and these points may therefore be marked off at once by considerations of symmetry.

Take a second plane VT2 and in a similar manner obtain the points 2 and 2_1. "Key" points are a, a_1 and b, b_1.

Obtain further points by choosing other similar planes and draw a fair freehand curve through them, as in the figure.

Note. Vertical section planes could also be used in this example.

EXAMPLES

(1), (2), (3) Determine the projections of the lines of intersection of the pairs of cylinders shown. The axes are at right angles to each other and are parallel to the V.P.

(4) Determine the projections of the line of intersection of the two given cylinders. *Hint*: use *inclined* section planes, taken parallel to the axis of the branch cylinder.

(5) Turn the plan of the cylinders in Ex. (3) through an angle of 30° outwards and determine the projections of the line of intersection.

Further example in Isometric Projection:—

(6) Make an isometric drawing showing the curve of intersection of the cylinders in Ex. (2).

Note.—These curves of intersection, together with the developments of the surfaces (dealt with in Chapter 17) are of importance in Plate Metal work.

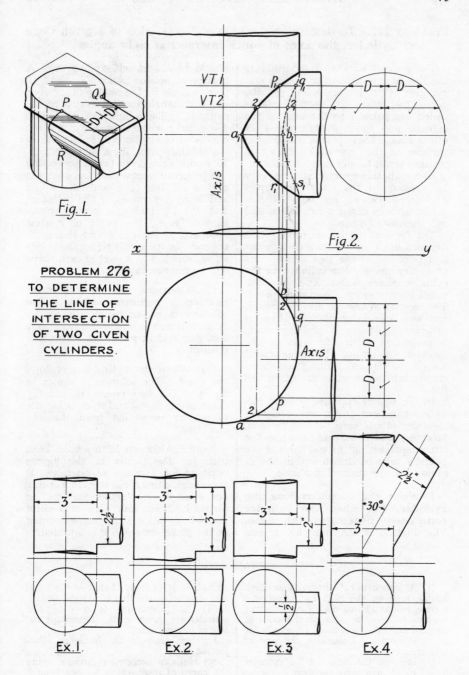

Fig. I.

VT I
VT 2

Axis

P_1 q_1
2 2
a_1 b_1
r_1 s_1

D — D

x ———————————— y

Fig. 2.

**PROBLEM 276.
TO DETERMINE
THE LINE OF
INTERSECTION
OF TWO GIVEN
CYLINDERS.**

b
2
q
Axis
D   D
2
P
a

Ex. 1.   Ex. 2.   Ex. 3.   Ex. 4.

3"   2½"
3"   3"
3"   2"
2½"   30°   3"

Problem 277. To determine the Line of Intersection of a given Cone and Cylinder, the Axes of which intersect at right angles.

The *general method* of determining the line of intersection of a cone and cylinder is to locate points on the curve by means of section planes which are parallel to the axis of the cylinder and which contain the apex of the cone. These planes will cut the surfaces of the cone and cylinder in straight lines, as will be seen later.

Although this general method may be applied to the simple examples considered below, *horizontal section planes* give an easier construction and will therefore be used.

The form of the line of intersection depends upon the relative dimensions and positions of the two solids. The cone may envelop the cylinder, or the cylinder envelop the cone, or both solids may envelop a common sphere. The resulting curves are dissimilar, and as a knowledge of their shape is of great assistance in plotting them, a projection along the axis of the cylinder, i.e. an end elevation, should always be drawn; this view shows clearly into which class the curve will fall.

In the figures opposite the end elevation has been drawn separately for clearness of reference. It will be seen, however, that in solving examples the circle representing the end view of the cylinder may be drawn within the $\triangle$ giving the elevation of the cone.

I. When the cone envelops the cylinder, i.e. when the cylinder completely penetrates the cone. The solution is shown in fig. 1, and

should be almost self-explanatory. A horizontal section plane VT gives, in plan, a circle for the section of the cone and a rectangle for the section of the cylinder. These intersect in points p, q, r, and s, which are on the plan of the curves of intersection; the elevations are obtained from them by projection. Key points are given (a) by projecting the points of intersection between the outline of the cone and the upper and lower generators of the cylinder; and (b) by taking a section plane through $m_2 n_2$, the point of nearest approach of the circle and the $\triangle$ in the end elevation, this giving the points m, m_1, and n, n_1. A part of each curve will be hidden, in plan.

II. When the cone and cylinder envelop a common sphere. The solution is shown partly in fig. 2. In elevation, the line of intersection is straight, and in plan (not shown), it is elliptical.

III. When the cylinder envelops the cone. The solution is shown in fig. 3. The key points are readily determined, m, m_1 and n, n_1 being obtained by projection from the end elevation.

Note. Although letters have been used for the points in the figures opposite for purposes of reference, it is preferable to use numbers in working out examples. Number the section planes in order, and give each point (or pair of points), the same number as the plane on which it is situated.

EXAMPLES

(1) A cylinder $1\frac{1}{2}''$ diam., axis horizontal, rests on the H.P. and penetrates a cone, height $2\frac{1}{2}''$, base $2\frac{1}{4}''$ diam., resting on the H.P. The axis of the cone is vertical and intersects that of the cylinder at right angles. Determine the line of intersection.

(2) Increase the diam. of the cylinder in (1) to $1\frac{3}{4}''$ and solve problem.

(3) Displace the cylinder in (1) until

its axis is $\frac{1}{4}''$ in front of that of the cone, and determine the line of intersection.

(4) The projections of a cone and cylinder are shown in fig. Determine the line of intersection.

Further example in Isometric Projection:—

(5) Make an isometric drawing showing the curve of intersection for one branch in Ex. (1).

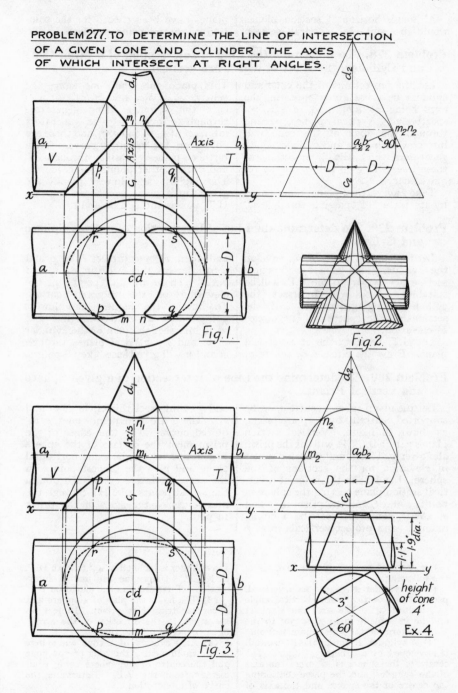

PROBLEM 277. TO DETERMINE THE LINE OF INTERSECTION OF A GIVEN CONE AND CYLINDER, THE AXES OF WHICH INTERSECT AT RIGHT ANGLES.

Fig. 1.

Fig. 2.

Fig. 3.

Ex. 4.

Although horizontal section planes would be suitable, vertical section planes have been chosen for the solution of the following problems.

Problem 278. To determine the Line of Intersection of a given Cone and Cylinder, the Axes of which are vertical.

Let the projections of the cone and cylinder be those given opposite, the axes being ab a_1b_1, and cd c_1d_1 respectively. Vertical planes passing through the apex of the cone will intersect both cone and cylinder along generators, and the intersection of these lines will give points on the required curve.

Take any vertical plane, represented by its trace HT, passing through ab.

This plane cuts the cone along the generators be b_1e_1, and bf b_1f_1; it also cuts the cylinder along generators through p and q. The generators intersect at p, p_1 and q, q_1, and these are points on the required curve. Take further planes passing through ab and obtain other points on the curve. The highest and lowest points are given by a plane so chosen that its H.T. passes through ab and cd.

Problem 279. To determine the Line of Intersection of a given Sphere and Cylinder.

Let the sphere, centre o, o_1, envelop the cylinder, axis cd c_1d_1. Vertical section planes par[l] to the V.P. will be suitable; these will intersect the cylinder along generators, and the sections of the sphere will be circles in elevation.

Let H.T. be the trace of a vertical plane. Draw the circle e_1f_1, centre o_1

and diam. ef, and project from p and q to intersect this circle in p_1, r_1 and q_1, s_1. These are four points on the elevation of the required curves. To obtain the highest and lowest points, draw a diam. through o and cd to intersect the plan of the cylinder in m and n; vertical planes through m and n will give these " key " points.

Problem 280. To determine the Line of Intersection of a given Sphere and vertical Prism.

Let the given sphere, centre o, o_1, be supported centrally by a square prism, as shown in figure. Vertical section planes par[l] to the V.P. will cut the prism along vertical lines, and will give circles in elevation for the sections of the sphere. Let HT be the trace of a vertical section plane cutting the sphere in points e and f, and the prism in p and q. With centre o_1 describe a circle, diam. ef, and project verticals from p

and q to intersect this circle in p_1 and q_1. These are two points on the required curve. The four edges of the prism meet the surface of the sphere in points lying in the same horizontal plane and give the highest points on the curve of intersection. To obtain the lowest points take HT through the mid-points of the sides of the prism in plan.

EXAMPLES

Use vertical section planes in each of the following:

(1) A cylinder 2" diam., axis vertical, penetrates a vertical cone, vertical angle 55°, altitude 4". The axes are $\frac{1}{2}$" apart and lie in a plane inclined at 30° to the V.P. Determine the line of intersection.

(2) A cylinder, 2" diam., axis horizontal, is enveloped by a sphere $3\frac{1}{2}$" diam. The centre of the sphere is ·6" from the axis of the cylinder, and the plane containing the centre of the sphere and the axis of

the cylinder is inclined at 45° to both H.P. and V.P. Determine the line of intersection. (First draw an end elevation.)

(3) The fig. is the plan of a sphere and a regular triangular prism. Project an elevation, in the direction of the arrow, of the curve of intersection.

(4) An elevation of a cone and sphere is shown in figure, the axis of the cone and the centre of the sphere being equidistant from the V.P. Determine the curve of intersection.

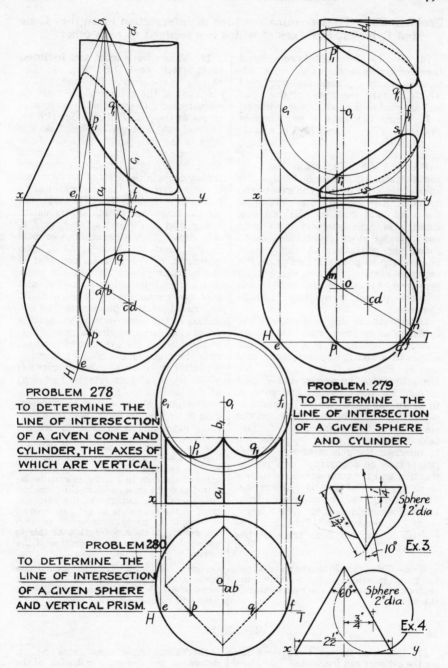

PROBLEM 278
TO DETERMINE THE
LINE OF INTERSECTION
OF A GIVEN CONE AND
CYLINDER, THE AXES OF
WHICH ARE VERTICAL.

PROBLEM. 279
TO DETERMINE THE
LINE OF INTERSECTION
OF A GIVEN SPHERE
AND CYLINDER.

PROBLEM 280
TO DETERMINE THE
LINE OF INTERSECTION
OF A GIVEN SPHERE
AND VERTICAL PRISM.

Sphere
2" dia

10° Ex. 3.

60° Sphere
2" dia.

Ex. 4.

Problem 281. To determine the Line of Intersection of a given Cone and Cylinder, the Axes of which are inclined to each other.

In the two examples considered here, one simple and one more advanced, section planes are employed which contain the apex of the cone and are par[l] to the axis of the cylinder.

I. When the axis of the cone is vertical. Figs. 1 and 2. Let ab a_1b_1 be the axis of the cone, and cd c_1d_1 that of thè cylinder, the latter being par[l] to the V.P. Determine the horizontal trace, ht, of a line through the apex of the cone par[l] to the axis of the cylinder. Any section plane containing the apex, and with its H.T. passing through ht, will intersect both cone and cylinder along generators; fig. 2 shows the generators in the section intersecting in points P and Q on the required curve. (*Note:* the second generator AH would be used if the cylinder passed completely through the cone.)

Determine the elliptical trace of the cylinder, as in Prob. 266. Draw any line HT1 through ht intersecting the ellipse in e and f, and the circle in g and h. Project e_0f_0 and g_0 and draw the generators through e_0 and f_0 par[l] to c_1d_1, for the cylinder, and from g_0 to a_1 for the cone. These intersect in p_1 and q_1, and give two points on the elevation of the required curve. Obtain their plans, p and q, by projecting to intersect the generator gab. Other points may be determined in a similar manner, and the complete curve drawn as in fig. 1. The plan is symmetrical about cd, and may be completed from the half which is plotted. The " key " point r, r_1 is given by HT2, taken tangential to the elliptical trace of the cylinder.

Note.—This example could also be solved by using spherical sections—taking the centre of the sphere at the intersection of the axes.

II. When both axes are inclined to the H.P. and V.P. Fig. 3. Let ab a_1b_1 be the axis of the cone, and cd c_1d_1 that of the cylinder: the axes do not intersect. Determine the horizontal traces of the cone and cylinder (Probs. 265 and 266), and also of a line from the apex of the cone par[l] to the axis of the cylinder. Section planes will intersect the cone and cylinder along generators if they contain the apex of the cone and if their horizontal traces contain ht.

Draw any line HT1 through ht, intersecting the elliptical traces in e, f, g, and h. Project e_0, f_0, g_0, and h_0 and draw three generators, two par[l] to c_1d_1, and one from g_0 to a_1—the generator a_1h_0 would be required if the cylinder passed through the cone. These lines intersect in p_1 and q_1, and give the elevations of two points on the curve. Other points may be obtained in the same way. The limiting point r_1 is given by the tangent, HT2, to the trace of the cylinder at j, which cuts the trace of the cone at k and l: a similar point on the dotted curve is given by a tangent line (not drawn) at the back of the plan.

The plan of the curve of intersection is omitted in fig. 3, and its determination is left as an excercise for the student.

Note 1.—Because the tangent line HT2 to the cylinder intersects the elliptical trace of the cone in k and l, it may be inferred that the cone envelops the cylinder; conversely, if a tangent to the elliptical trace of the cone were to cut the trace of the cylinder, the cylinder would envelop the cone. A clear conception of this is essential for the correct projection of the curve.

Note 2.—If the H.T. of the cylinder is inaccessible it may be more convenient to use the V.T. Prob. 282 on the following page will make the method clear.

EXAMPLES

(1) Determine the curve of intersection for the cylinder and frustum of cone given in figure. The axes intersect and the plane containing them is par[l] to V.P.

(2) Taking dimensions from fig. 3, determine the elevation *and plan* of the line of intersection of the given cone and cylinder.

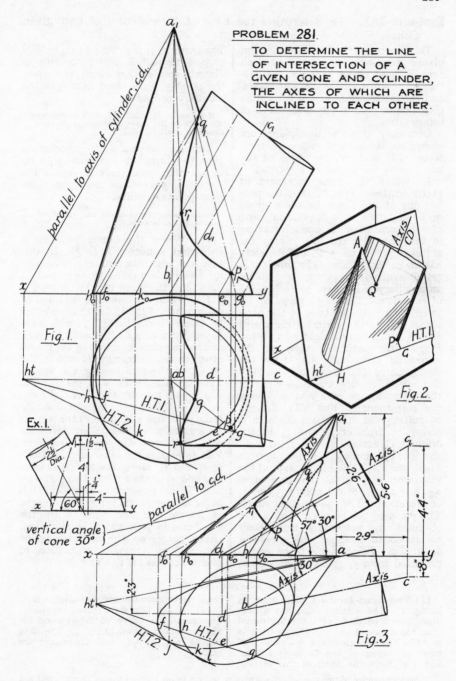

PROBLEM 281.

TO DETERMINE THE LINE
OF INTERSECTION OF A
GIVEN CONE AND CYLINDER,
THE AXES OF WHICH ARE
INCLINED TO EACH OTHER.

Fig. 1.

Fig. 2.

Fig. 3.

Ex. 1.

parallel to axis of cylinder, cd₁

vertical angle of cone 30°

parallel to c₁d₁

Problem 282. To determine the Line of Intersection of two given Cones.

The general method is to use section planes containing the apex of each cone. These planes will cut the surface of each cone along generators, and the intersections of the generators give points on the required curve of intersection.

Obviously, if all section planes are to contain the apex of each cone, their traces will contain the traces of the line joining the apices of the cones.

I. When the axes intersect at right angles. Fig. 1. Let the projections of the cones be those given in figure, axes ab a_1b_1 vertical, and cd c_1d_1 horizontal. A difficulty arises in this case, for the line joining the apices is par[1] to the V.P. and therefore has no vertical trace. The problem is readily solved, however, by taking an auxiliary V.P. perp. to each plane of reference. Take x_1y_1 perp. to xy, to coincide with the base of the horizontal cone, and determine the traces ht and vt, as shown clearly in figure. Project the auxiliary view of the horizontal cone only, i.e. draw the circle, centre c_2d_2. Draw any line HT1 through ht, cutting the base circle of the vertical cone in e and m; draw the corresponding line VT1 through vt, cutting the other base circle in f_2 and g_2. Transfer f_2 and g_2 to the original elevation, giving f_1 and g_1, and join c_1f_1 and c_1g_1. Project e_1 from e and join a_1e_1. The intersection of the generator a_1e_1 with the generators c_1g_1 and c_1f_1 gives the points 1_1 and 2_1 on the elevation of the required curve. The plans of the points, 1 and 2, are obtained by projecting from their elevations to intersect the generator ae, in plan. Points 5 and 6 are readily projected from 5_1 and 6_1 in elevation.

The point 4, 4_1 is given by taking VT2 to pass through k_2 and proceeding as above; this section plane also gives the point 3, 3_1, derived from l_2 in the auxiliary projection.

If the traces HT1 and VT1 do not meet within the limits of the paper, Problem 209, page 212, should be used.

The smaller curve of intersection may be plotted in the same way by using other generators in which the section planes cut the cones, e.g. those from points m and n.

Note.—This example could also be solved by using spherical sections.

II. When the cones envelop a common sphere. Fig. 2. In this case the elevations of the curves of intersection are given by two straight lines, and the plans by two ellipses.

III. When the axes are inclined to each other. Fig. 3. Let the two cones be those shown in fig. 3, with axes ab a_1b_1 and cd c_1d_1 (not intersecting), and having circular traces on the planes of reference. Determine ht and vt, the traces of the line joining the apices. Take any plane HT1, VT1, the traces of which pass through ht and vt and meet in xy. HT1 intersects one circular trace in m and n, and gives the section generators ma and na; their elevations m_1a_1 and n_1a_1 are easily projected. VT1 intersects the other circular trace in r_1 and s_1, and gives the generators r_1c_1 and s_1c_1, their plans being rc and sc. These generators intersect each other in four points lying on the curve of intersection and denoted by the number 1 in plan and 1_1 in elevation. Other points on the curves may be obtained by taking a succession of planes such as HT1, VT1.*

EXAMPLES

(1) Two cones intersect each other as in fig. 1. The vertical cone has a base diam. and altitude of $2\frac{3}{4}''$; the horizontal cone has a base diam. of $2\frac{1}{2}''$ and altitude of $2 \cdot 6''$. The axes intersect in a point $1 \cdot 1''$ from the apex of the horizontal cone, and $1 \cdot 4''$ from the apex of the vertical

cone. Determine the curve of intersection.

(2) Using the data given opposite, determine the curve of intersection for the two cones shown in fig. 3. The axes are ab a_1b_1, cd c_1d_1, and the traces are circular.

* *Note.*—Examples of this type are very confusing and the points must be plotted systematically and numbered consecutively if the correct solution is to be obtained. Extreme accuracy in drawing is essential

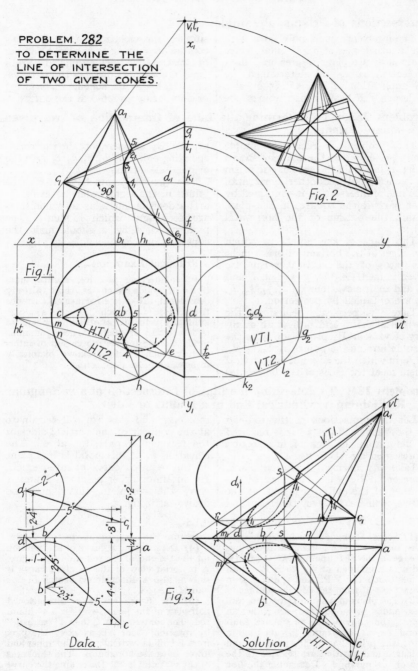

PROBLEM. 282
TO DETERMINE THE
LINE OF INTERSECTION
OF TWO GIVEN CONES.

Fig. 1.

Fig. 2.

Fig. 3.

Data.

Solution

Intersections of Prisms, Pyramids, etc.

The intersections of polyhedra with one another are straight lines, since their faces are plane figures, and their projections may be determined by the application of Probs. 235 or 239. Although the basic constructions are simple, the solutions may be very complicated owing to the large number of construction lines employed. Problems must therefore be treated systematically, and a scheme of lettering or numbering adopted at the outset.

Problem 283. To determine the Line of Intersection of two given square Prisms.

This problem is relatively simple when presented in the manner shown in fig. 1, in which the axis of one prism is vertical and the other horizontal. The line of intersection is contained by that part of the square, in the plan, within the outline of the horizontal prism.

The plans of the points in which the edges of the horizontal prism meet the faces of the vertical prism are given immediately by g, h, j, k, l, and m, and their elevations g_1, h_1, j_1, k_1, l_1, m_1 are obtained by projection.

Take x_1y_1 perp. to the axis of the horizontal prism and project an auxiliary elevation. By projection from the plan determine the points a_2, b_2, c_2, d_2, e_2, f_2, in which the edges of the vertical prism meet the faces of the horizontal prism. By transfer, obtain the corresponding points a_1, b_1, c_1, d_1, e_1, f_1 in the original elevation.

The various points have now to be joined in the right order, as in fig. 1, and a decision made as to which lines are seen and which hidden. The pictorial view, fig. 2, should make the construction clear.

Practical Examples.

In engineering practice, abrupt junctions of solids, such as those already considered, are of less frequent occurrence than those in which the solids merge gradually into one another by means of circular radii or fillets. These junctions produce some interesting curves of intersection, such as those shown pictorially in fig. 4 and Ex. 4 opposite.

Problem 284. To determine the Line of Intersection of a rectangular Rod joining a cylindrical End in a Radius or Fillet.

Let the projections of the rod-end be those given in fig. 3; it is required to determine the line of intersection represented by AB in fig. 4.

Take any horizontal section plane, represented by VT. The plan of the contour of the cylindrical end, at the radius, will be an arc, centre o and rad. o_1a_1, and this arc will terminate at the contour of the vertical sides of the rectangular part, i.e. at p. The elevation p_1 of this point is one point on the required curve of intersection. By obtaining other points in a similar way, the complete curve may be drawn, as in fig. 3.

EXAMPLES

(1) Solve Prob. 283 using the data given in fig. 2. The horizontal prism has an edge in the H.P. inclined at 30° to xy, and a face inclined at 60° to H.P. Both prisms touch the V.P., and the corner in contact is 1·3″ from the edge in contact.

(2) The plans of a square pyramid (full lines) and a prism (dotted) are shown in figure; the ends of the prism are equilateral $\triangle$s. The pyramid has its base in H.P.; the prism has a face in H.P., and its length and position are referred to the plan of the pyramid. Determine the line of intersection of the two solids.

(3) Using the dimensions given in fig. 3, determine the curve of intersection shown.

(4) Determine the curve of intersection for the tee-end of the rod shown in figure. A pictorial view of the required curve is shown, also a dimensioned sketch of one half of the rod.

(5) The figure shows a dimensioned half-plan of the forked-end for a 2″ diam. rod. The contour from C to D is " turned " by rotation about AB as axis; the part from E to F is vertical; and the upper and lower sides are horizontal. The thickness of the material is 2″. Determine the curve of intersection between the curved contour and the flat sides.

285

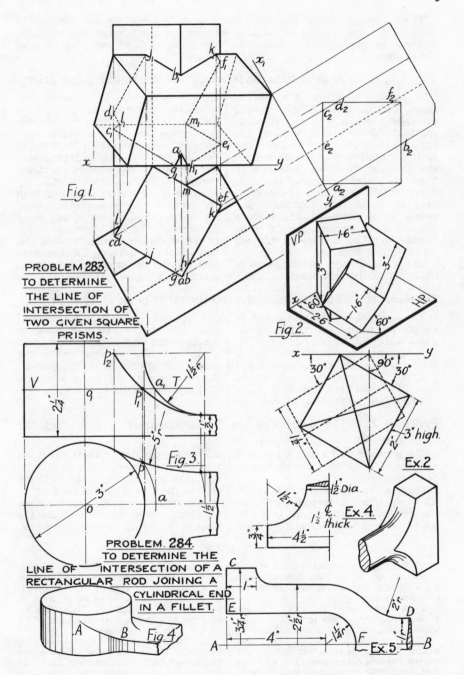

Fig 1.

PROBLEM 283.
TO DETERMINE
THE LINE OF
INTERSECTION OF
TWO GIVEN SQUARE
PRISMS.

Fig.2.

VP.
HP.

Ex.2

Fig.3.

PROBLEM. 284.
TO DETERMINE THE
LINE OF INTERSECTION OF A
RECTANGULAR ROD JOINING A
CYLINDRICAL END
IN A FILLET.

Fig.4

Ex.4

Ex.5.

When a surface can be laid out on a plane it is said to be developable and the figure obtained is called its development.

In dealing with the developments of prisms, pyramids, cones and cylinders, the ends or bases will not be considered.

Problem 285. To determine the Development of a right regular Prism.

A pentagonal prism is shown in figure with a circular hole passing centrally through it. The development of the surface consists of five similar rectangles in contact and their dimensions may be taken from the plan and elevation. One hole in the development is circular; the other (that containing an edge of the solid) may be obtained by plotting a few points on the curve—e.g. the distance $p_0 2$ in the development is equal to $p2$ in plan, and P on the curve is obtained by projection from p_1.

The various edges in the development should be numbered to corre-

spond with the edges in the projections.

The **shortest distance** between any two points on a surface is given by the line joining their positions in the development; they may usually be joined in two directions (or more) and that giving the shortest distance must be taken. Consider the two points r, r_1 and s, s_1. Transfer the distances $r4$ and $s1$ from the plan to the development, giving $r_0 4$ and $s_0 1$, and obtain R and S by projection. The line joining R and S is the shortest distance between the two points r, r_1 and s, s_1. The construction for the elevation of this line is given in the figure.

Problem 286. To determine the Development of an oblique Prism.

The figure shows the projections of an oblique prism with regular hexagonal ends, length of edge D. The development consists of six equal-sided par^{ms}. Number the corners and from each draw a line perp. to the axis. With dividers set to the distance D step off the points 1, 6, 5 . . . in the

development. From these points draw lines par[1] to the axis and equal in length to the slant edge, and complete as shown. The construction for the projections of a line giving the shortest distance between the points r, r_1 and s, s_1 is shown clearly in figure.

Problem 287. To determine the Development of a right regular Pyramid.

The projections of a square pyramid and the V.T. of an inclined section plane are given.

Produce the base line and mark off $EF = a1$. Erect a perp. EA equal to the altitude of the pyramid. Join AF; AF is the true length of a slant edge.

To obtain the development draw the four similar isosceles $\triangle$s A14, A43, A32, A21, each base being equal to the edge of the base of the pyramid.

The method of projecting the lines in which the section plane cuts the faces is clearly indicated in the figure.

EXAMPLES

(1) Draw the development of an oblique prism on a regular pentagonal base. Edge of base $1\frac{1}{2}''$, length of axis $2''$, inclination of axis $45°$.

(2) A cube $2''$ edge has a circular hole $1''$ diam. drilled centrally through opposite edges. Develop the complete surface.

(3) A workshop consists of three similar " bays " as shown in figure. An electric cable is taken, along the wall and roof, from A to B and then from B to C. Determine its shortest length and draw

the plan of its path. Scale: $1'' = 20$ feet.

(4) A flue is in the form of the frustum of a pyramid, the plan of which is given. A lightning conductor is taken from A to B around the outside. Determine its shortest length and draw the plan of its path. Scale: $1'' = 6$ feet.

(5) Taking dimensions from Prob. 287, determine the development of the pyramidal surface beneath the section plane V.T.

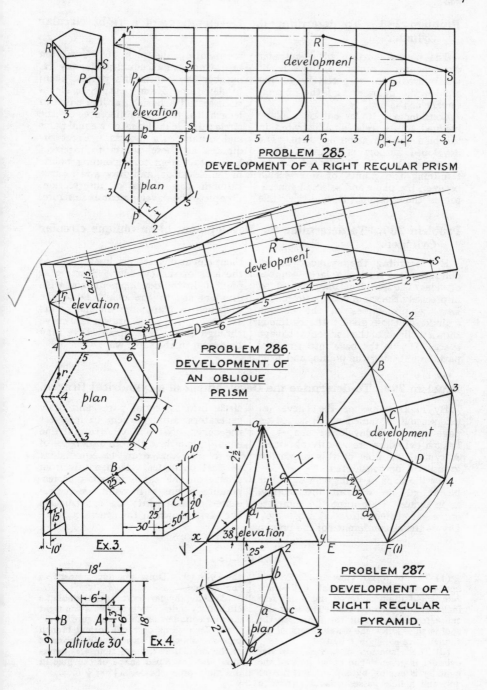

PROBLEM 285.
DEVELOPMENT OF A RIGHT REGULAR PRISM

PROBLEM 286.
DEVELOPMENT OF AN OBLIQUE PRISM

Ex.3.

Ex.4.
altitude 30'

PROBLEM 287.
DEVELOPMENT OF A RIGHT REGULAR PYRAMID.

Problem 288. To determine the Development of a right circular Cylinder.

The curved surface of a cylinder develops into a rectangle, the sides of which are equal respectively to the axis of the cylinder and the circumference of the end.

Consider a cylinder cut by a plane inclined to its axis, as illustrated in the plan and elevation opposite. The developed surface is bounded by a curve which may be plotted by determining the points of intersection between the plane and selected generators of the cylinder. Divide the circle

representing the plan of the section into a number of equal parts, say 12, number the points and project them in elevation. From point 1, in elevation, set off a horizontal line equal in length to the circumference of the circle; divide it also into 12 equal parts and number the points. Erect ordinates at these points to intersect horizontals from corresponding points in the elevation, and draw a fair curve through the points of intersection. Complete the development as in figure.

Problem 289. To determine the Development of an oblique circular Cylinder.

By regarding the cylinder as an oblique prism having a large number of sides, the problem may be solved in precisely the same way as Prob. 286. Let the projections of the oblique cylinder be those given: the xy line is shown inclined to the horizontal to save space. Divide the base into 12 equal parts, numbering the points, and from

their elevations set off lines perp. to the axis of the cylinder. Starting at point 1, in the development, and with dividers set to the chord D of the divisions in plan, step off the points 12, 11, 10, 9 . . . and draw a fair curve through them. The upper curve may be drawn in the same way.

Problem 290. To determine the Development of a cylindrical Branch.

By making use of the curves of intersection, obtained as described on page 274, the developed shapes of branch cylinders are readily obtained. A typical example of this important group will be considered.

Let the axes of two given cylinders be at rt. ∠s without intersecting, as in figure. Determine the complete curve of intersection (Prob. 276). Divide the circumference of the branch

circle into 12 equal parts and draw generators through them to intersect the curve. Set out a horizontal line equal in length to the circumference of the branch and draw 12 equidistant vertical ordinates. Horizontals from points on the curve intersect corresponding ordinates in points on the required development: the construction is clearly shown in the drawing.

EXAMPLES

(1) The pictorial view in Prob. 288 opposite shows a vertical cylindrical funnel, 24″ diam., intersecting a flat roof inclined at 45° to its axis. The least distance from the top of the funnel to the roof is 18″. Draw the development of the funnel. Scale: $\frac{3}{4}$″ = 1 foot.

(2) The elevation of a right circular cylinder is given. The upper part of the cylinder is cut off by a plane inclined at 30°, and a hole passes centrally through

the cylinder. Draw the development of the surface.

(3) Draw the development of the oblique circular cylinder which connects the right circular cylinders in the given figure.

(4) Two views of a cylindrical branch pipe are shown. Draw the development of the branch at the junction, and determine the developed shape of the hole in the main pipe. Scale: $1\frac{1}{2}$″ = $\frac{1}{2}$ foot.

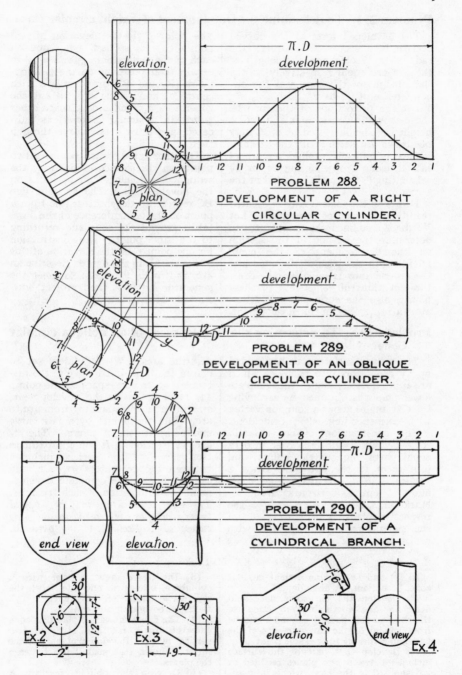

elevation.

development.

$\pi.D$

PROBLEM 288.

DEVELOPMENT OF A RIGHT
CIRCULAR CYLINDER.

plan

elevation

development.

PROBLEM 289.

DEVELOPMENT OF AN OBLIQUE
CIRCULAR CYLINDER.

plan

development.

$\pi.D$

end view

elevation.

PROBLEM 290.

DEVELOPMENT OF A
CYLINDRICAL BRANCH.

Ex.2.

Ex.3.

Ex.4.

elevation

end view

Problem 291. To determine the Development of a right circular Cone.

The developed form of the curved surface is a sector of a circle; the rad. of the sector and the length of the arc are equal respectively to the slant height of the cone and the circumference of its base.

Let the projections of the cone be those shown in figure, apex a, a_1, slant height L. The development a_1BC may be drawn by setting off the circumference graphically around an arc of rad. L, using Probs. 28 and 27, or by calculating θ, the angle included in the sector.

Points on the surface may be located readily by means of generators. Let VT be an inclined section plane: to determine the outline of the section on the development of the cone. Divide both the base of the cone and the arc BC into 12 equal parts; draw the elevations of generators to these points, also the corresponding generators $a_1 1$, $a_1 2$, &c., in the development.

The point p_1 is the elevation of two points on the section, lying one ·on each of the generators $a_1 5$ and $a_1 9$; to obtain their positions on the development (P$_1$ and P$_2$) draw an arc with centre a_1 and radius $a_1 p_2$ to cut the generators $a_1 5$ and $a_1 9$. Locate other points in a similar manner, as indicated, and draw a fair curve through them.

The shortest distance between two points on the surface is given by the straight line joining their positions on the development. The straight line BC represents the shortest line from a point on the circumference of the base, taken around the cone and returning to the same point. The construction for one point r_1 on the elevation of this line is shown in the figure: describe an arc, rad. $a_1 R_1$ (or $a_1 R_2$) to meet the generator $a_1 1$ in r_2, and project horizontally to intersect $a_1 4$.

Problem 292. To determine the Development of an oblique circular Cone.

The curved surface may be divided up by generators into triangles which are approximately plane; the complete development is obtained by arranging these triangles about a common vertex with corresponding edges coincident.

The construction is conveniently set out thus: Divide the base into 12 equal parts, number the points, and join them to the apex a. Draw a vertical line $a_2 a_3$ = altitude of cone, and set out a horizontal from a_3. Mark off $a_3 1$, $a_3 12$, $a_3 11$ . . . = the respective plans $a 1$, $a 12$, $a 11$ . . . of the generators. Join $a_2 1$ and with centre a_2 and radii $a_2 1$, $a_2 12$, $a_2 11$ . . .

describe arcs. With dividers set to one of the base divisions, e.g. the distance 1—2, in plan, space off the points 12, 11, 10 . . . on the development, starting at 1 and spanning from arc to arc. Join the points by a fair curve and complete as in figure. The $\triangle$ $a_2 1$, 12 (sectioned) is the approximate true shape of the surface included between the generators $a 1$, $a_1 1$, and $a 12$, $a_1 12$, and the complete development is a summation of such triangles. *The same construction may be applied to any oblique cone, whatever the base curve; also, of course, to any pyramid.*

EXAMPLES

In (1) and (2) assume a right circular cone, base diam. 3″, altitude $3\frac{1}{2}$″.

(1) Develop the curved surface and determine the length and projections of the shortest line passing around the cone from a point on the circumference of the base and returning to the same point.

(2) Develop that part of the surface included between two planes inclined at 10° and 50° to the base, which intersect in a tangent to the base circle.

(3) The figure shows a cone with a cylindrical branch on one side only; the shortest generator of the branch is $1\frac{1}{2}$″. Develop the branch and cone.

(4) The figure shows the plan of a cone and the traces HT and H$_1$T$_1$ of two vertical planes. Develop that part of the surface, shown sectioned, lying between the planes.

(5) Develop the oblique frustum A shown in figure.

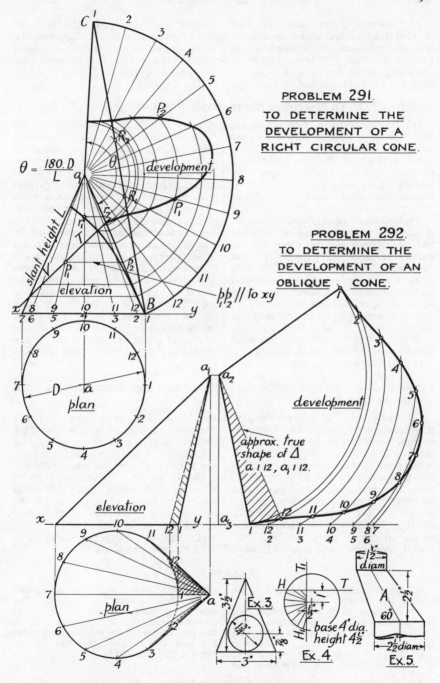

PROBLEM 291.
TO DETERMINE THE
DEVELOPMENT OF A
RIGHT CIRCULAR CONE.

$\theta = \dfrac{180.D}{L}$

development

slant height L.

elevation

bb_{12} // to xy

plan

PROBLEM 292.
TO DETERMINE THE
DEVELOPMENT OF AN
OBLIQUE CONE.

development

approx. true
shape of Δ
a 1 12, a_1 1 12.

elevation

plan

Ex. 3.

$3\frac{1}{2}$"

$\frac{7}{8}$"

3"

$1\frac{3}{4}$"

Ex. 4.

base 4" dia.
height $4\frac{1}{2}$"

Ex. 5.

$\frac{1}{2}$"
diam

A

60°

$2\frac{1}{2}$"

$2\frac{1}{2}$ diam.

Any surface of revolution may be divided into either zones or lunes * and an approximate development of these obtained. The more numerous the parts chosen the closer will be the degree of approximation which the sum of the developed surfaces bears to the given surface. In the constructions given below, relatively large parts of the surfaces will be dealt with to avoid a confusion of lines.

Problem 293. To determine the approximate Development of a Sphere.

(a) In Zones. Fig. 1 shows a part elevation of a sphere with its upper hemispherical surface divided into three zones, A, B, and C, the distance between the points 1-2, 2-3, 3-4, being equal. By joining the points, zones A and B become the projections of frusta of right circular cones; and the spherical cap C becomes a cone of small altitude. A surface approximating to that of the hemisphere will be given by the curved surfaces of these frusta and the cone, developed as in Prob. 291. In fig. 1 only one half of each developed surface is shown to save space; the strips A_1 and B_1, and the sector C_1 give the development of one half the surfaces of A, B, and C, and consequently represent one-fourth the total development of the whole sphere.

(b) In Lunes. The elevation of a lune representing one-twelfth of the total surface of the sphere is shown (shaded) in fig. 2; its end elevation is given by the thickened semi-circle. Divide this semi-circle into six equal parts, number them, and project points a and c in elevation. With centre o draw the arcs ab and cd. Rectify the semi-circle, using Prob. 26, mark off its length along a centre line CL, and divide the line into six equal parts. Erect ordinates at the points of division and transfer to them the distances ab, cd, and ef from the elevation, giving a_1b_1, c_1d_1, and e_1f_1, &c. The figure given by joining these points represents an approximate development of the selected lune, and twelve of these figures will enable the spherical surface to be built up.

Problem 294. To determine the approximate Development of a Tore or Anchor Ring.

(a) In Lunes. Fig. 3 shows the part projections of a given ring. Consider a lune contained between planes inclined to each other at 30°, as shown in the plan. Divide the circular section into twelve equal parts, drop projectors from the points into the plan, and describe tangent concentric arcs with centre o. Set off a centre line CL, equal in length to the circumference of the circular section, divide it into twelve equal parts and erect ordinates, numbering them as in figure. Transfer distances such as ab to the respective ordinates, giving points b_1 and b_2 on the outline of the curve, and complete the development in the manner shown.

(b) In Zones. The surface may be divided into zones and the parts treated as frusta of cones, as for the sphere. The part elevations of two such frusta are shown dotted in fig. 3, and their development should need no further explanation.

EXAMPLES

(1) Develop, in zones, one-fourth of the surface of a sphere 3″ diam.

(2) The mean diam. of an anchor ring is 5″ and the diam. of the section is $1\frac{3}{4}$″. Develop, in lunes, one-twelfth of its surface.

(3) Three semi-circular hoops are used to form the frame of a lampshade, the plan of which is represented by six equally spaced radii to a 12″ diam. circle. The shade is covered with silk, all horizontal sections being regular hexagons. Draw the projections of the shade, one-quarter full size, and develop, half size, one of the six curved surfaces.

* A Zone is the surface included between two planes perp. to the axis; a Lune that included between two planes which contain the axis.

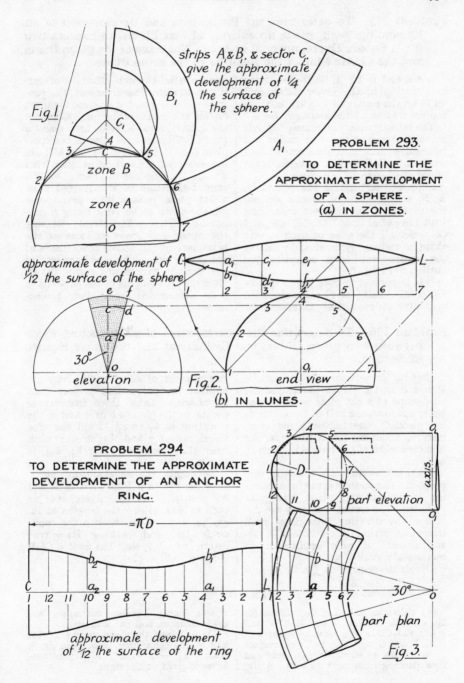

strips A_1 & B_1 & sector C_1 give the approximate development of ¼ the surface of the sphere.

Fig. 1

C_1

B_1

A_1

4
3   c 5
zone B
2   6
zone A
1   7
0

PROBLEM 293.

TO DETERMINE THE APPROXIMATE DEVELOPMENT OF A SPHERE.
(a) IN ZONES.

approximate development of
1/12 the surface of the sphere

C L
a_1 c_1 e_1
b_1 d_1 f_1
1   2   3   4   5   6   7

e f
c d
a b
30°   o
elevation   Fig. 2   end view

3   4
2   5
1   6
7

q_1

(b) IN LUNES.

PROBLEM 294
TO DETERMINE THE APPROXIMATE DEVELOPMENT OF AN ANCHOR RING.

3   4   5
2   6
1   D   7
12   11   10   9   8   axis
part elevation
o_1

$= \pi D$

b_2 b_1
C a_2 a_1 L
1 12 11 10 9 8 7 6 5 4 3 2 1   2 3 4 5 6 7
a
30°   o

approximate development
of 1/12 the surface of the ring

part plan

Fig. 3.

Problem 295. To determine the Projections and Development of an Expanding Bend, made up entirely of Flat Plates, to connect two given Square Ducts, unequal in Area. 'The Axes of the given Ducts and the Centre Line of the Bend lie in the same Plane.

Let A and B, fig. 1, be the positions in elevation of the given ducts, and let C be the centre of the arc forming the centre-line of the required bend.

The rate of reduction in area depends on the purpose of the duct and it will be assumed here that the areas at four sections D, E, F, and G, on equidistant radial lines from C. are given.

Suppose the bend at the points D, E, F, and G to be of square section, the planes of the squares coinciding with the radial lines CD, CE, CF, and CG. Set off the edge of each square along a radius, symmetrically about the centre-line; then complete the outline as in fig. 1.

It is important to note that **a quadrilateral cannot be plane unless its opposite sides either intersect or** are parallel to each other. Further, if two straight lines intersect, the projections of the point of intersection must lie on the same projector. Consider the quadrilateral MNPO, figs. 1 and 2 (fig. 2 is an isometric view of the complete bend): the edges OP and MN are not par[1] and, as will be evident from a plan, they do not intersect. MNPO cannot therefore be constructed from a flat plate (unless it is permissible for the plate to be bent along a diagonal). The quadrilaterals forming the upper and lower surfaces of the bend satisfy the above condition and are each plane. The developments of the various figures are easily drawn after the true lengths of the edges have been determined from their projections.

Problem 296. To determine the Development of a Connecting Piece between two given Ducts, one Cylindrical and the other Square in Section.

Let the projections of the ducts be P and R in fig. 3. To determine the true shape of a flat plate which, when bent, will form one half of the connecting piece Q, the halves being symmetrical about the centre line *afde*. An isometric view of the halves is given in fig. 4.

There are several ways of solving this problem. The method given here makes use of conical and plane triangular surfaces. The conical surfaces are developed by dividing them into small triangular strips, a method of " triangulation " of wide application in metal plate work.

Divide the semi-circle *fge*, fig. 3, into any number of equal parts—here 12; number the points and project their elevations. Draw lines from these points, in the plan to *b* or *c*, and in the elevation to b_1 or c_1. Find the true length of $b7$, b_{17}, and of each other such line; the construction for $b4$, b_{14}, is shown.

To obtain the development, fig. 5, draw the rt. $\angle$d $\triangle$ FAB; AB = *ab* and AF = $a_1 f_1$. Set off the succession of $\triangle$s such as B12, taking the lengths of 12, 23, 34, etc., as one-twelfth the semicircle *fge*, and making B2 = true length of $b2$, b_{12}, etc. The final $\triangle$ CDE is again right-angled.

EXAMPLES

(1) Take the following dimensions for fig. 1: sides of ducts A and B, 1' 6" and 3' 0"; rad. CB, 4' 0"; sides of squares at D, E, F, and G, 2·6', 2·2', 1·9', 1·65' respectively. Draw an elevation, plan, and end view (looking from right to left). Scale: 1" = 1 foot. Develop the upper and lower surfaces, and one side surface.

(2) Obtain the development of one half of the connecting piece Q in fig. 3. Scale: $1\frac{1}{2}$" = 1 foot. A dimensioned figure is given on the right.

295

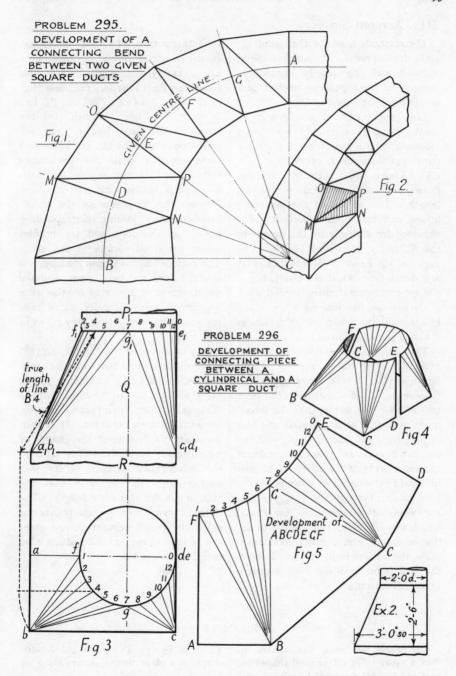

PROBLEM 295.
DEVELOPMENT OF A
CONNECTING BEND
BETWEEN TWO GIVEN
SQUARE DUCTS.

Fig. 1

Fig 2.

PROBLEM 296

DEVELOPMENT OF
CONNECTING PIECE
BETWEEN A
CYLINDRICAL AND A
SQUARE DUCT

true length
of line
B 4

Fig 3

Fig 4

Development of
ABCDECF

Fig 5

Ex. 2.
2'-0"d.
3'-0"so

211a. Aerofoil Surfaces.

The methods used for the definition and development of aerofoil wing surfaces will be briefly considered because of their general interest and possible use for other surfaces.

The proportions of many aerofoil shapes have been standardized, and values from one example are given on page 342 and used opposite. The *Chord Line* is sub-divided by ordinates drawn at percentage distances along its length. The values of the ordinates above and below the chord line are obtained by multiplying the figures in the Table by the *Chord Length*. The curve at the *Leading Edge* is a circular arc struck from a defined point, fig. 3, the arc extending slightly beyond what is defined as the leading edge. The line through the mid-points of ordinates is called the *Mean Line*.

The numerals describing standard aerofoils define the shapes. The first figure (2) in NACA.23021 gives the *Camber* as 2% of the chord; the next two figures (30) imply that the maximum camber is at an ordinate one half of 30% from the leading edge; and the last two figures (21) give the maximum thickness, 21% of the chord. A st. line inclined to the chord at an angle, fig. 1, here passing through the 35% point, is the *Wing Reference Line* or *Plane*, which is parallel to the normal direction of flight. Fig. 2 shows, to a larger scale, the outline in fig. 1 turned about the 35% point until the wing reference plane is horizontal.

Construction of Wing Sections.

In the example shown, incomplete, in figs. 2 and 3, the length of the *Root Chord* has been taken as 120" and that of the *Tip Chord* as 40". The 35% line is common to both. Multiplying the values in the Table by 120" and 40" gives the lengths of the ordinates and these, set off to scale, give the shapes shown. It is customary to use ordinates perp. to the wing reference line, through the % points on the chord. Straight lines joining corresponding points on the root and tip profiles define the ruled surface of the wing. The surface shaded is bounded by two straight lines and two curves, and the development of any area on the wing presents no great difficulty, as we have the positions of any boundary point in two views.

The incidence of the wing tip section may differ from that of the root section, as in fig. 3 where the angles are 3° and 5°. Again, the aerofoil at the tip usually differs from that at the root, for aerodynamical reasons. It is then necessary to determine the shapes of intermediate aerofoils. This requires the sub-division of each of the percentage lines in the ratio that the station divides the wing length. The dotted curve drawn through points one third along each percentage line gives the aerofoil shape one third of the wing length from the tip.

EXAMPLE

The Table on page 342 relates to NACA.23021. Set off aerofoil shapes for root and tip, taking chord lengths of 140" and 45", and angles of incidence of 5° and 3°, as in fig. 3. Determine the aerofoil shape on a plane three quarters along tip to root.

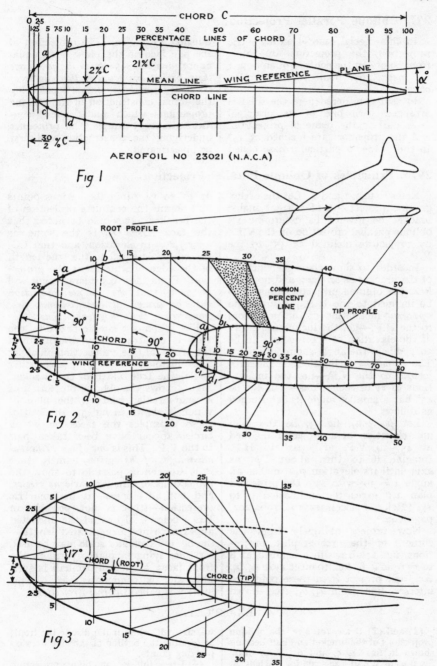

CHORD C

0 25 | 1·25 | 5 | 75 | 10 | 15 | 20 | 25 | 30 | 35 | 40 | 50 | 60 | 70 | 80 | 90 | 95 | 100

PERCENTAGE LINES OF CHORD

21% C

2% C

MEAN LINE   WING REFERENCE   PLANE

CHORD LINE

$\alpha°$

$\frac{30}{2}$ % C

AEROFOIL NO 23021 (N.A.C.A)

Fig 1

ROOT PROFILE

90°

CHORD

90°

WING REFERENCE

COMMON PER CENT LINE

TIP PROFILE

90°

Fig 2

17°

CHORD (ROOT)

3°

CHORD (TIP)

Fig 3

297. Oblique Parallel Projection.

In this special case of axonometric projection, the projectors are taken from points on the contour of an object to meet the plane of projection *obliquely*.

In that considered here the solid is arranged with one of its principal faces par[l] to the plane of projection, and the projectors are inclined at 45° to this plane; a particular case called

Cavalier Projection. The projection of the par[l] face on the plane is an equal figure, *and lines perp. to the face have their projections equal to the lengths of the lines themselves.* Before considering the actual construction of the oblique projection, which under these conditions is very simple, the principles underlying the construction will first be investigated.

297a. Principles of Oblique Parallel Projection.

Refer to figs. 1 and 2. Let the orthographic projections of a right-angled bracket be given: to determine its oblique parallel projection on the V.P., by projectors inclined at 45° to the V.P.

In order to show the upper surface of the bracket as an *area*, and not as a *line*, the oblique projectors must also be inclined to the H.P. Assume the *apparent* inclination of the projectors to the H.P.—i.e. the inclination to xy of the elevation of a projector—to be 30°; this is the angle commonly used. The *apparent* inclination of the projectors to the V.P.—i.e. the inclinations to xy of their plans—will not be 45° but a smaller angle β, determined as follows.

Let pa p_1a_1, fig. 3, be the plan and elevation of a projector inclined at 45° to V.P. and par[l] to H.P. Imagine it to turn about Pp_1 as axis until its elevation p_1a_3 makes an angle $\alpha = 30°$ with xy; the projected plan a_2p gives the inclination β to xy. Refer if necessary to Prob. 202, page 202.

Now project obliquely from the corners of the orthographic projections, fig. 1; draw lines inclined at β to xy from a, b, &c., to meet xy in a_0, b_0, &c., and project from these points to intersect lines from a_1, b_1, &c., drawn

at 30° to xy. Join the various points and obtain the complete projection of the bracket. It should be noted that the face ABEFDC is the same as $a_1b_1e_1f_1d_1c_1$ in elevation, and that CG, the width, is equal to the true length of the edge of which it is the projection. From these considerations it will be seen that *the oblique projection may be drawn at once without first constructing the orthographic views*.

Fig. 4 shows a simple bracket: the view is constructed by drawing first the principal face, and setting off along lines inclined at 30° (or any other angle) the true width of the bracket. Fig. 6 shows the oblique projection of a semi-circular bearing, the plan of which is shown in fig. 5. In both of these examples the faces on which circles appear have been taken par[l] to the V.P. This is *one of the principal advantages of the system*, namely, that it is frequently possible to show the projections of circular parts as *circles*, and not as ellipses—as in isometric projection—(this is not possible in Ex. 3 below). To avoid the distorted appearance and exaggerated length— see fig. 6—a smaller scale may be used for lines lying in planes perp. to the front face: a scale of one-half is suitable and is commonly used. This system is called *Cabinet Projection.*

EXAMPLES

(1) and (2) Draw, full size, the oblique projections of the bracket and half bearing shown in figs. 4, 5, and 6. Also draw fig. 6 using a half-size scale for the length.

(3) Two views of a bracket are given. Draw an oblique projection of the bracket,

full size, with the circular disc to the front. Determine the outline of the side arcs by plotting points.

(4) Draw, full size, an oblique projection of the overhung crank and disc shown.

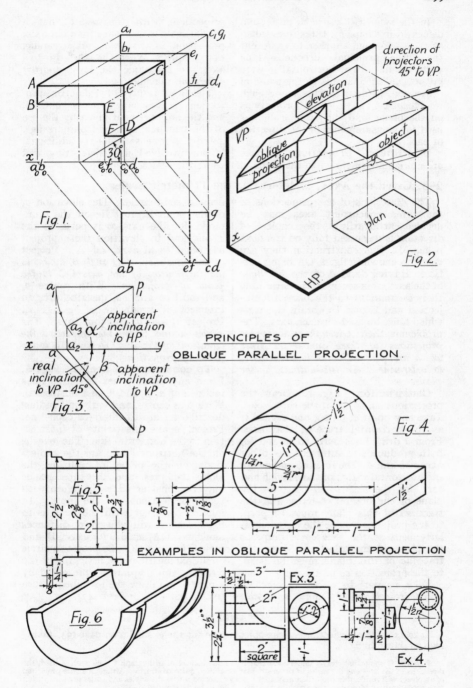

Fig 1.

Fig. 2

direction of projectors 45° to V.P.

V.P.

elevation

oblique projection

object

plan

HP

apparent inclination to H.P.

apparent inclination to V.P.

real inclination to V.P - 45°

Fig. 3.

PRINCIPLES OF
OBLIQUE PARALLEL PROJECTION.

Fig. 4

Fig 5

EXAMPLES IN OBLIQUE PARALLEL PROJECTION

Fig 6

Ex.3.

2 square

Ex. 4.

In the *isometric system* of projection discussed in Chap. 13, three lines taken at rt. ∠s to tone another to represent the three principal directions (one vertical and two horizontal), were given *equal* inclinations to the plane of projection. Equal lengths along each line, or axis, gave equal projections on the plane, and the same scale was used for measurements in the direction of each axis.

In **trimetric projection** the axes are not equally inclined to the plane of projection, with the result that a special scale is necessary for each axis: projection is, however, orthographic, not oblique.* When the scales for two axes are the same the term *dimetric projection* may be used.

The appearance of isometric projections leaves much to be desired, and the improvement given by the use of the trimetric system (compare figs. 4 and 5) compensates for the additional labour involved in constructing and using the scales.

298. Given the Axes, to determine the Trimetric Scales.

The directions of the projections of the three concurrent axes may be decided arbitrarily; the choice of direction is discussed fully on the next page. In this construction they are taken as shown in fig. 2, OA being vertical. Having decided on the *directions* of the axes, it is necessary to determine their *inclinations* to the plane of projection and thence to obtain the *scale* which must be used for each axis. The problem is, then: *Given the orthographic projections of three concurrent lines which are perpendicular to each other, to determine their inclinations to the plane.*

Construction. Fig. 1. Draw the projections *oa, ob, oc* of the three axes. Take any point *a*, on *oa*, and regard it as the horizontal trace of that axis. From *a* draw lines perp. to *co* and *bo*, both produced, to intersect the other axes in *b* and *c*. The lines *ab* and *ac* are the horizontal traces of the △s *oab* and *oac*: for the axis *oc* is perp. to the plane of the other two axes, and the trace *ab* of this plane must be perp. to the projection *oc* of the axis; similarly because the axis *ob* is perp. to the plane of the other two axes, the trace *ac* of this plane must be perp. to the projection *ob* of the axis.

Take *xy* par¹ to *co* and project b_1 and c_1 from *b* and *c*. On b_1c_1 as diam.,

draw a semi-circle. The elevation o_1 of the point *o* must lie on the semi-circle, for the axis *oc* is perp. to the △ *oab*, and in elevation these project as the sides of a rt. ∠d △. Project o_1 and join o_1c_1. The angle θ, $o_1c_1b_1$, is *the inclination of the axis OC to the plane of projection*. With centre *o*, and radii *ob* and *oa*, describe arcs to intersect *oc* produced in b_2 and a_2. Project a_3 and b_3 and join to o_1. *The angles α and β are the respective inclinations of the axes OA and OB to the plane of projection.*

To construct the Scales. Fig. 3. From any point *p* set off three lines inclined at α, β, and θ, to a horizontal. With *p* as centre, and with unit radius, describe an arc to intersect each line. Project *p* and the points of intersection to the horizontal line. The lengths on the horizontal line are the projections, on the plane, of unit lengths along the axes, and the three scales may be constructed as in figure with these projections as the unit. In making drawings with the axes as in fig. 2, scale A will be used for distances along axis OA, scale B for axis OB, and scale C for axis OC. To project a circle on an axometric plane first project the circumscribing square—represented by a par^m—and draw the inscribed ellipse to this par^m (as in Prob. 47).

EXAMPLES

Make axometric projections of the objects used for the examples on page 193, taking axes as shown in fig. 2.

* *Note.*—It should be clearly understood that if oblique instead of orthographic projection is used, the directions of the axes *and* the ratios of the scales may be chosen quite arbitrarily. Although a true projection of an object will result, this projection may appear very distorted unless the angles and scales are kept within reasonable limits. Moreover, the absolute scale will have to be determined.

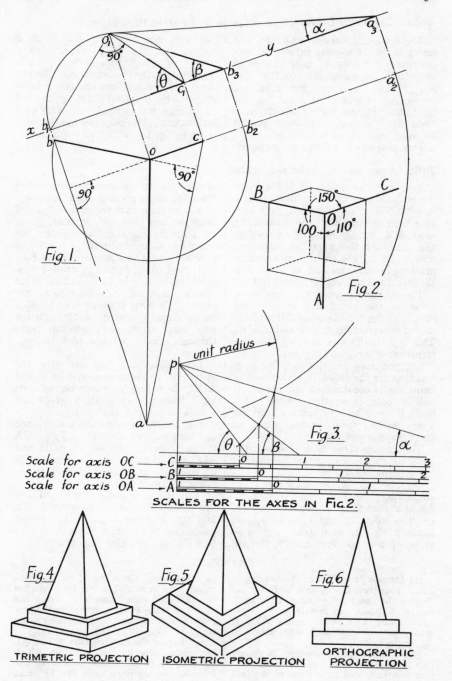

Fig. 1.

Fig. 2.

Fig. 3.

unit radius

Scale for axis OC
Scale for axis OB
Scale for axis OA

SCALES FOR THE AXES IN Fig. 2.

Fig. 4

Fig. 5

Fig. 6

TRIMETRIC PROJECTION

ISOMETRIC PROJECTION

ORTHOGRAPHIC
PROJECTION

298a. Choice of Direction of Axes in Trimetric Projection.

The choice of axes is arbitrary and offers scope for a great variety of projections. Yet only a limited range of projections is acceptable to the eye. Fig. 1 opposite shows five trimetric projections of a unit cube, and marked on them are the values which have been used for the apparent angles between concurrent edges.

The projection in fig. 1a approxi-mates to the isometric (in which a diagonal of the solid projects as a point). The difference is barely sufficient to justify the variation. In the other four projections the effect is to lengthen the diagonal (dd in fig. 1b) and to change its inclination. The projection in fig. 2 on page 301 lies between those in fig. 1c and fig. 1d opposite.

298b. Axes to suit selected scales.

The true inclinations of the axes to the plane of projection can in each case be obtained as described on page 282. It will be obvious that the greater the angle BOC, fig. 1e, the more nearly does the scale for the axis OA approach unity. In (e), the error introduced by marking actual, instead of scale, distances off along OA is very small indeed.

The foregoing leads to a consideration of the possibility of selecting simple scales for the axes OB and OC and of determining $\angle$AOB and $\angle$AOC. This is virtually a reversal of the construction of fig. 1 on page 301.

Let the scale for the axis OC be $\frac{1}{2}$, and that for the axis OB, $\frac{7}{8}$, these scales being easily constructed and used. As will be seen later, the selection of a scale for one axis imposes a limiting value on the scale for the other (in orthographic projection).

The construction is as follows, the lettering of fig. 2 opposite being that of fig. 1, page 301.

On a horizontal line, draw any semi-circle $b_1o_1c_1$. In this semi-circle set off c_1o_1 at angle θ to c_1b_1 such that $\cos\theta = \frac{1}{2}$; i.e. $\theta = 60°$. Through o_1 draw o_1b_3 at angle β to c_1b_1 such that $\cos\beta = \frac{7}{8}$; although $\beta = 29°$, very nearly, the use of the ratio in the construction rather than the angle gives greater accuracy. It will be clear that the sum of θ and β must be less than 90°, so that, given the scale for one axis, the scale for the other has an upper limit.

Draw any line b_2oc parallel to b_1c_1, and through b_1, o_1 and c_1 draw perps. intersecting b_2oc in b_2, o, and c. With centre o and rad. o_2b_3 describe an arc cutting the perp. through b_1 in b. Join bo, produce it, and draw through c a line perp. to it to meet the perp. through b in a point a (not shown). Join ao.

The angles aob and aoc give the values of the required angles AOB and AOC which are, very nearly, 94° and 103° (accurate calculation gives the values as 94° and 102° 37').

Fig. 3 shows a unit cube constructed to these values; OB : OC :: $\frac{7}{8}$: $\frac{1}{2}$; OA is taken as 1; the angles AOB and AOC are 94° and 103°. The appearance of the cube is to be preferred to an isometric projection.

The trimetric projection of solids of more complicated shapes is not difficult, and fig. 4 shows a partly completed projection of a regular dodecahedron—(see Ex. 3). See also page 304.

EXAMPLES

(1) Draw a projection of the bearing of Ex. 6, page 197, taking axes as in fig. 1c. Repeat the construction using axes as in fig. 1e. Express your preference.

(2) Using the axes and scales of fig. 3, draw a projection of the bracket shown in fig. 6, page 197.

(3) A regular dodecahedron is shown partly projected in fig. 4. The easy construction indicated makes use of the fact that the pentagonal faces may be regarded as tilted about the edges of an inscribed cube; e.g. the shaded face in fig. 4 touches the inscribed cube along ab. Using the axes and scales of fig. 3, complete the projection of a regular dodecahedron having edges $1\frac{1}{2}''$ long.

(4) Repeat the construction of Ex. (3), but using irregular pentagons in which the lengths of the edges parallel to the faces of the inscribed cube are $1\frac{1}{2}''$, and of the other edges 1·146.

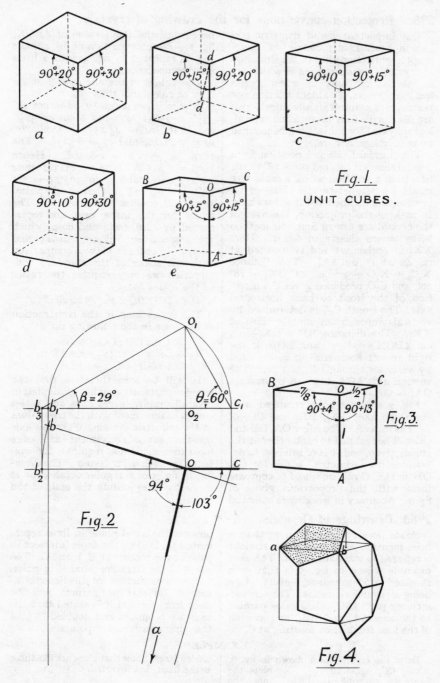

Fig. 1.

UNIT CUBES.

Fig. 2

Fig. 3.

Fig. 4.

298c. Projection conventions for the drawing of crystals.

One important use of trimetric projection (called clinometric or clinographic projection) lies in the drawings of crystals, the structures of which are of interest not only to mineralogists but to engineers. The identification and portrayal of nature's architectural forms are dealt with fully in works on crystallography and the subject can be touched upon only lightly here.

The standard selected position for a crystal shows it rotated some 18° to the left, from a position where a main face would be parallel to the plane of reference, and inclined about 9° forwards. In making the projection, the axes of the crystal are drawn first, the method being shown clearly in fig. 1. Here XX_1 is horizontal and is bisected at rt. ∠s at O by AA_1. The distance $X_1C_2 = X_1O \div 3$, i.e. $\angle C_2OX_1 = 18°\ 26'$, and C_2O produced gives the direction of the front to back horizontal axis. The length C_1C is determined by verticals through the middle third of XX_1. The distance $XB = OX \div 27$, i.e. $\angle BOX = 2°\ 7\frac{1}{2}'$, and BOB_1 is the right to left horizontal axis, bounded by verticals through X and X_1. The vertical axis AOA_1 is given by making $OA_1 = OA = OC_2$.

Fig. 2 shows a cube arranged with the three axes AA_1, BB_1 and CC_1 of fig. 1 as its axes. The edges OA, OB and OC will be inclined to each other at the angles given, and it is of interest to determine the 'trimetric' scales for OA, OB and OC (Prob. 298) and to compare these with the proportions given in fig. 1. Accuracy in working is essential

because of the small value of ∠BOX, fig. 1. The construction of fig. 1, page 301 is repeated in fig. 4 with a little more compactness.

The values of the angles θ, β and α may be calculated thus:
If oc is taken as unity, $od = \cos 20°\ 33\frac{1}{2}' = ·93631$; $oa = od \div \cos 87°\ 52\frac{1}{2}'$, so that $oa = 25·2513$; similarly, $oe = 7·98446$, and $ob = 8·5276$. The perp. $oo_1 = \sqrt{oe} = 2·82566$. Hence $\tanθ = 2·82566$, $\tanβ = ·33136$, $\tanα = ·1119$, so that $θ = 70°\ 30\frac{1}{2}'$, $β = 18°\ 20'$, $α = 6°\ 23\frac{1}{2}'$. Careful graphical work will confirm these values. The scales for the three axes are represented by $\cosθ$, $\cosβ$, and $\cosα$, which are respectively ·33367, ·94924 and ·99378. In order to make a comparison with the scales of the standard construction we may express the ratios of the scales as:
$$OA : OB : OC :: 2·9783 : 2·8448 : 1.$$

The scales used in the construction of fig. 1 are in the following ratio:
$$OA : OB : OC :: \sqrt{10} :$$
$$(3 \div \cos 2°7\frac{1}{2}') : (1 \div \cos 18°26').$$
i.e. $OA : OB : OC :: 3 : 2·848 : 1.$

It will be seen therefore that the simple construction in fig. 1, adopted in crystallography, gives axes and scales closely in agreement with the true values in the trimetric system. Where an axis construction is convenient and edge measurement is not required, the system has much in its favour. The drawing in fig. 3 of a regular octahedron is given by lines joining the ends of the axes.

298d. Drawings of Crystals.

Space does not permit more than a brief mention of this, and the student is referred to standard works. The easy example shown in fig. 5 is that of a rhombic dodecahedron, every face being a regular rhombus. The circumscribing par^m DEFG has sides parallel to the axes AA_1 and BB_1, and as each of the four front faces meeting at C_1 in-

tersects the axial plane in lines represented by DEFG, the faces intersect in pairs at the corners D, E, F and G. The edges of the faces are parallel in pairs.

The construction of the icositetrahedron (crystal of garnet) will be clear from fig. 6 if it is stated that the axes are produced and doubled to give the outer construction points.

EXAMPLE

Draw the crystal form shown in fig. 6, taking CC_1 equal to $1\frac{1}{2}''$. Complete the figure by showing, in dotted lines, the hidden faces. Show that these are like those at the front, but inverted.

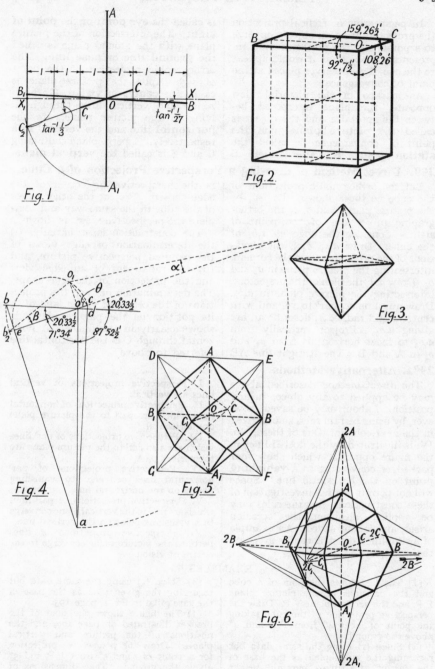

Fig. 1.

Fig. 2.

Fig. 3.

Fig. 4.

Fig. 5.

Fig. 6.

In perspective or conical projection the projectors from an object converge to a point, and the projected view represents the object as it would appear to the eye of an observer placed at the point of convergence.

As marked in the pictorial view opposite, the plane interposed between the spectator and the object is called the **picture plane**, and the point of convergence is called the **station point, S**; alternatively S is called the **eye point** or the **point of sight**. The intersection of the picture plane with the ground plane is called the **ground line** or **base line**. The orthographic projection of the point of sight upon the picture plane is called the **centre of vision, C**; horizontal and vertical lines through C, lying in the picture plane, are the **horizontal line** and the **vertical line** respectively. A perp. plane containing C and S is called the **vertical plane**.

299. Direct Method of drawing a Perspective Projection of a Cube.

Let the orthographic projections of the cube be those shown. Let oo_1 be the picture plane, and s, s_1 the orthographic projections of the point of sight. Consider the edge ab a_1b_1 of the cube. Draw a_1s_1, b_1s_1, the elevations of the projectors from the corners, intersecting the picture plane in a_4 and b_4. Draw abs, the plan of the projectors, intersecting the picture plane in a_2b_2. Draw any line oc_2 par^1 to xy, and with centre o and rad. oa_2b_2 describe an arc giving a_3b_3. Project vertically from a_3b_3 to meet horizontals from a_4 and b_4 in A and B. The straight line AB is the perspective projection of the edge chosen. Project the other edges of the cube in the same way and complete the perspective view, as shown.

The construction is, essentially, (i) the determination of edge views of the required perspective picture, and (ii) the rabatment of an edge view and the projection of the true shape. The determination of c, c_1, the projections of the centre of vision, and of C, its position on the picture plane, is shown clearly in the figure. A horizontal through C is the horizontal line referred to above.

299a. Alternative Methods.

The direct method described above may be applied to any object in any position. Labour may be saved, however, by using certain rules summarized in the next column. One of these, No. IV, is illustrated by the dotted lines in the figure opposite, which show four par^1 edges converging to a **vanishing point** on the horizontal line. Space will not permit of a full investigation of these constructions, but the rules may be applied whenever the occasion arises, and labour saved. An example of a simplified construction is given on the next page.

I. Perspective projections of vertical lines are vertical.

II. Perspective projections of horizontal lines which are par^1 to the picture plane are horizontal.

III. Perspective projections of par^1 lines which are also par^1 to the picture plane are themselves par^1.

IV. Perspective projections of par^1 horizontal lines converge to vanishing points on the horizontal line.

V. Perspective projections of par^1 lines which are par^1 to the vertical plane converge to a vanishing point on the vertical line.

VI. Perspective projections of lines perp. to the picture plane converge to the centre of vision.

EXAMPLES

(1) The fig. shows the plan of a cube and the positions of the picture plane P.P. and the vertical plane V.P. Draw the perspective projection of the cube taking the point of sight 4" from P.P and 4" above the ground.

(2) Solve (1) using the same data but regarding the given plan as the base of the frustum of a square pyramid, height $3\frac{1}{2}$", upper edge $1\frac{1}{2}$", surmounted by a pyramidal top, 1" high.

(3) Solve (1) using the same data but regarding the given plan as the base of the gate pillar in Ex. 1, page 195.

(4) The figure shows the plan of the bracket illustrated on page 169, and the positions of the picture and vertical planes. Draw the perspective projection for a point of sight 5" from P.P and 3" above the ground. Take dimensions of the bracket from Ex. 1, page 169.

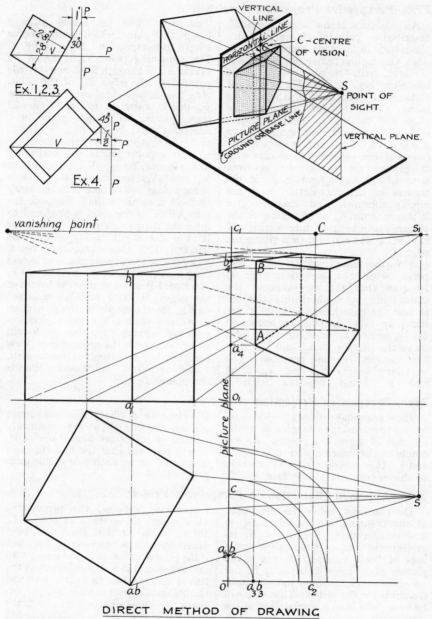

Ex. 1, 2, 3.

Ex. 4.

VERTICAL
LINE

HORIZONTAL LINE

C – CENTRE
OF VISION.

PICTURE PLANE

GROUND OR BASE LINE

S
POINT OF
SIGHT.

VERTICAL PLANE.

vanishing point

picture plane

DIRECT METHOD OF DRAWING
A PERSPECTIVE PROJECTION.

299b. Perspective Projection simplified.

An inspection of the "direct method" described on page 306 shows that simplifications in the construction are possible. One results from a rearrangement of views, with the plan, fig. 2, above the elevation, fig. 4. These show a rectangular block ABCD, arranged as in fig. 1 with respect to the picture plane and the station point S, projections s, s_1. By drawing rays from s to a, b, c and d in fig. 2, and projecting the points of intersection of these rays with the picture plane to meet the rays from s_1 to $a_1b_1c_1d_1$ and $e_1f_1g_1h_1$ in fig. 4, the corners of the perspective view are readily obtained. Only one projector is shown, from b_0, in the trace of the picture plane in fig. 2, intersecting s_1b_1 in b, fig. 4, and s_1f_1 in f, thus giving the edge bf.

A further simplification is to dispense with the elevation, fig. 4. Imagine that the elevation of the station point s_1, the horizontal through it, and the base line, all in fig. 4, are lifted up into the position shown in fig. 3, where s_1 and the horizontal lie in the picture plane, and the base line assumes its new position. The vanishing points V_{ab} and V_{bc} must now be found together with a

"measure" line, i.e. a line lying in the picture plane, which shows its true length in perspective. Refer to fig. 1. The vanishing point V_{bc} is found by taking a ray through S (shown dotted) parallel to the edge BC of the solid. *Any line through the observer's eye taken parallel to a line, or system of parallel lines, pierces the picture plane in a vanishing point for this line, or system*; and for a system of horizontal lines, the vanishing point lies in the horizontal line. Hence, fig. 3, sV_{bc} is parallel to bc, and sV_{ab} parallel to ab. Before these points can be used it is necessary to have a measure line. Suppose the face ABFE of the solid is produced to intersect the picture plane. It will cut it in a line which will give the true length of BF. Hence project ab, fig. 2, to intersect the picture plane in m and drop a perp. to meet the base line in F. Mark off FB equal to the true length of the edge FB. If F and B are joined to V_{ab}, the intercepts on the projectors give the points a, b, e and f. The remaining corners can now be found, using V_{bc}, and the perspective view completed. This combination of plan, vanishing points, and measure lines is commonly used.

299c. Perspective Projection of curves.

These are usually drawn freehand, referring to enclosing rectangles, or by the location of a few key points. An example of the latter is shown in figs. 2 and 3. It is supposed that a circle is to be shown centrally on the face ABFE.

By using an auxiliary semi-circle, and the measure line BF, points are quickly located on projectors from the edge ab. It will be noted that the half-chord pq is marked off on each side of the mid-point of BF.

299d. Apparent Distortion in Perspective Projection.

The plan and, below, the elevation, of two columns, are shown in fig. 5. The station point S is given by its projections s, s_1. Perspective drawings of these columns on the given picture plane show a disparity in size between them, the one more distant from S being the larger. That this will be so is clear from a comparison of the lengths of the intercepts A and B. Similarly, if the circles in the plan

represented spheres, the perspective view would show these as spheroids. In fig. 5 the station point has been taken over-close to the columns; but if the perspective drawing is viewed with one eye placed at S, the obliquity of the line of vision serves as a corrective and the columns appear in their true proportions. To avoid exaggerated effects the position of S should be central and not nearer than about the width of the view.

EXAMPLE

Refer to Ex. 1, page 194. Draw perspective views of the pillar from a station point 5 ft. high, 6 ft. in front, (a) centrally placed, (b) displaced 4 ft. to the left.

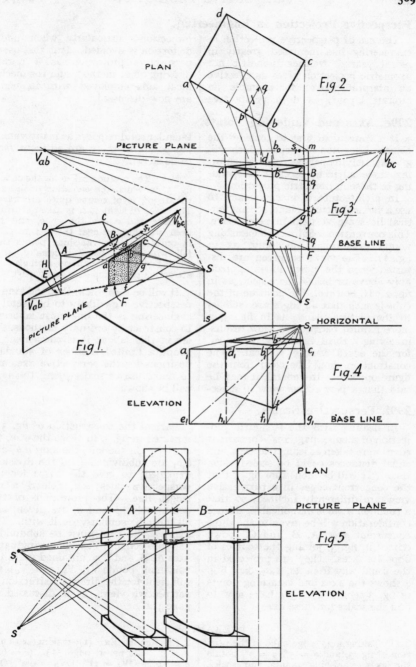

Perspective Projection in Engineering.

The use of perspective projection in engineering has increased greatly in recent years. Neither isometric nor axometric projection gives as effective an interpretation of, for example, an aircraft layout, as does perspective projection—particularly when undue distortion is avoided. It is this use of perspective projection as a normal drawing office method, and the mechanical aids employed with it, which are now discussed.

299e. Axes and Vanishing Points.

It is convenient first to consider the choice of position of a cube which will give a satisfying perspective view showing three adjacent faces, corresponding to those in isometric projection.

In architecture it is customary to use a vertical picture plane, so that vertical lines are vertical in perspective. This convention requires two vanishing points (sometimes one only, as in fig. 1). The views so given are distorted when the viewpoint is appreciably above or below the object, as in fig. 2. If, however, the position of the cube is such that all edges are inclined to the picture plane, as in fig. 3, a more natural perspective view results, involving a third vanishing point, V_A, for the edges usually vertical. The construction should be clear from the figure opposite. In general, it may be said that a perspective view of a rectangular solid requires as many vanishing points as there are adjacent faces to be shown.

Note. The position taken for the cube in fig. 3, in which the elevation of fig. 2 is tilted 30°, is of course quite arbitrary.* The point of sight s, s_1 is taken centrally opposite and above the cube, and has been brought over-near the picture plane in order that the vanishing points shall fall within the page; normally, s, s_1 would be taken farther away, giving a more attractive projection—see fig. 1, page 313.

It will be seen, then, that the kind of perspective view likely to be useful in engineering is the one most laborious to construct by ordinary methods. To lessen this labour it is convenient to adopt a limited number of standard positions for the perspective axes and to make use of scales along them, as will be shown.

299f. Perspective Scales.

In dealing with the isometric projection of a cube, page 194, concurrent edges were taken as isometric axes, and equal distances could be scaled along them. It will be seen from fig. 3 that the concurrent edges in a perspective view are differently inclined, so that a common scale is inapplicable. Hence consideration will be given to the three concurrent *axes*, A, B, and C, shown dotted in fig. 3, joining the centres of opposite faces. They are projected in the usual way from aa, a_1a_1, &c. Fig. 4 shows the axes and vanishing points of fig. 3 separately. We have now to find the scales for these axes.

Part of the construction of fig. 3 is repeated in fig. 5 to twice the size, for clarity. On the left, the axes a_1a_1 and b_1b_1 are subdivided (and the divisions extended) to give the bases for the perspective scales; a_1a_1 is clearly a true length line. The projections of two points only, p and q, are given, and all must be similarly dealt with. The axis c_1c_1 likewise must be subdivided and projected. The three graduated axes A, B, and C so obtained (with the vanishing points V_A, V_B and V_O) are sufficient for the direct construction of perspective views, as is discussed on page 312.

EXAMPLES

(1) Taking a $2\frac{1}{2}''$ edge cube in the position of fig. 3, and $x = 5''$, $y = 5''$, obtain the axes, vanishing points and scales. Check with figs. 4 and 5.

(2) Repeat Ex. (1) making $x = 7\frac{1}{2}''$, $y = 4''$. See fig. 1, page 313. ($a = 61°$; $\beta = 50°$; $OV_A = 11''$, $OV_B = 9.2''$, $OV_O = 12''$.)

* Instead of tilting the cube 30° the picture plane could be inclined 30°, giving the same result.

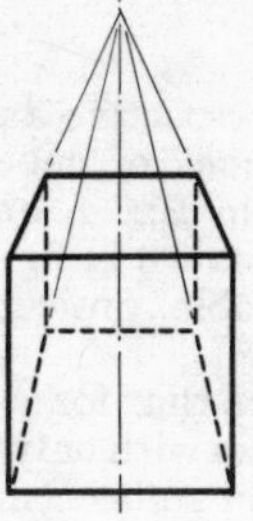

Fig. 1.

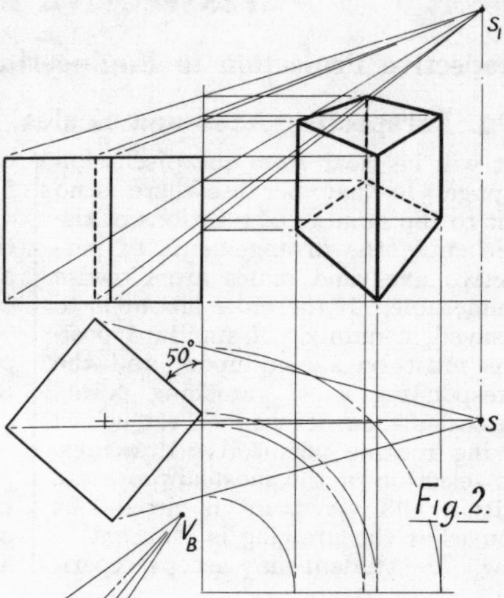

Fig. 2.

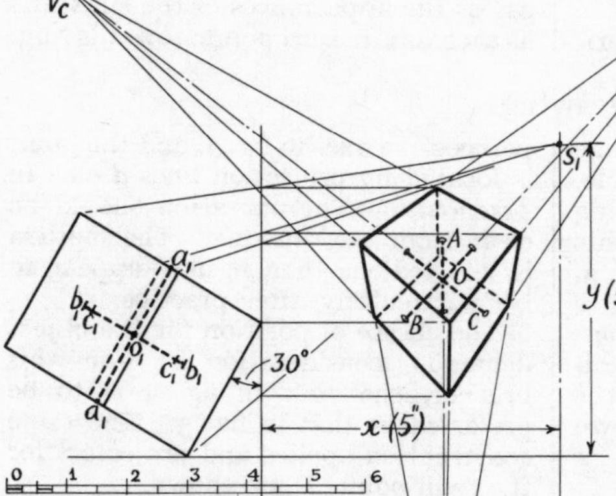

50°

s

V_B

V_C

s_1

A

O

B

C

a_1

$b_1 c_1$

O_1

$c_1 b_1$

a_1

30°

$y(5")$

$x(5")$

0 1 2 3 4 5 6
SCALE : INCHES.

Fig. 3.

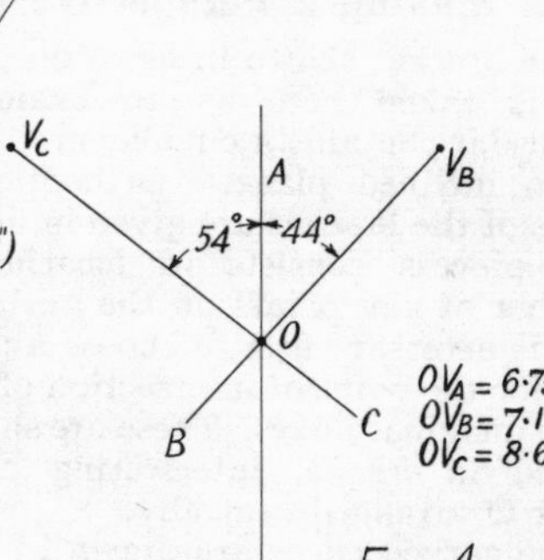

b_1

c

a

O

d

c

b

s

V_A

V_C

A

V_B

54° 44°

O

C

B

V_A

$OV_A = 6.75"$
$OV_B = 7.1"$
$OV_C = 8.65"$

Fig. 4.

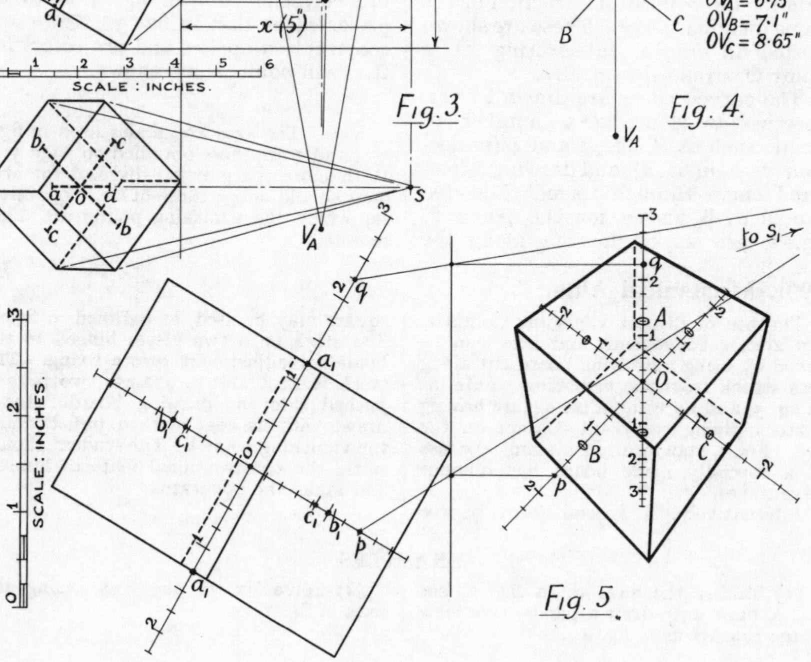

3

2

q

to s_1

q

3

2

A

1

O

$b_1 c_1$

O

$c_1 b_1$

p

a_1

B

C

p

3

SCALE : INCHES.
0 1 2 3

Fig. 5.

Perspective Projection in Engineering (Contd.)

299g. Perspective Axes and Scales.

It will be clear from the discussion on page 310 that, because there is no limit to the number of positions of the basic cube, the arrangements of perspective axes and scales are likewise innumerable. If therefore labour is to be saved, a number of standard positions must be agreed upon, and the corresponding axes, vanishing points and scales then determined for use in making routine perspective drawings. The selection of the most appropriate position and viewpoint is dependent upon what the drawing is required to show. The student may accept experi-

mentally the perspective axes resulting from the positions of cube and point of sight given in Ex. 2, on page 310; these axes are shown in fig. 1, and give a very acceptable projection of the cube.

It may well be that for some articles, perspective views with only two vanishing points would suffice; and for these one of the scales would be uniform.

There is no need, of course, for any particular axis to be the vertical one; the student should turn fig. 1 and assess the appearances of the six views as each axis in turn points towards him.

299h. Making a Perspective Drawing.

The bracket shown in fig. 6 on page 197 is taken here as an example, because it contains a circular arc lying in an inclined plane. Orthographic views of the bracket are given in fig. 2. The process consists in locating a number of key points on the outline.

It is necessary first to choose a position for the point of intersection of the three imaginary axes. These are shown dotted in fig. 2, intersecting at a point O arranged centrally.

The curved edges are drawn in perspective by locating a number of points such as P, figs. 3 and 4 (projections p, p_1 in fig. 2), and drawing a freehand curve through them. The distances α, β and γ for the point P, fig. 2, are set off *to scale* along the

perspective axes in fig. 3, and the point is located by projection lines *drawn in perspective*. The operation should be clear from the drawing. The method is less tedious than it may appear to be, particularly after practice.

The choice of position for the object demands consideration. For this bracket, the view in fig. 4 is to be preferred to that in fig. 3. The same construction applies and projectors for the twin points P are shown.

Note. The axes and scales used in figs. 3, 4 and 5 are those obtained on page 311. With more space at his disposal the student should adopt those of fig. 1 opposite, for which the vanishing points are more remote.

299i. Mechanical Aids.

The use of distant vanishing points is not always convenient, and time can be saved by using a drawing board cut along arcs struck from the vanishing points, as in fig. 5, and by using a tee square having a stock giving two-point contact on the arc. For a standardized system, the use of a specially made board has obvious advantages.

Alternatively, a special form of tee

square may be used, as outlined in fig. 6. The stock is in two pieces hinged to the blade for adjustment before fixing. The stock is kept always against two pegs P pushed into the drawing board. Lines drawn with the edge marked, pass through the vanishing point V. The student should settle the constructional features himself and make the apparatus.

EXAMPLES

(1) Taking the axes as in fig. 1 (see Ex. 2, page 310), draw a perspective view of the bracket as in fig. 4.

(2) Solve Ex. 6, page 198, using the axes of fig. 1.

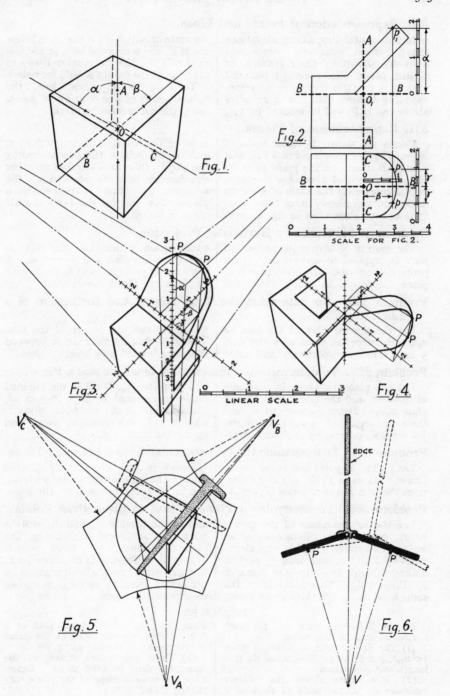

Fig.1.

Fig.2.

SCALE FOR FIG.2.

0   1   2   3   4

Fig.3.

LINEAR SCALE

0   1   2   3

Fig.4.

EDGE

Fig.5.

Fig.6.

300. Representation of Points and Lines.

In Horizontal Projection, elevations are dispensed with and points and lines are shown by their *indexed*, or *figured*, *plans*. Fig. 1 shows a pictorial view of a triangle and fig. 2 its corresponding figured plan; B is 12 units above the H.P. and is denoted by b_{+12}

(or more usually b_{12}); A is 4 units below the H.P. and is denoted by a_{-4}; and C is in the H.P. The conventional way of marking these points should be noted.

To describe a curved line, the figured plans of a succession of points along the curve must be given.

301. Representation of Planes.

Planes are shown by their *scales of slope*. Fig. 3 shows a plane VTH, and a line AB drawn on the plane perp. to H.T. The figured plan of a line such as AB represents the scale of greatest slope and completely fixes the plane. It is always shown, as in fig. 4, with

another thicker line drawn par[1] to it and on the left of the line ascending the slope; this thick line has no other significance than to show that the adjacent par[1] line *ss* represents a plane. The unit for the vertical scale must be stated.

Solution of Problems by Horizontal Projection.

The system of horizontal projection may be applied to solve most of the problems on the straight line and plane, although in many cases the

construction ultimately amounts to a solution by plan and elevation. A few typical problems will be dealt with to illustrate the methods used.

Problem 302. To determine the True Length and Inclination of a Line.

Let the figured plan of the line be a_5b_2. Erect perps. to ab at a and b of 5 and 2 units respectively and join:

this gives the true length of the line. The angle θ is the inclination between the given and " true length " lines.

Problem 303. To determine the Intersection of a Line and a Plane.

Let the plane be given by its scale of slope *ss*, and the line by its figured plan $a_{16}b_3$. Draw xy par[1] to *ss* and erect ordinates of 5 and 10 units. Join these points by VT and draw

TH perp. to xy. VTH is the inclined plane represented by *ss*. Project cd, the elevation of ab: the point of intersection e is the required point and its figured plan is $p_{8.5}$.

Problem 304. To determine the Perpendicular from a Point to a Plane.

Let a_{16} be the point and *ss* the plane. Draw VTH as in Prob. 303 and determine the position of c. Draw cf perp. to

VT and project its plan aq (dotted lines). The figured plan of the foot is $q_{5.5}$, the index being taken from the scale.

Problem 305. To determine a Plane to contain three given Points.

Let the figured plans of the points be a_{17}, b_3, c_{10}. Regarding $a_{17}b_3$ as an xy line determine c and e, the elevations of a and b, and draw fg par[1] to ab and distant 10 units—the index of c. Project d_{10}. This point is on the same level as c_{10}, the third given point,

and a line joining c_{10} and d_{10} will be horizontal and will also lie in the plane containing the three points. The scale of slope *ss* is therefore perp. to $d_{10}c_{10}$ and the scale itself is given by projecting from a_{17} and c_{10}, as shown clearly in the figure.

EXAMPLES

In the following take ·1″ for both horizontal and vertical units.

(1) The figured plan of a line is given by $a_{14}b_{-3}$, 2·8″ long. Determine its true length and inclination.

(2) The figure shows the relative positions of *ss*, the scale of slope of a

plane, and $a_{20}b_4$ the indexed plan of a line. Determine the index of the point of intersection of the line and plane.

(3) Three points are shown in the figure by their indexed plans. Determine the scale of slope of the plane containing them.

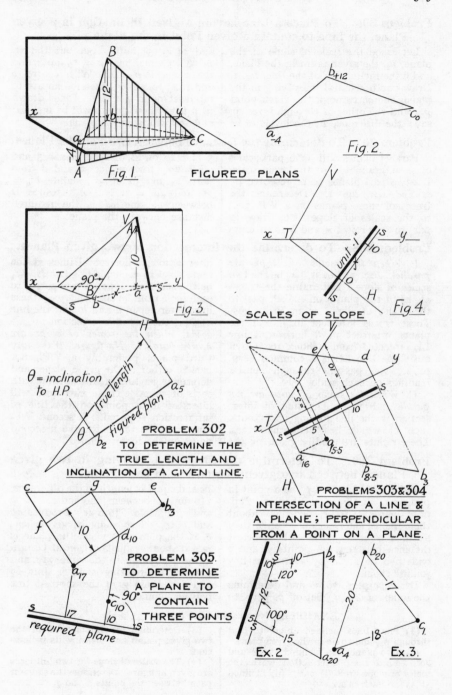

FIGURED PLANS

Fig. 1

Fig. 2

Fig. 3

Fig. 4

SCALES OF SLOPE

θ = inclination to H.P.

true length

figured plan

PROBLEM 302
TO DETERMINE THE TRUE LENGTH AND INCLINATION OF A GIVEN LINE.

PROBLEMS 303 & 304
INTERSECTION OF A LINE & A PLANE; PERPENDICULAR FROM A POINT ON A PLANE.

PROBLEM 305.
TO DETERMINE A PLANE TO CONTAIN THREE POINTS

required plane

Ex. 2

Ex. 3.

Problem 306. To Place a Line having a given Inclination in a given Plane, the Line to contain a given Point in the Plane.

Let ss be the scale of slope of the plane, p_{25} the given point in the plane, and θ the inclination of the line, fig. 1. Draw two horizontal lines lying in the plane, one ab through the given point p_{25}, and the other cd through level 10, say: the difference in the levels of ab

and $cd = 15$ units. Set out the rt. $\angle$d $\triangle$ Ppe, making P$p = 15$ units, and the angle P$ep = \theta$. With centre p and rad. pe, i.e. R, describe an arc to intersect cd in q and r. Lines drawn to p from these points will be inclined at θ and will lie in the plane.

Problem 307. To determine the Distance between two Parallel Planes.

Parallel planes will have par^l scales of slope graduated to the same scale. Let two par^l planes be represented by ss and s_1s_1, fig. 2. Determine the traces of these planes on a V.P. par^l to the scales of slope: i.e. draw xy par^l to ss, project a and b from units

5 and 20 on ss, making $a_0a = 5$ and $b_0b = 20$; then project c and d from units 15 and 30 on s_1s_1, making $cc_0 = 15$ and $dd_0 = 30$. The distance D between ab and cd is the required distance between the planes.

Problem 308. To determine the Intersection of two given Planes.

I. *When the scales of slope are parallel. Fig. 3.* Let ss, s_1s_1 be the two scales of slope. Determine the traces ab, cd of the planes on a V.P. par^l to the scales of slope, as in Prob. 307. These traces intersect in l, and the planes intersect in a horizontal line LL projected from l, index 16. Obviously the line LL is a common perp. to ss, s_1s_1 passing through similar readings on each scale.

II. *When the scales of slope are not parallel. Fig. 4.* The line of intersection is the line joining two points, each of which lies in both planes. These points are readily given by the

intersection of horizontal lines at the same level in each plane. Let ss, s_1s_1 be the scales of slope. Draw perps. to each scale from levels 0 and 300; these lines intersect in a and b, and the line $a_{300}b_0$ is the line of intersection.

III. *When the scales of slope are nearly parallel. No figure.* Take any third plane, preferably a V.P., not par^l to either of the given planes and determine its line of intersection with each of them. These two lines will intersect in a point on the line of intersection required. A second V.P. must then be taken to give a second point on the required line.

Problem 309. To determine the Shortest Line lying in two given Planes, between two given Points, one in each Plane.

Refer to fig. 4. Let p_0 be a point in plane s_1s_1, and q_{10} a point in ss. Suppose the planes to be opened out about their line of intersection ab until they lie in a common plane: the shortest distance between the points p and q will then be given by a straight line joining them.

Draw pc perp. to ab and determine the index of c, 240. Set off pe perp. to

pc and equal in length to the difference, 240 units, between the indices of p and c, to scale. Join ce. Produce cp and mark off ce along it, giving cm; m is the plan of p when the plane is raised about ab until horizontal. Obtain n, for the point q, in the same way, and join mn, intersecting ab in o. Join po and qo: these give the shortest line between p and q.

EXAMPLES (vertical unit for (1) and (3), 0·1″)

(1) Place lines inclined at 20° to pass through p_{17} and to lie in the given plane.

(2) Two planes are inclined at 30° and 60°. Find their intersection when the scales of slope are: (i) par^l; (ii) inclined at 45° to each other.

(3) Determine the intersection of the two given planes and measure its inclination.

(4) The scales of slope for two hill faces are given in figure. Determine the shortest path between the points p and q.

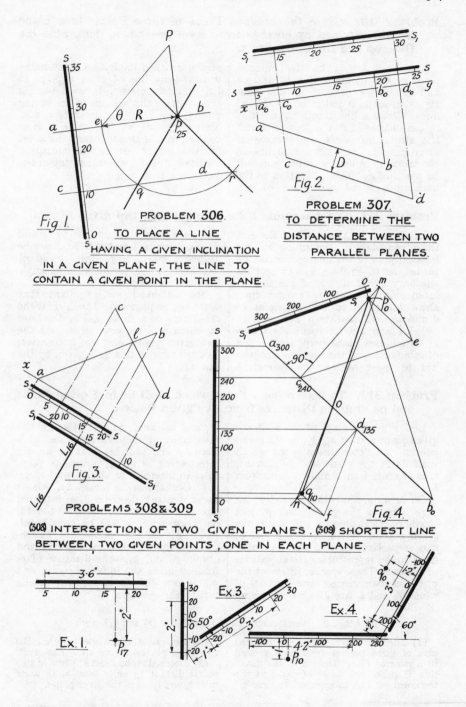

Fig. 1.

PROBLEM 306.
TO PLACE A LINE
HAVING A GIVEN INCLINATION
IN A GIVEN PLANE, THE LINE TO
CONTAIN A GIVEN POINT IN THE PLANE.

Fig. 2.

PROBLEM 307.
TO DETERMINE THE
DISTANCE BETWEEN TWO
PARALLEL PLANES.

Fig. 3.

PROBLEMS 308 & 309

Fig. 4.

(308) INTERSECTION OF TWO GIVEN PLANES. (309) SHORTEST LINE
BETWEEN TWO GIVEN POINTS, ONE IN EACH PLANE.

Ex. 1.

Ex. 3.

Ex. 4.

Problem 310. Given the Indexed Plans of three Points in a plane. Strata, obtained by borings from level ground, to determine the Outcrop and Dip of the Strata.

Let $a_{-62}b_{-150}c_{-205}$ be the indexed plans of the points. Join ba and ca and determine e_0 and f_0—the construction for e_0 is indicated by dotted lines. Draw a line through e_0f_0: this is the *outcrop*, i.e. the line in which the strata intersects the surface of level ground. Any horizontal line in the strata is called the *strike*, and will be par[1] to f_0e_0. The inclination to the horizontal, θ, of the plane of the strata is called its *dip*. To determine θ, draw any line of dip mn perp. to f_0e_0; set off from mn, to scale along a line of strike through one of the boring points, say c_{-205}, an ordinate $= 205$, giving p; join pm. The angle $pmn = \theta$, the dip. If the apparent thickness of the strata or seam, measured vertically, is L, its actual thickness, D, is $=$ L $\cos\theta$.

Problem 311. To determine the Angle between two given Planes.

Let ss, s_1s_1 be the scales of slope of the given planes. Determine ab, the indexed plan of the intersection of the planes, as in Prob. 308. Project an auxiliary elevation, cb, of the line of intersection, on ab as ground line. Draw VT perp. to bc, cutting bc in d and ab in T, and draw TH perp. to ab; VT and TH represent the traces of an inclined plane perp. to the line of intersection of the planes. Produce TH to meet horizontals through b (i.e. lines perp. to ss and s_1s_1) in e and f. With centre T and rad. Td, describe an arc to cut ab in g. Join ge and gf. The angle egf (θ) is the required angle between the planes.

The method adopted is similar to that described on page 242. The $\triangle$ egf is the rabatment of a $\triangle$ formed by the lines in which a plane perp. to ab cuts the two given planes and the horizontal: an edge view of this $\triangle$ is given by the line dT.

Problem 312. To determine a Point which shall lie in a given Plane and be at given Distances from two given Points.

Let ss be the scale of slope of the given plane and let a_{15}, b_{22} be the given points. To determine a point which shall lie in the plane and be distant, say, 12 units from a and 15 units from b.

On ss as ground line draw VT, an edge view of the plane, and project auxiliary elevations c and d of the given points. With centres c and d, and respective radii 12 and 15 units, draw circles representing the elevations of spheres. These circles intersect the plane in two circles, the centres of which f_1 and e_1 are given by drawing df_1 and ce_1 perp. to VT. Obtain the rabatment of these circles: i.e. with centre o and rad. oe_1 describe an arc intersecting ss in e_0; draw e_0e perp. to ss to intersect ae, drawn par[1] to ss, in e. Obtain f in the same way. About these centres describe the rabatted circles, which intersect in two points: only one point g will be considered. Raise g into the plane VT, by reversing the above construction, and obtain g_1 and p_{12}—the distance of g_1 from ss gives the index of p, 12, and p_{12} is one of the two points which satisfy the question.

EXAMPLES (vertical unit for (1), 0·05″, for (2) and (3), 0·1″)

(1) The figure represents the indexed plan of three borings from level ground to a mineral vein. Determine the direction of strike, the angle of dip, and the thickness per foot of apparent thickness.

(2) Determine the angle between the two given planes, ss and s_1s_1. Ans. 124°.

(3) Determine the indexed plans of two points distant 15 units from a, 18 units from b, and lying in the given plane.

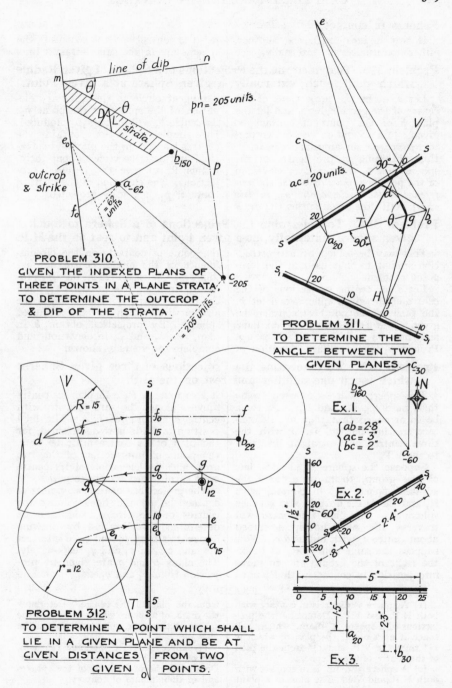

line of dip

$pn = 205$ units.

θ'

D

strata

θ

outcrop & strike

e_0

f_0

b_{150}

P

a_{-62}

= 62 units

c_{-205}

= 205 units

PROBLEM 310.
GIVEN THE INDEXED PLANS OF
THREE POINTS IN A PLANE STRATA
TO DETERMINE THE OUTCROP
& DIP OF THE STRATA.

V

c

$90°$

s

$ac = 20$ units.

10

d

20

T

θ

g

b_0

s

a_{20}

$90°$

s_1

20

H

f

10

0

s_1

-10

PROBLEM 311.
TO DETERMINE THE
ANGLE BETWEEN TWO
GIVEN PLANES.

c_{30}

b_{-160}

N

Ex.1.

$\begin{cases} ab = 2\cdot8" \\ ac = 3" \\ bc = 2" \end{cases}$

a_{-60}

V

$R = 15$

f_0

f

f_1

15

d

b_{22}

s

g

g_0

g_1

P_{12}

c

10

e_1

e

e_0

$r = 12$

a_{15}

T

s

5

S

0

PROBLEM 312.
TO DETERMINE A POINT WHICH SHALL
LIE IN A GIVEN PLANE AND BE AT
GIVEN DISTANCES FROM TWO
GIVEN POINTS.

s

60

40

20

0

-20

-20

$60°$

s

s

s

40

20

0

-20

$2\cdot4"$

$\cdot8"$

$1\frac{1}{2}"$

Ex.2.

$5"$

0   5   10   15   20   25

$1\cdot5"$

$2\cdot3"$

a_{20}

b_{30}

Ex.3.

Spheres in contact.

If two spheres touch one another, either externally or internally, the point of contact and the centres of the spheres are in the same straight line.

Problem 313. To determine the Projections of a Sphere of given Radius which shall touch, externally, a given Sphere at a given Point.

Let a, a_1 be the projections of the centre of the given sphere, and let the plan p of the point of contact be given. Join ap and on x_1y_1 drawn par^l to ap project an auxiliary elevation, thus obtaining a_2 the centre of the sphere, and p_2 the auxiliary elevation of the point (p_2 *must lie on the circumference of the circle, centre a_2, in this view*). With the given radius (R) draw a tangent circle to the auxiliary elevation of the sphere at the point p_2; its centre b_2 will lie on a_2p_2 produced. By simple projection obtain the centres b and b_1 in the plan and elevation, and describe circles about them, radius R, to give the required projections. The elevation of the point of contact, p_1, is readily projected.

Problem 314. To determine the Projections of a Sphere to touch a given sphere, externally, at a given Point and to rest on the H.P.

This may be solved by a construction similar to that used in the foregoing problem.

Let a, a_1 be the projections of the centre of the given sphere, and let p, the plan of the point of contact, be given. Join ap and draw an auxiliary elevation on x_1y_1 taken par^l to ap. Project p_2, the auxiliary elevation of the point of contact, on the circumference of the circle, centre a_2, and join a_2p_2. With centre b_2, on a_2p_2 produced, draw a circle to touch x_1y_1 and the circle centre a_2; this circle is the auxiliary elevation of the required sphere. By projection obtain b in plan, and b_1 and p_1 in elevation, and complete the views as shown.

Problem 315. To determine the Projections of three given Spheres which touch one another and rest on the H.P.

The spheres will be referred to by their radii R_1, R_2, and R_3. Let them be so arranged that they are in contact, resting on the H.P., with the three centres in a straight line par^l to the V.P.

Suppose the sphere R_3, at the left of the group, to move around the adjacent sphere R_1, in contact with R_1 and the H.P., until it also touches sphere R_2. In plan its centre will traverse the arc shown, described about centre a, with radius a_0c_0. Now suppose the same sphere R_3 to be on the right of the group, and to move into position in contact with R_2 until it also touches R_1. In plan its centre moves along the arc drawn with centre b and radius b_0c_0. The final position of R_3 is now determined, for the plan of its centre must be at c, the point of intersection of the two arcs, and the elevation of its centre is at c_1, obtained by projection.

The projections of the two points of contact lie on the lines joining the centres of the circles, and may be determined in elevation by drawing horizontals from p_2 and q_2 to intersect a_1c_1 and b_1c_1 in p_1 and q_1 respectively. The plans p and q are given by projection from p_1 and q_1.

EXAMPLES

(1) A sphere 3″ diam. is in contact with both H.P. and V.P. Determine the projections of a sphere, 2″ diam., which shall touch it at a point, the plan of which is $2\frac{1}{2}$″ from the V.P. and $1\frac{1}{4}$″ from the plan of the centre of the 3″ sphere.

(2) A sphere 3″ diam. is in contact with both H.P. and V.P. The plan of a point on its surface is 2″ from the V.P. and 1·4″ from the plan of the centre. Determine the diam. of a sphere which will touch the sphere at the given point and rest on the H.P.

(3) Three spheres 4″, $2\frac{1}{2}$″, and $1\frac{1}{2}$″ diam. rest on the H.P. and touch one another. Determine the projections of the spheres and of their points of contact.

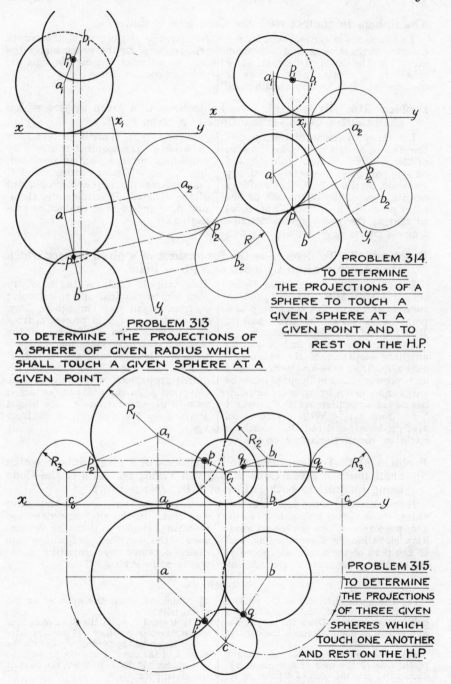

PROBLEM 313
TO DETERMINE THE PROJECTIONS OF
A SPHERE OF GIVEN RADIUS WHICH
SHALL TOUCH A GIVEN SPHERE AT A
GIVEN POINT.

PROBLEM 314.
TO DETERMINE
THE PROJECTIONS OF A
SPHERE TO TOUCH A
GIVEN SPHERE AT A
GIVEN POINT AND TO
REST ON THE H.P.

PROBLEM 315.
TO DETERMINE
THE PROJECTIONS
OF THREE GIVEN
SPHERES WHICH
TOUCH ONE ANOTHER
AND REST ON THE H.P.

The Sphere in contact with the Cone and Cylinder.

Two surfaces in contact will have a common tangent plane and a common normal at the point of contact. If a sphere is in contact with a cone or cylinder, the line joining the point of contact and the centre of the sphere will be perp. to the generator of the cone or cylinder passing through the point of contact.

Problem 316. To determine the Projections of a given Sphere which shall touch a given vertical Cone at a given Point.

Let the projections of the apex of the cone be a, a_1 and let p be the plan of the point of contact. Join ap, and on x_1y_1 par^1 to ap project an auxiliary elevation of the cone. The auxiliary elevation p_2 of the point falls on the outer generator a_2p_2. With rad. equal to that of the given sphere describe a circle, centre c_2, to touch the gener-ator a_2p_2 at p_2. This is the auxiliary elevation of the required sphere. By projection obtain c, on ap produced, and c_1; draw circles centres c and c_1 representing the required projections of the sphere. Determine the elevation p_1 of the point of contact by projection.

Problem 317. To determine the Projections of a given Sphere which shall touch a given Cylinder at a given Point.

Let the axis of the cylinder be inclined to both H.P. and V.P., projections aa and a_1a_1. Let the plan p of a point on the surface be given, and let R be the rad. of the given sphere.

Take x_1y_1 par^1 to aa and project an auxiliary elevation of the cylinder—axis a_2a_2. Then take x_2y_2 perp. to a_2a_2, and project an auxiliary plan; the axis is now given by a_3, and the point lies on the circumference at p_3, distant "d" from x_2y_2. With centre c_3 on a_3p_3 produced, and rad. R, describe a circle to touch the circle centre a_3. This tangent circle is the auxiliary plan of the required sphere. Project p_2 from p and p_3. Through p_2 draw p_2c_2 perp. to a_2a_2 and project c_2 from c_3. Determine centres c and c_1, in the original plan and elevation, by projection. Describe circles with rad. R and centres c and c_1, giving the required projections of the sphere. The elevation p_1 of the point of contact is readily projected from p; its height above $xy = $ the distance of p_2 from x_1y_1.

Problem 318. To determine the Projections of a given Sphere which shall touch a given Cone at a given Point, the Axis of the Cone being inclined to both H.P. and V.P. (No figure.)

If the plan only of the point is given, obtain its elevation by means of Prob. 250, page 250. Then project an auxiliary elevation on a ground line par^1 to the plan of the axis and, secondly, an auxiliary plan on a ground line perp. to the axis in the auxiliary elevation. In the auxiliary views the problem is similar to that for a vertical cone; the required projections are easily obtained by projecting backwards, as in Prob. 317.

EXAMPLES

(1) The given figure shows the plan of a cone, axis vertical, and the plan of a point on its surface. Draw the projections of a sphere 2″ diam. to touch the cone at the given point.

(2) In the given figure ab a_1b_1 are the projections of the axis of a cylinder $1\frac{1}{2}″$ diam., and p is the plan of a point on its surface. Draw the projections of a sphere $2\frac{1}{2}″$ diam. to touch the cylinder at the given point.

(3) Regard b, b_1 as the apex of a cone which envelops a sphere $2\frac{1}{4}″$ diam., centre a, a_1. Let p be the plan of a point on the surface of the cone. Draw the projections of a sphere $2\frac{1}{4}″$ diam. to touch the cone at the given point.

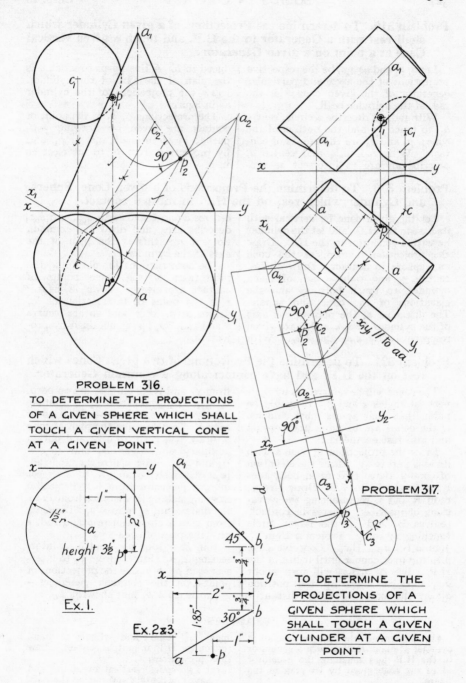

PROBLEM 316.

TO DETERMINE THE PROJECTIONS OF A GIVEN SPHERE WHICH SHALL TOUCH A GIVEN VERTICAL CONE AT A GIVEN POINT.

Ex. I.

Ex. 2 & 3.

PROBLEM 317.

TO DETERMINE THE PROJECTIONS OF A GIVEN SPHERE WHICH SHALL TOUCH A GIVEN CYLINDER AT A GIVEN POINT.

Problem 319. To determine the Projections of a given Cylinder which shall rest with a Generator in the H.P. and touch a given vertical Cone at a point on a given Generator.

Let a, a_1 and ac a_1c_1 be the respective projections of the apex and particular generator of the given cone; let the rad. of the cylinder be R.

With rad. R, describe a circle, centre b_1, to touch xy and the outline of the cone; p_2 and b_0 are the points of contact. With rad. a_0b_0 and centre a, describe an arc to intersect ac pro-

duced in b. A line perp. to ab at b is the plan of the axis of the cylinder. Draw the projections of the cylinder as in figure.

The projections p, p_1 of the point of contact are given by drawing p_2p_1 par^1 to xy to intersect a_1c_1 in p_1, and by projecting from p_1 to intersect ac in p.

Problem 320. To determine the Projections of a given Cone, Sphere, and Cylinder which rest on the H.P. in mutual contact.

Let the given cone be vertical with its base in the H.P., and let the cylinder lie with a generator in the H.P. Draw the projections shown, of the cone and sphere in contact. Draw circles, centres c_2 and radii = rad. of given cylinder, to touch both xy and the elevations of the cone and sphere. The distances of the plan of the axis of the cylinder from a and b are given respectively by a_0c_0 and b_0c_0. With

centres a and b and radii a_0c_0 and b_0c_0 describe arcs, and draw the common tangent cc: this is the plan of the axis of the cylinder.

The constructions for obtaining the projections of the three points of contact are clearly shown in figure: r is the point of intersection of bc, drawn perp. to cc, and an arc, centre b and rad. b_0r_0; r_1 is obtained by projection.

Problem 321. To determine the Projections of two given Cones which rest on the H.P. and have contact along a common Generator.

The cones will be referred to by their vertical angles α and β. It should be noted that their apices must coincide if the cones are to have line contact and also rest on the H.P.

Draw the projections of cone α with its axis par^1 to xy. Draw the elevation of cone β above that of α, both axes lying in the same V.P. Draw circles, radii R and r, representing the elevations of inscribed spheres in contact, centres B_1 and c_1, the larger circle touching xy in c_0. Project c from c_1. Join a_1B_1, a_1c_1, B_1c_1. Conceive cone β to roll over cone α until it lies in the H.P.: the $\triangle$ $a_1B_1c_1$ will turn about a_1c_1 until it takes up the position $a_1b_1c_1$, the point b_1 being distant r

from xy and lying on B_1b_1 drawn perp. to a_1c_1. Tangents from a_1 to a circle, rad. r, centre b_1, give the elevation of cone β. To determine the plan of cone β, draw x_1y_1 perp. to a_1c_1, and obtain auxiliary projections B_2 and a_2c_2 of B_1 and a_1c_1. With centre a_2c_2 and rad. $(a_2c_2)B_2$, describe an arc to intersect in b_2 a projector from b_1. Project b from b_1, making its distance from xy = the distance of b_2 from x_1y_1. Tangents from a to a circle centre b and rad. r give the plan of cone β.

Join bc. Join p_2c_0 and b_1c_1, intersecting in p_1. Project from p_1 to intersect bc in p: p, p_1 are the projections of points on the common generator—given by ap a_1p_1 (not shown).

EXAMPLES

(1) Determine the projections of a cylinder 2″ diam. resting with a generator in the H.P. and touching the generator ab of the cone given by its plan in the figure.

(2) A cone, base $2\frac{1}{4}$″ diam., height 3″,

sphere $2\frac{1}{2}$″ diam., and cylinder 1″ diam. rest on the H.P. in mutual contact. Draw their projections.

(3) Two cones, vertical angles 40° and 20°, have line contact and rest on the H.P. Determine their projections.

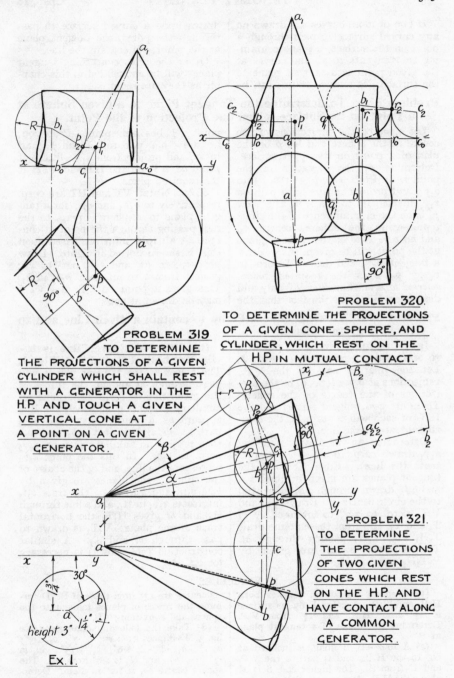

PROBLEM 319.
TO DETERMINE
THE PROJECTIONS OF A GIVEN
CYLINDER WHICH SHALL REST
WITH A GENERATOR IN THE
H.P. AND TOUCH A GIVEN
VERTICAL CONE AT
A POINT ON A GIVEN
GENERATOR.

PROBLEM 320.
TO DETERMINE THE PROJECTIONS
OF A GIVEN CONE, SPHERE, AND
CYLINDER, WHICH REST ON THE
H.P. IN MUTUAL CONTACT.

PROBLEM 321.
TO DETERMINE
THE PROJECTIONS
OF TWO GIVEN
CONES WHICH REST
ON THE H.P. AND
HAVE CONTACT ALONG
A COMMON
GENERATOR.

height 3"

Ex. 1.

If two or more curves are drawn on any curved surface to pass through a point on the surface, a plane containing the tangents to all the curves at the given point is a tangent plane to the surface. If a straight line can be drawn upon a curved surface to pass through the point, the tangent plane at the point will contain the line.

Other theorems concerning tangent planes will be enunciated in this chapter as they are required.

Problem 322. To determine the Tangent Plane to a given Sphere at a Point on its Surface, given one Projection of the Point.

Let o, o_1 be the projections of the centre of the sphere and let p be the plan of a point on its upper surface. Determine p_1 the elevation of the point as in Prob. 182, page 186. Join o_1p_1, and project an auxiliary plan on x_1y_1 taken par[l] to o_1p_1: in this view p_2 is on the circumference of the circle representing the sphere, centre o_2, and an edge view of the tangent plane at the point will be given. At p_2 draw a tangent p_2V, and draw VT perp. to x_1y_1; p_2VT is the required plane, referred to x_1y_1. Consider VT only and determine HT from the fact that the point p, p_1 lies in the plane—Prob. 209; i.e. draw p_1a_1 par[l] to VT and pa par[l] to xy, and project from a_1 to intersect pa in a: a is on the HT and VTH is the required plane.

322a. Note: VT and HT are perp. respectively to o_1p_1 and op; for a tangent plane to a sphere is perp. to the rad. passing through the point of contact. A **simple solution,** based upon this, is shown dotted in figure. Draw pb perp. to po, and p_1b_1 par[l] to xy; project from b to intersect p_1b_1 in b_1. This gives a point on VT, and VTH may be drawn at once.

Problem 323. To determine a Plane to contain a given Line and to touch a given Sphere.

(a) When the given line is parallel to one plane of projection. Fig. 2.
Let the line be par[l] to the V.P., projections ab a_1b_1; let o, o_1 be the projections of the centre of the sphere. There are two tangent planes and the H.T. of each will contain ht, the horizontal trace of the line.

Determine an auxiliary plan on x_1y_1 drawn perp. to a_1b_1—$(a_2b_2$ represents the line). Edge views of the tangent planes are given by lines h_1t_1 and h_2t_2, drawn through a_2b_2 tangential to the circle centre o_2. Draw V_2T_2 and V_1T_1 perp. to x_1y_1 to intersect xy in T_2 and T_1, and from these points draw T_2H_2 and T_1H_1 passing through ht. The required traces are given by $V_1T_1H_1$ and $V_2T_2H_2$.

(b) When the given line is inclined to both H.P. and V.P. Fig. 3.
The traces of the tangent planes will pass through ht and vt, the traces of the given line ab a_1b_1. Determine an auxiliary projection in which the line appears as a point: first take x_1y_1 par[l] to ab and project a_2b_2 and o_2; then project an auxiliary plan on x_2y_2 perp. to a_2b_2. In this the point a_3b_3 represents the line, and o_3 the centre of the sphere; the planes are given by h_1t_1, v_1t_1, and h_2t_2, v_2t_2. The trace v_1t_1 intersects x_1y_1 in H_1, and a line through H_1 and ht gives H_1T_1 the horizontal trace of one plane. V_1T_1 is drawn to pass through vt and T_1. A similar construction (not shown) is necessary for v_2t_2, h_2t_2.

EXAMPLES

(1) A sphere $2\frac{3}{4}''$ diam. touches both H.P. and V.P. A point P on its surface is $2\frac{1}{4}''$ from H.P. and $1\frac{3}{4}''$ from V.P. Determine the traces of a tangent plane at P.

(2) A line AB, $2''$ long, is inclined at 60° to the H.P., and is par[l] to the V.P. and $1''$ from it. The higher end B is $2''$ above the H.P. A sphere $2''$ diam. touches both H.P. and V.P. and the projectors of its centre are $1\frac{3}{4}''$ from those of B. Determine the traces of planes tangent to the sphere and containing AB.

(3) Take the following dimensions for fig. 3. Distances from xy: o, $\cdot7''$; o_1, $1\cdot3''$; a, $1''$; a_1, $\cdot8''$; b, $1\cdot6''$; b_1, $1\cdot75''$. aa_1 is $\cdot7''$ from oo_1, and bb_1 is $1\cdot6''$ from aa_1. The sphere centre oo_1 is $1\cdot2''$ in diam. Determine the traces of planes tangent to the sphere and containing the line.

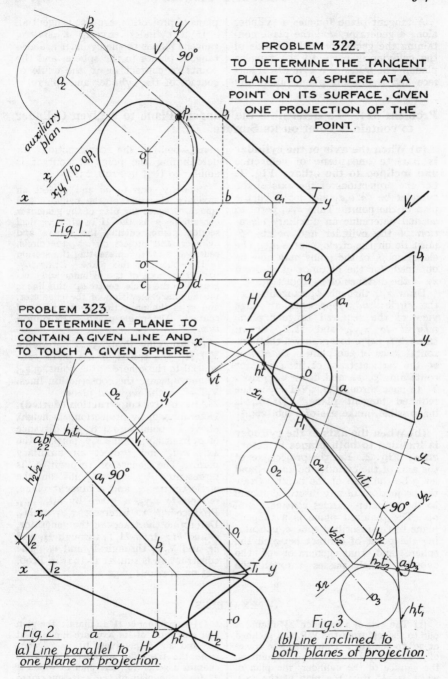

PROBLEM 322.
TO DETERMINE THE TANGENT PLANE TO A SPHERE AT A POINT ON ITS SURFACE, GIVEN ONE PROJECTION OF THE POINT.

auxiliary plan

x_1y_1 // to op_1

Fig 1.

PROBLEM 323.
TO DETERMINE A PLANE TO CONTAIN A GIVEN LINE AND TO TOUCH A GIVEN SPHERE.

Fig 2
(a) Line parallel to one plane of projection.

Fig. 3.
(b) Line inclined to both planes of projection.

A tangent plane touches a cylinder along a generator, and the plane containing the generator and the axis of the cylinder is perp. to the tangent plane. The *traces* of a cylindrical surface and a tangent plane, upon any plane of projection, are also tangential.

If a cylinder envelop a sphere, tangent planes to the cylinder are also tangent planes to the sphere, and the points of contact lie in the circle of contact of the cylinder and sphere.

Problem 324. To determine the Tangent Plane to a given Cylinder to contain a Point on its Surface.

(a) When the axis of the cylinder is par^l to one plane of reference and inclined to the other. Fig. 1. Let the projections of the axis of the cylinder be aa a_1a_1 and let p be the plan of the point. Take x_1y_1 perp. to aa and determine an auxiliary elevation of the cylinder and point; p_2 must lie on the circle, centre a_2. The elevation p_1 of the point may now be obtained, for the distance of p_1 from $xy =$ the distance of p_2 from x_1y_1.

Draw v_1t_1 through p_2 tangential to the circle: v_1t_1 represents an edge view of the required tangent plane *referred to* x_1y_1, and H_1T_1 drawn through t_1 perp. to x_1y_1 gives the horizontal trace of the plane. Determine vt, the vertical trace of the generator containing p, p_1, and draw V_1T_1 from T_1 to pass through vt. $V_1T_1H_1$ is the required tangent plane. . (If T_1 is inaccessible, make use of Prob. 209.)

(b) When the axis of the cylinder is inclined to both planes of reference. Fig. 2. Let the projections of the axis of the cylinder be aa a_1a_1 and let p be the plan of the point. Draw the projections of any inscribed sphere to the cylinder—circles centres o and o_1. The required plane is a tangent plane to the inscribed sphere, containing the point of contact between the sphere and that generator of the cylinder which passes through the given point; the construction for determining the point of contact is similar to that in Prob. 251:—

Take x_1y_1 par^l to aa and project an auxiliary elevation of the cylinder and sphere. Regard the plan of the generator containing p as the H.T. of a vertical section plane, cutting the sphere and cylinder, and project on x_1y_1 the circle, centre o_2, rad. r, representing the section of the sphere by this plane. Draw p_2q_2 par^l to the axis of the cylinder and tangential to the circle, centre o_2; this line is the auxiliary elevation of the generator, and q_2 is the projection of the point of contact. Determine q and q_1 by projection.

By applying Prob. 322a determine $V_1T_1H_1$, the traces of the plane tangential to the sphere at the point, q, q_1; qb and q_1b_1 are the construction lines. $V_1T_1H_1$ is the required plane.

Alternative construction (dotted). Determine p_1 by projection, its height above xy being equal to the distance of p_2 from x_1y_1. Take x_2y_2 perp. to the axis a_2a_2 and project an auxiliary plan. The generator through p_2 is represented by p_3, a point on the circle centre o_3, and a tangent plane, *referred to* x_2y_2, is given by h_1t_1, v_1t_1. Produce v_1t_1 to intersect x_1y_1 in H_1. Determine the traces of the generator, vt and ht; draw H_1T_1 through H_1 and ht, and V_1T_1 through T_1 and vt. The construction is similar to that in Prob. 323 on previous page.

EXAMPLES

(1) The axis of a cylinder, $2\frac{1}{4}''$ diam., is par^l to and $1\frac{1}{2}''$ from the V.P. and inclined at $30°$ to the H.P. Determine the traces of a tangent plane to contain a point on the surface of the cylinder, the plan of which is $1\frac{1}{8}''$ from the plan of the axis, measured away from xy.

(2) A cylinder is $2\frac{1}{2}''$ in diam.; the plan and elevation of its axis are inclined at $30°$ and $40°$ respectively to xy. Determine the traces of a tangent plane to contain a generator, the plan of which is $\frac{3}{4}''$ from the plan of the axis—measured away from xy.

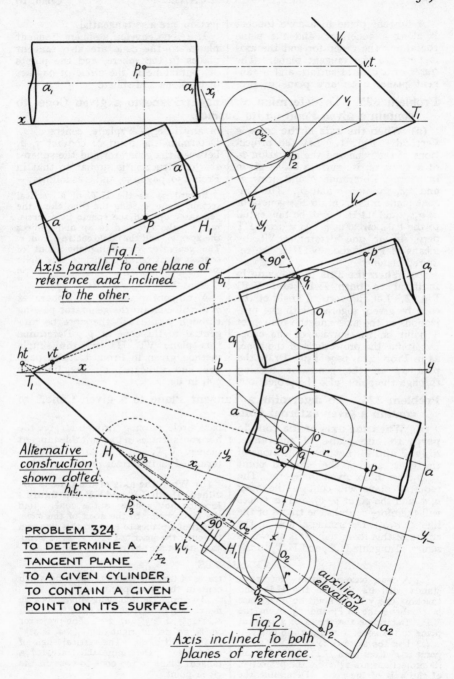

Fig. 1.
Axis parallel to one plane of
reference and inclined
to the other.

Alternative
construction
shown dotted

PROBLEM 324.
TO DETERMINE A
TANGENT PLANE
TO A GIVEN CYLINDER,
TO CONTAIN A GIVEN
POINT ON ITS SURFACE.

auxiliary elevation

Fig. 2.
Axis inclined to both
planes of reference.

A tangent plane to a cone touches it along a generator, and the plane containing the generator and the axis is perp. to the tangent plane. The *traces* of a conical surface and a tangent plane, upon any plane of pro-jection, are also tangential.

If a cone envelop a sphere, tangent planes to the cone are also tangent planes to the sphere, and the points of contact lie in the circle of contact of the cone and sphere.

Problem 325. To determine a Tangent Plane to a given Cone, to contain a given Point on its Surface.

(a) When the axis of the cone is vertical. Fig. 1. Let the projec-tions of the cone and the elevation p_1 of a point on its surface be as given in figure. Determine p and draw ac and a_1c_1 through p and p_1. The tan-gent plane must contain the generator ac a_1c_1, and its H.T. will be tangential to the base circle at c. At c draw HT perp. to ac and determine VT by means of Prob. 209. VTH is the re-quired plane.

(b) When the axis of the cone is inclined to both H.P. and V.P. Fig. 2. Let the vertical angle of the cone be given, together with the pro-jections of the axis and the plan p of a point on the surface of the cone. Determine the projections of the cone, as in Prob. 249, page 250, Draw the plan ap of the generator passing through the point, also the projections of any inscribed sphere, centre o, o_1, Determine the point of contact q, q_1 between the generator and the sphere: this construction is similar to that in Prob. 250, page 250, and is as follows:—

Regard ap as the H.T. of a vertical section plane, cutting the cone along the generator ap and the sphere in a circle, rad. r, take x_1y_1 par^1 to ap and project the apex a_2, and the circle centre o_2 rad. r. The generator containing the point of contact must be tangential to this circle and is given by a_2p_2 touching the circle in q_2. Then project q and q_1.

A tangent plane to the sphere at q, q_1 will contain the generator passing through p and will therefore be tan-gential to the cone at p. Determine this plane VT, HT by the simple method given in Prob. 322a on page **326** and indicated by the lines qb q_1b_1 in fig. 2.

Problem 326. To determine a Tangent Plane to a given Cone,* to contain a given external Point.

(a) When the axis of the cone is perp. to one plane of reference. Figs. 3 and 4. Let the projections of the cone, and of the external point p, p_1, be those given in figure. The required plane will contain the apex a, a_1 and the point p, p_1, and its traces will therefore contain the traces of the line ap a_1p_1. Determine ht and vt, the traces of this line; only ht is shown in figure. Tangents H_1T_1 and H_2T_2 to the base circle, passing through ht, give the horizontal traces of two suitable tangent planes. V_1T_1 and V_2T_2 are readily de-termined as in former problems.

(b) When the axis of the cone is in-clined to both H.P. and V.P. (No figure.) Inscribe any sphere to the cone. Join the given point to the apex of the cone. Determine planes to contain this line and to touch the inscribed sphere, as in Prob. 323: these are the required planes.

EXAMPLES

(1) A cone, base $2\frac{1}{4}''$ diam., height $3''$, stands with its base on the H.P. and touching the V.P. A point on its surface is $1''$ above the H.P. and $1\frac{1}{2}''$ from the V.P. Determine the traces of a tangent plane to the cone, to contain the point.

(2) Use the same cone as in Ex. 1. A point is $1''$ from both H.P. and V.P., and its projections are $1\frac{3}{4}''$ from the projectors of the axis of the cone. Determine the traces of tangent planes to the cone which contain the given point.

(3) Apply the following dimensions to fig. 2. Distances from xy: p, $1\cdot8''$; o, $1\cdot6''$; o_1, $1\cdot25''$; a, $2\cdot5''$; a_1, $2\cdot7''$. The projector oo_1 is $\cdot65''$ to the right of pp_1, aa_1 is $1\cdot5''$ to the right of oo_1. The vertical angle of cone = $30°$. Determine the traces of a tangent plane to the cone to contain the given point.

* A cylinder may be treated as a cone with its apex infinitely distant.

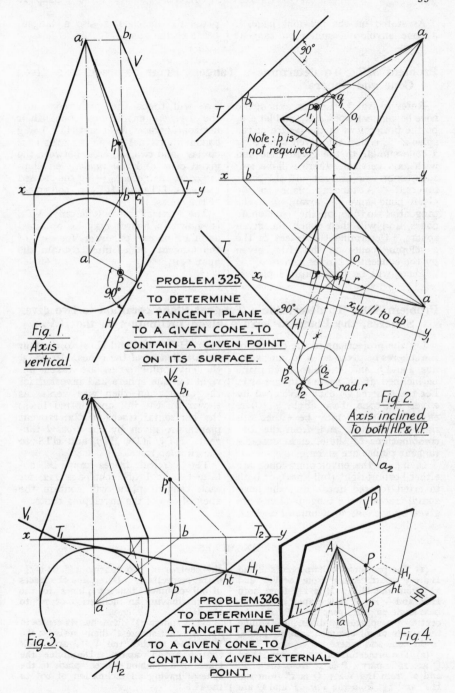

Fig. 1.
Axis vertical

Note: *p* is not required.

PROBLEM 325.
TO DETERMINE
A TANGENT PLANE
TO A GIVEN CONE, TO
CONTAIN A GIVEN POINT
ON ITS SURFACE.

x,y ∥ to ap

rad. r

Fig. 2.
Axis inclined to both H.P & V.P.

Fig. 3.

PROBLEM 326.
TO DETERMINE
A TANGENT PLANE
TO A GIVEN CONE, TO
CONTAIN A GIVEN EXTERNAL
POINT.

Fig. 4.

As stated on the previous page, if a cone envelop a sphere, a tangent plane to the cone is also a tangent plane to the sphere.

Problem 327. To determine a Tangent Plane common to a given Cone and Sphere.

Refer to fig. 1. Let the axis of the cone be vertical, apex a, a_1, and let o, o_1 be the projections of the centre of the sphere. Envelop the given sphere by a cone similar to the given cone and with axis vertical: there will be two such cones, one upright and the other inverted. A tangent plane to the given cone may be arranged to be tangential to one of the enveloping cones, and will then touch the given sphere. Determine the traces of the enveloping cones on the H.P., given by the concentric circles, centre o.

There will be four suitable planes:

two will touch the given cone and the *upright* enveloping cone, their horizontal traces, HT1 and HT2, being common external tangents to the base circles; and two will cross between the given cone and the *inverted* enveloping cone, their traces being the crossed tangents, HT3 and HT4, to the traces of the cones.

The vertical traces (one only, VT4, is shown in figure) may be obtained as in Prob. 209, page 212, for each of the tangent planes must contain the apex a, a_1.

Problem 328. To determine a Tangent Plane common to two given Spheres, the Plane to have a given Inclination to the H.P.

Let the projections of the centres of the spheres be given by p, p_1 and q, q_1, figs. 2 and 3, and let the required plane be inclined at an angle θ to the H.P. Let the spheres be each enveloped by a vertical cone, base angle θ. The enveloping cones may be either upright or inverted, and, from the four combinations of these, eight suitable tangent planes are given.

In fig. 2, the enveloping cones are either both upright (full lines), or both inverted (dotted lines), and the horizontal traces of the tangent planes are given by the *outside* common tangents,

HT1, HT2, HT3 and HT4, to the four circular traces of the cones. In fig. 3, the enveloping cones are taken upright for one sphere and inverted for the other, and then vice versa, as shown by the full and dotted lines; the horizontal traces of the tangent planes are given by the *crossed* tangents, HT5, HT6, HT7, and HT8, to the circular traces.

The vertical traces may be obtained as in Prob. 209, page 212, for each tangent plane will contain the apex of one of the enveloping cones.

EXAMPLES

(1) The projectors of two points A and B are 3″ apart. A is 3″ from the V.P. and 2½″ from the H.P.; B is 1¾″ from both H.P. and V.P. A is the apex of a cone, base angle 70°, axis *perp. to V.P.* B is the centre of a sphere 1½″ diam. Determine the traces of all common tangent planes to the cone and sphere.

(2) The projectors of two points P and Q are 2½″ apart. P is 1″ from the H.P. and 2″ from the V.P.; Q is 1″ from the H.P. and 1¾″ from the V.P. P and Q are

the centres of two spheres 1½″ and 1¼″ diam. respectively. Determine the traces of all common tangent planes to the spheres, having an inclination of 70° to the H.P.

(3) A sphere 2½″ diam. has its centre in *xy*. Another sphere 2″ diam. rests on the H.P. and touches the V.P., and their centres are 3″ apart. Determine the traces of a common tangent plane to the spheres having an inclination of 60° to the H.P.

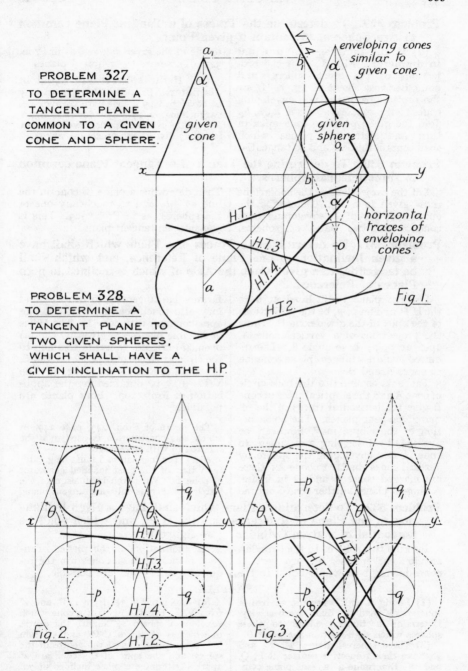

PROBLEM 327.
TO DETERMINE A
TANGENT PLANE
COMMON TO A GIVEN
CONE AND SPHERE.

given
cone

enveloping cones
similar to
given cone.

given
sphere

horizontal
traces of
enveloping
cones.

Fig. I.

PROBLEM 328.
TO DETERMINE A
TANGENT PLANE TO
TWO GIVEN SPHERES,
WHICH SHALL HAVE A
GIVEN INCLINATION TO THE H.P.

Fig. 2.

Fig. 3.

Problem 329. To determine the Traces of a Tangent Plane common to two Spheres, to contain a given Point.

(a) If the spheres are unequal in diameter. Fig. 1. Let the projections of the given spheres and point be those shown in fig. 1. Draw the projections of the two enveloping cones to the spheres, apices a, a_1, b, b_1. Join the given point to each apex in turn and determine planes which shall contain this line and be tangential to one of the given spheres, as in Prob. 323. These are the required planes.

(b) If the spheres are equal in diameter. Fig. 2. The spheres may be enveloped by a cone with its apex between them, or by a cylinder as in fig. 2. Regard the cylinder as a cone with its apex infinitely distant and proceed as at (a).

Problem 330. To determine the Traces of a Tangent Plane common to three unequal Spheres.

Let the projections of the spheres be those given in fig. 3. Determine the apices a, a_1, b, b_1 of the enveloping cones common to two pairs of spheres. Then determine a plane to contain the line ab a_1b_1 and to touch any one of the spheres, as in Prob. 323. This is one suitable tangent plane.

Problem 331. To determine the Traces of a Plane which shall have a given Inclination to one Plane of Reference, and which shall be tangential to a given Cone the Axis of which is inclined to both Planes of Reference.

Let the plane be inclined at θ to the H.P. and let p, p_1 be the projections of the apex of the given cone B. Draw the projections of a vertical cone A, apex at p, p_1, base angle θ. The required plane is a tangent plane common to cones A and B.

Tangents common to the base circle of cone A and the elliptical trace of cone B give the horizontal traces of the required tangent planes. The construction of the ellipse may, however, be avoided, by inscribing any sphere to cone B, and enveloping the sphere by vertical cones similar to cone A; these are marked cones C and D in figure. A tangent plane to either cone C or cone D may be tangential to cone B, and may also touch cone A, which has an apex common with cone B. The H.T.s of suitable planes are given by the (a) open, (b) crossed, tangents common to the circular traces of (a) cones A and C, (b) cones A and D—as in fig. 4. The V.T.s may be obtained by the application of Prob. 209. Four planes are possible.

The figure of Prob. 247, page 249, in which this construction is used, may be referred to with advantage.

Note. The inclination θ must be greater than that of the least inclined generator of cone B. When the inclination of the latter $= \theta$ there is only one tangent plane.

Problem 332. To determine a Plane which shall have a given Inclination to one Plane of Reference, say the H.P., and which shall be tangential to a given Cylinder. (No figure.)*

Inscribe two spheres to the cylinder, envelop the spheres by two vertical cones having base angles equal to the given inclination of the plane. Tangent planes to these cones are the required planes—four are possible.

EXAMPLES

(1) Points A and B are the centres of spheres $1\frac{1}{2}''$ and $\frac{3}{4}''$ diam. respectively. Determine a tangent plane to these spheres which shall contain the point C.

(2) Three spheres, diams. $2''$, $1''$, and $\frac{3}{4}''$, have their respective centres at P, Q, and R. Determine a tangent plane common to the spheres.

(3) A point P is $3\frac{1}{2}''$ from H.P. and $2''$ from V.P. P is the apex of a cone which envelops a sphere $1\frac{1}{2}''$ diam., the centre of which is $1''$ above H.P., and $3''$ from V.P.; the projectors of the centre of the sphere and the apex of the cone are $2''$ apart. Determine a plane inclined at $70°$ to H.P. to touch this cone tangentially.

* The inclination of the plane cannot be less than that of the axis of the cylinder.

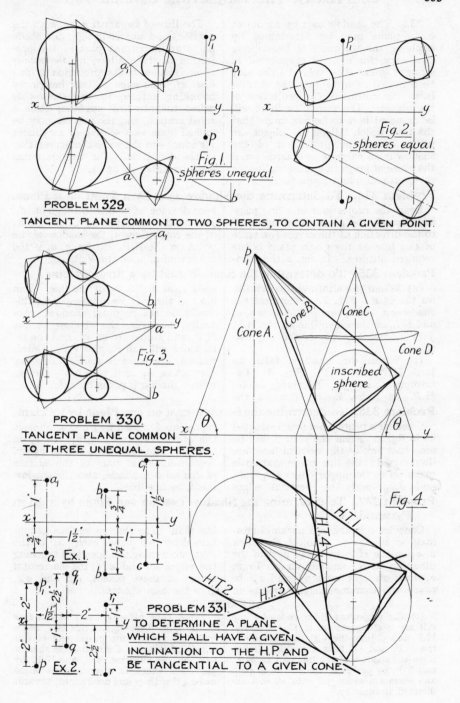

Fig. 1.
spheres unequal.

Fig 2.
spheres equal.

PROBLEM 329.
TANGENT PLANE COMMON TO TWO SPHERES, TO CONTAIN A GIVEN POINT.

Fig. 3.

PROBLEM 330.
TANGENT PLANE COMMON
TO THREE UNEQUAL SPHERES.

Cone A.
Cone B
Cone C
Cone D

inscribed
sphere

Fig 4.

H.T.1
H.T.4
H.T.2
H.T.3

Ex. 1.

Ex. 2.

PROBLEM 331.
TO DETERMINE A PLANE
WHICH SHALL HAVE A GIVEN
INCLINATION TO THE H.P. AND
BE TANGENTIAL TO A GIVEN CONE.

333. The shadow cast by an object on a plane may be determined by applying the theorems of Descriptive Geometry, for light is propagated in straight lines. The rays of light may be regarded as projectors, the shadow being the outline of the projection of the object. The source of light will be assumed here to be so remote that the rays which fall on the object are par[1]; hence the projection of the shadow is an exercise in oblique (or, if the plane of projection is normal to the rays, orthographic) projection.

The **line of separation** between the portions of a surface in light and shade plays an important part in the projection of the shadow, for *the outline of the shadow is the projection of this line of separation.* The imaginary bounding surface to the shadow is called the **shadow surface;** it is a ruled surface, and the shadow may be looked upon as a section of a cylinder (or cone)—in the widest interpretation of the definition. The various terms are illustrated in fig. 1.

Problem 334. To determine the Shadow cast by a Point on a Plane.

Draw the projections of a line passing through the given point and par[1] to the direction of the rays. The trace of this line on the given plane is the required shadow. In fig. 2 the horizontal trace of a line through a, a_1, par[1] to the projected directions R, R_1 of the rays, gives a_s the shadow of the point on the H.P. In fig. 3, a_s is the shadow of a, a_1 on the V.P.

Problem 335. To determine the Shadow cast by a Straight Line.

(a) When the shadow lies wholly on the H.P. Fig. 2. Determine the shadows a_s and b_s cast by the ends a, a_1 and b, b_1 of the given line on the H.P. Join a_s and b_s; $a_s b_s$ is the required shadow.

(b) When the shadow falls on both H.P. and V.P. Fig. 3. Determine a_2 and b_s, the traces *on the H.P.* of rays passing through the

ends a, a_1 and b, b_1 of the line. Join $a_2 b_s$: if the V.P. were transparent this would be the required shadow. Actually, however, the shadow on the H.P. will stop at c, on xy, and the remainder will fall on the V.P. Determine a_s, the trace *on the V.P.* of a ray through a, a_1, and join $a_s c$. The required shadow is given by $a_s c b_s$.

Problem 336. To determine the Shadow cast on one Plane by a Prism.

Consider a pentagonal prism situated as in fig. 4. From the plan it will be seen that two of the vertical faces are illuminated; the line of separation is made up of the upper edges $b_1 c_1$, $c_1 d_1$, and $d_1 e_1$, and the vertical edges

represented in plan by b and e. Project the upper corners b_1, c_1, d_1, and e_1 on the H.P. and join, as shown in figure. Section-line that part of the surface of the solid in shade, also the shadow itself, by fine lines.

Problem 337. To determine the Shadow cast on one Plane by a given Pyramid.

Consider a hexagonal pyramid situated as in fig. 5. Project the apex a, a_1 on the H.P. by a ray par[1] to the direction R, R_1, and obtain a_s. From a_s draw straight lines $a_s b$ and $a_s d$ to touch the extreme angular points in

the plan. The shadow surface is contained by two planes each par[1] to the rays, containing the apex and touching the edges ab and ad; the horizontal traces of these planes, $a_s b$ and $a_s d$, give the cast shadow.

EXAMPLES

(1) The projectors of the ends of a line AB are $1.4''$ apart; A is $1''$ above the H.P. and $3''$ from the V.P., B is $2\frac{1}{2}''$ above the H.P. and $.9''$ from the V.P. Determine the shadow of the line cast on H.P. and V.P. by par[1] rays which in both plan and elevation make $30°$ with xy and are directed towards xy.

(2) A cube $2\frac{1}{2}''$ edge rests with a face in the H.P. One edge of this face makes $50°$ with xy and the nearer end of this edge is $1''$ from xy. Determine the shadow cast by the cube on H.P. and V.P. by par[1] rays which in plan and elevation make $45°$ with xy and are directed towards xy.

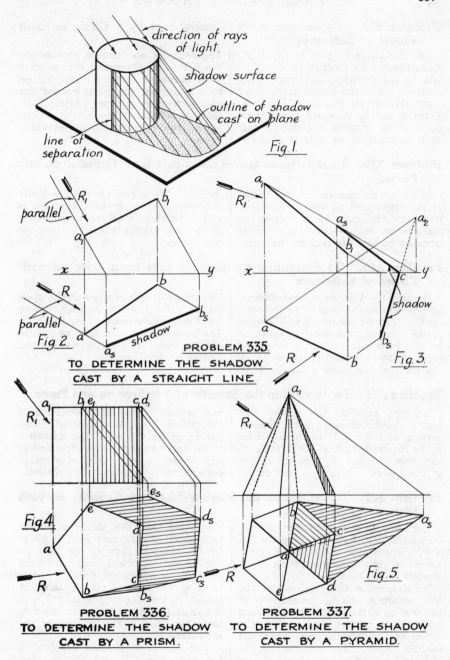

direction of rays of light.

shadow surface

outline of shadow cast on plane

line of separation

Fig.1

R_1

parallel

b_1

a_1

x y

b

R

parallel

Fig.2 a a_s shadow b_s

PROBLEM 335.

TO DETERMINE THE SHADOW
CAST BY A STRAIGHT LINE

a_1

R_1

a_s

b_1

a_2

x y

c

a shadow

b_s

R b *Fig.3*

a_1 b_1 e_1 $c_1 d_1$

R_1

e_s

Fig.4 e d_s

a d

R b b_s c c_s

PROBLEM 336.

TO DETERMINE THE SHADOW
CAST BY A PRISM.

a_1

R_1

a_s

b c

a

R *Fig.5*

e d

PROBLEM 337.

TO DETERMINE THE SHADOW
CAST BY A PYRAMID.

Problem 338.   To determine the Shadow cast by a Circle on both Planes of Reference.

Let the circle be par[l] to the H.P., as in figure. Its shadow on the H.P. will be a segment of an equal circle, centre o_s; o_s is the horizontal trace of a ray through the centre of the circle. From a_s and b_s, in which the shadow circle meets xy, draw $a_s a$ and $b_s b$ par[l] to R to intersect the plan in a and b.

The segment abc casts a shadow on the V.P. To determine this shadow —an elliptical curve on the V.P.— take a number of points on the arc abc, such as p, p_1, and project their shadows on the V.P.—e.g. p_s. The *tangent* ray, r, in plan is an important one.

Problem 339.   To determine the Shadow cast by a Cylinder on one Plane.

With o_s, the shadow of the centre of the upper end, as centre, and rad. equal to the rad. of the cylinder, describe an arc. Draw the tangents common to this arc and to the circle

centre o. These give the outline of the shadow; the generators through a and b, together with the upper semi-circle acb, constitute the line of separation.

Problem 340.   To determine the Shadow cast by a Cone on both Planes of Reference.

Assume the V.P. to be transparent, and determine the shadow of the cone on the H.P. by obtaining the shadow of the apex, a_2, and drawing tangents to the base circle. These tangents intersect xy in b_s and c_s, which mark the limit of the shadow on the H.P.

Determine a_s, the shadow of the apex on the V.P. Join $a_s b_s$ and $a_s c_s$: these are the shadow outlines on the V.P. The lines of separation and the surface of the solid in shadow are shown clearly in the figure.

Problem 341.   To determine the Shadow of a Sphere on one Plane.

Tangent rays to the sphere will form a circular cylinder as the shadow surface, and the H.T. of this cylinder is the required shadow. Determine o_s, the trace of a ray through the centre of the sphere. Take $x_1 y_1$ par[l] to oo_s

and project an auxiliary elevation of the sphere. Using this view determine the elliptical trace of the cylinder, as indicated in the figure. The line of separation is the circle of contact between the sphere and the cylinder.

Problem 342.   To determine the Shadow cast by a Cylinder on both Planes of Reference.

Let the axis of the cylinder, ab $a_1 b_1$ be par[l] to the H.P. and perp. to the V.P. Suppose the planes to be transparent in turn, thus permitting a full shadow to fall on each.

To determine the shadow on the V.P. obtain a_s and b_s, the shadows of the centres of the ends. Describe arcs about these centres with radii = the radius of the cylinder, and draw the outline shown. The shadow on the H.P. will consist of two elliptical arcs and two straight lines, $c_s d_s$ and $e_s f_s$.

The latter are the shadows of the generators cd $c_1 d_1$, and ef $e_1 f_1$, given by the tangent rays r_1 and r_2. To obtain the elliptical arcs plot the shadows of points on the circumferences of the ends of the cylinder: it should be noted that the part $d_s h_s$ is the shadow of the circular arc dh $d_1 h_1$ and that $c_s k_s e_s$ is the shadow of the semi-circle cke $c_1 k_1 e_1$.

The line of separation consists of the generators CD and EF and the semi-circles CKE and DHF.

EXAMPLES.   *See page 340*

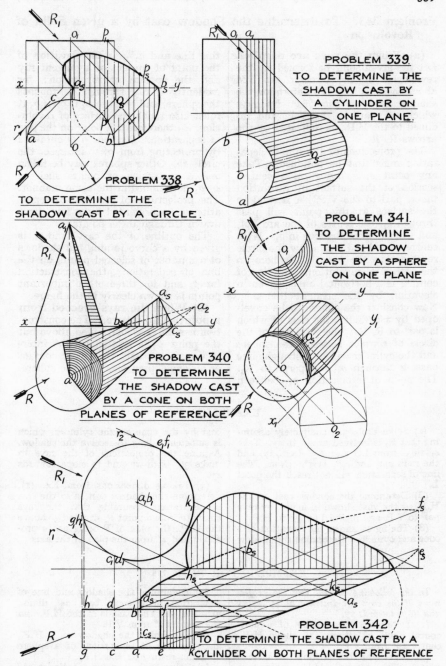

PROBLEM 339.
TO DETERMINE THE
SHADOW CAST BY
A CYLINDER ON
ONE PLANE.

PROBLEM 338.
TO DETERMINE THE
SHADOW CAST BY A CIRCLE.

PROBLEM 341.
TO DETERMINE
THE SHADOW
CAST BY A SPHERE
ON ONE PLANE

PROBLEM 340.
TO DETERMINE
THE SHADOW CAST
BY A CONE ON BOTH
PLANES OF REFERENCE

PROBLEM 342.
TO DETERMINE THE SHADOW CAST BY A
CYLINDER ON BOTH PLANES OF REFERENCE

Problem 343. To determine the Shadow cast by a given Solid of Revolution.

(a) When the rays are parallel. Let the given solid be formed by the revolution of the circular arc LM about the axis LO. To determine the shadow cast on the H.P. by rays which are par[1] to the V.P. and inclined to the H.P., as indicated by the arrows R, R_1.

The projections of the line of separation must first be obtained. Take any point a, a_1 lying on a central section of the surface of revolution taken par[1] to the V.P. A normal to the surface at this point will pass through c_1, the centre of the arc LM, and will intersect the axis in o_1. With centre o_1 and rad. o_1a_1 draw the circle representing the inscribed sphere to the surface of revolution. The line of contact is a horizontal circle, given in elevation by the straight line a_1b_1. Now consider the sphere to be enveloped by a cylinder, the axis of which is par[1] to R, R_1: the elevation of the circle of contact between the sphere and the cylinder will be a straight line passing through o_1 and perp. to R_1. The point of intersection p_1, between

this line and a_1b_1, is the elevation of the point of contact of a tangent ray and the surface of revolution; for evidently a ray through p_1 touches the sphere, and because p_1 is on a_1b_1 it must also touch the surface of revolution. p_1 therefore is a point on the line of separation. Its plan p is cbtained by projecting from p_1 to intersect the circle ab. Other spheres may be taken and a succession of points such as p, p_1 determined in the same manner. The projections of the line of separation are represented by a fair curve drawn through these points.

The outline of the cast shadow is given by a curve joining the shadows of a number of selected points on the line of separation; the construction for p_s and for three other important points is shown clearly in the figure.

(b) When the rays proceed from a point. (No figure.) The construction is similar to that given above but the point p_1 is located by the intersection of a_1b_1 with the line of contact of an enveloping cone and the sphere.

EXAMPLES

(1) Solve Prob. 343 completely assuming that R_1 is inclined at 45° to xy. Take c_1 1·65" from LO and 1·3" from xy, and the radii c_1a_1 and c_1b_1 as 3". (Note. The line of separation will not reach the point L.)

(2) Determine the shadow cast on the H.P. by the vase shown in figure, by rays par[1] to V.P. and inclined at 45° to H.P.

(3) The figure shows the elevation of a cone and cylinder. Determine the shadow

cast by the cone on the cylinder, which is sufficiently long to receive the shadow. Assume the projections of the rays to make 30° with xy and to slope towards xy.

(4) Taking dimensions from Ex. (1), determine the shadow cast, also the line of separation, assuming that the rays emanate from a point P, situated 5" above the H.P., the plan of P being on ba produced and 2" from the plan of the axis.

EXAMPLES FOR PAGE 338

In the following assume that the projections of the rays are inclined at 45° to xy and slope towards xy.

(1) Determine the shadow of a horizontal circle 2½" diam., 1" above H.P., centre 1½" from V.P.

(2) Determine the shadow of a horizontal ellipse, 1" above H.P.; major axis 3" long, par[1] to and 1¼" from V.P.; minor axis 2".

(3) Determine the shadow and line of separation of a cone, base 2¼" diam., altitude 3". The base is in the H.P. and the apex is 2" from V.P.

(4) Determine the shadow on a H.P., also the line of separation, of a sphere 2" diam. resting on the H.P.

(5) Solve Prob. 342 for a cylinder 2½" diam., axis 1¼" long, nearest face 1" from V.P., and axis 1½" from H.P.

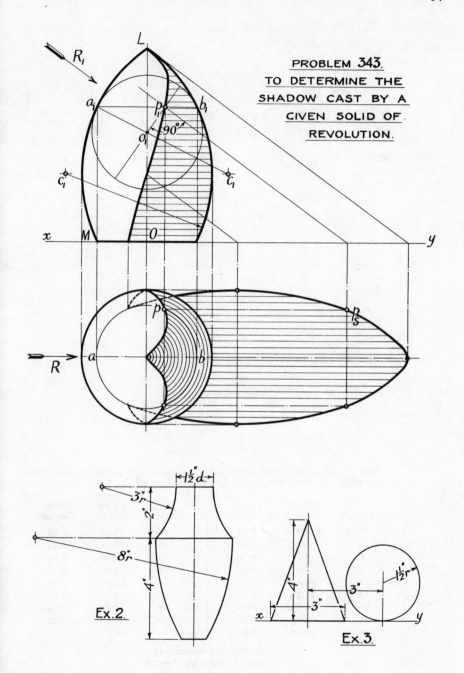

341

PROBLEM 343.
TO DETERMINE THE
SHADOW CAST BY A
GIVEN SOLID OF
REVOLUTION.

Ex.2.

Ex.3.

The following Table relates to the Aerofoil Surfaces dealt with on page 296. It gives Stations and Ordinates in % of aerofoil chord, for N.A.C.A.* 23021.

Station	Upper Surface	Lower Surface	Station	Upper Surface	Lower Surface
1·25	4·87	2·08	40	11·49	8·83
2·5	6·14	3·14	50	10·40	8·14
5·0	7·93	4·52	60	8·90	7·07
7·5	9·13	5·55	70	7·09	5·72
10	10·03	6·32	80	5·05	4·13
15	11·19	7·51	90	2·76	2·30
20	11·80	8·30	95	1·53	1·30
25	12·05	8·76	100	0·22	0·22
30	12·06	8·95			

Leading edge radius 4·85
Slope of radius tan^{-1} 0·305

* National Advisory Council for Aeronautics (U.S.A.).

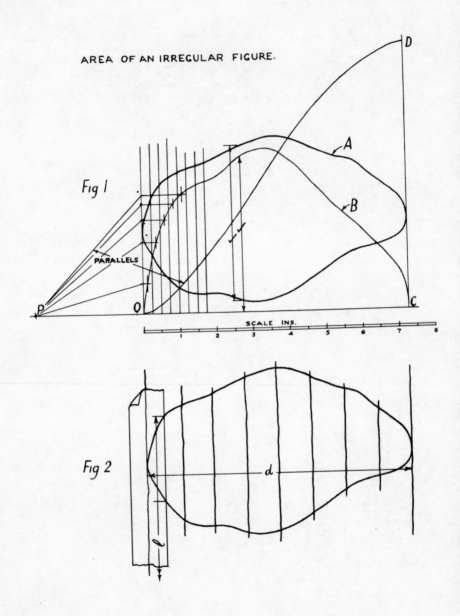

AREA OF AN IRREGULAR FIGURE.

Fig 1

PARALLELS

SCALE INS.

Fig 2

Two methods are shown above, one accurate but lengthy, the other rough but speedy and often sufficiently accurate.

The area within A in fig. 1 is subdivided by ordinates and the intercepts are transferred to give curve B. This is then integrated as on page 77. The area is given by OP × CD; and for the example shown was 3 × 7·04, i.e. 21·12 sq. in.

In the approximate method, fig. 2, we draw verticals *freehand* at the limits of the area. Bisect the distance d between them *by eye*, and then bisect the halves and quarters, again by eye and freehand. Eight strips of nearly equal widths are thus obtained. Do not draw the mid-ordinates but sum these using the edge of a strip of paper, as shown on the left, giving a total length l. Then the required area is l × d ÷ 8. Surprisingly close results can be obtained with a little care and some practice. In the example d was 7·2 and l 24·3, giving an area of 20·87 sq. in.

INDEX

The Numbers refer to pages